GRAVITATION, QUANTA AND THE UNIVERSE

Albert Einstein

Gravitation, Quanta and the Universe

Proceedings of the Einstein Centenary Symposium held at Ahmedabad, India,
29 *January* — 3 *February* 1979

EDITED BY
A.R. Prasanna
J.V. Narlikar
C.V. Vishveshwara

WILEY EASTERN LIMITED
New Delhi Bangalore Bombay Calcutta

Cover photograph of Einstein
Courtesy: International Communication Agency, New Delhi

WILEY EASTERN LIMITED
4835/24 Ansari Road, Daryaganj, New Delhi 110002
4654/21 Daryaganj, New Delhi 110002
6 Shri B.P. Wadia Road, Basavangudi, Bangalore 560004
Abid House, Dr Bhadkamkar Marg, Bombay 400007
40/8 Ballygunge Circular Road, Calcutta 700019

ISBN 0 85226 720 7

Published by Ravi Acharya for Wiley Eastern Limited, 4835/24 Ansari Road, Daryaganj, New Delhi 110002 and printed by Abhay Rastogi at Prabhat Press, 20/1 Nauchandi Grounds, Meerut 250002. Printed in India.

Foreword

"*Physics is essentially an intuitive and concrete science. Mathematics is only a means for expressing the laws that govern phenomena*".

"*The essence of a man like me lies just in what he thinks—not in what he does or suffers*".

Albert Einstein

We held a symposium at Ahmedabad during 29 January—3 February 1979 to celebrate the centenary of the birth of Albert Einstein. "Let every man be respected as an individual and no man idolized", so remarked Einstein once. Yet we idolize him and for very good reasons. By his deep thoughts and rigorous mathematical calculations Einstein bared some of the important secrets of the Universe. By conceptualizing the real world in terms of scientific theories, he showed the heights to which a single human can rise. And by his several theories which have so far proven to be a valid description of nature, he changed the meaning attached to the word "reality". Matter itself may behave differently, but a scientific description, accurate enough to describe "reality", could be conceived—and this he so demonstrated. His work, so to say, changed the world and continues to have profound influence. The many areas of research covered at the symposium are in fact those that were either founded by him or deeply influenced by his work. George Bernard Shaw once remarked "Ptolemy made a universe which lasted 1400 years; Newton also made a universe, which lasted 300 years. Einstein made a universe and I can't tell you how long that universe will last". Today Einstein's universe is more relevant than ever and that is why we do idolize him and celebrate the centenary of his birth.

It is indeed a great pleasure and privilege for me to be involved in some way with these celebrations. I would like to take this opportunity to thank the organizers of the symposium for their dedicated efforts. These are reflected in the list of eminent participants and by the fact that the texts of most of the papers presented are here. And in particular, I would like to record my thanks to one person, A.R. Prasanna, who worked assiduously and with an intellectual zeal, and to the host institutions which cooperated in organizing these celebrations.

DEVENDRA LAL
Director
Physical Research Laboratory
Ahmedabad

Editorial Preface

The Indian Association for General Relativity and Gravitation at its annual meeting in January 1978 resolved to celebrate Albert Einstein's birth centenary in early 1979 in a befitting manner by holding an international symposium. The Association invited research institutes like the Physical Research Laboratory, the Tata Institute of Fundamental Research and the Raman Research Institute to co-sponsor the organisation of the symposium. This proposal was enthusiastically supported by these institutes and later the Space Applications Centre, the Gujarat University and the Indian Institute of Science also joined in. A national organising committee consisting of G. Bandyopadhyay, P.K. Iyengar, J.V. Narlikar, S.P. Pandya, A.R. Prasanna, A.K. Ray Chaudhuri, P.C. Vaidya, C.V. Vishveshwara and Yash Pal, was formed. The Physical Research Laboratory came forward to host the symposium at Ahmedabad and generous grants in aid came from various national bodies such as the Indian National Science Academy, the University Grants Commission, the Council of Scientific and Industrial Research and the Department of Atomic Energy as well as from some of the sponsoring institutions. The symposium was held from 29 January till 3 February 1979 at Ahmedabad, with over two hundred participants, of whom fifteen were from abroad. We feel that the ambition of making the symposium worthy of the occasion was fulfilled, in every possible way. To quote from the concluding remarks of one of the participants, "Einstein as a modest man would have least wanted all the attention paid to him in his centennial year. But surely he would be happy to see physicists paying attention to his thoughts and ideas, and I think, he would be pleased that, besides all the commemorative gatherings there were also *Scientific* Symposia such as the one we have had here at Ahmedabad". The proceedings of that Symposium are being recorded in this book.

As may be seen from the book, the symposium reflected the true spirit of Einstein's own concept of the unity of nature by bringing together scholars in different aspects of physics like relativity, astrophysics, field theory and foundations of quantum mechanics and statistical mechanics. The scope of the symposium was such that participants from these various disciplines, particularly the younger ones (graduate and Ph. D. students) could interact with experts as well as among themselves. While preparing the proceedings care was taken to see that the invited and the review lectures were well documented with the discussions that followed the lectures. Broadly, the lectures presented here may be classified under the sections: (I) Foundations and Mathematical Aspects of Relativity, (II) Cosmology and Astrophysics, (III) Quantised Fields, Gauge Theories and the Foundations of Quantum Mechanics, and (IV) Statistical Physics and Photoelectric Effect. Further, the nature of these lectures spanned the whole spectrum from the popular expository review to the technical coverage of the most recent developments.

Starting with the considerations of the philosophical foundations of relativity and the creative genius of Einstein, as emphasised by V.V. Narlikar in his inaugural address, the first section develops onto a 'biased' review of relativity since Einstein, by E.T. Newman who has broadly outlined the new developments in understanding Einstein's equations using methods of modern differential geometry. As Newman has pointed out, in spite of many developments there has been no breakthrough in attempts to quantise gravity. Perhaps, Hawking's gravitational instantons and Pernose's theory of twistors seem promising lines of pursuit. Of all the rival theories of gravitation, general relativity has come out to be the best. And with the new developments in space research, as has been pointed out by R. Cowsik, it is now possible to make more accurate tests of these theories.

Considering the various important problems of the mathematical aspects of general relativity as envisaged today, M. Walker, D. Brill, A. Papapetrou, D.M. Chitre and E.T. Newman have covered a wide variety of topics like the asymptotic structure of space-time, positive mass conjecture, gravitational radiation, exact solutions and the theory of $\mathcal{H}$-spaces. Each of these lectures is self contained and gives a good glimpse into the state of art today.

The section on cosmology covers two important aspects: one is a review by J.V. Narlikar on cosmological theories starting with the models of Einstein and de Sitter coming upto the present day theories and their individual status with respect to the various types of observational evidence, such as the M—Z relation and the 3°K microwave background radiation. His treatment is entirely classical as against that of E. Gunzig who considers a novel approach to quantum cosmology, wherein the origin of the Universe is associated with spontaneous symmetry breaking through particle production. The particle production mechanism is interpreted as a phase transition in which the edge of the universe is considered as the boundary between two phases.

Coming to astrophysical aspects, we first have J. Perry on quasars as probes in understanding the cosmos, wherein she presents a new model to account for the anamoly of absorption red shifts of quasars. General relativistic aspects of astrophysics which is the main thrust in theoretical research today constitutes an important section. J. Pachner describing a model for spherically symmetric collapse has given a new method of numerical integration which compares favourably with some analytical results obtained earlier. C.V. Vishveshwara has reviewed the physics of the black holes at a nontechnical level, touching upon its relevance to astrophysics and mentioning some of the unsolved problems in the field. One of the very important and widely studied aspects of astrophysics is the energy generating mechanisms of high energy sources like quasars and X-ray sources. Many of these models make use of plasma accretion onto compact objects like neutron stars and black holes. A.R. Prasanna has presented a case for studying the structure and stability of such plasma discs around black holes in general relativistic framework by discussing possible stable orbits for charged particles very close to the event horizon and has he further presented some preliminary results concerning such a study.

The section on quantum physics begins with a study of quantised fields in the presence of gravity, i.e. in a curved background. N.D. Hari Dass discusses the Hawking radiation of black holes, by using the Schwinger formalism. As he points out it is indeed a tribute to Einstein himself that attempts to reconcile thermodynamic concepts with black hole phenomena lead to the inclusion of quantum behaviour in general relativity. With a large influx of elementary particle physicists into the study of gravitation, quantum gravity is becoming a widely studied field. In fact the last few years have seen a tremendous activity in the field of supergravity, a brief review of which may be found in the article of M. Sohnius who points out the occurence of anticommutators leading beyond the usual Lie groups for particle symmetries as the new characteristic feature of supersymmetric theories. It is well known that during the later part of his life Einstein was mostly occupied with the problem of unification of physical theories. As can be seen from the review by G. Rajasekharan of gauge theories, particle physicists seem to have met with partial success in this direction as far as the weak and the electromagnetic interactions are concerned. As he has conjectured, supersymmetry might show a new direction in the grand unification of all interactions.

Concerning the foundations of quantum mechanics it is well known that Einstein never agreed with the philosophy underlying the quantum uncertainty principle and was of the opinion that Nature is basically deterministic. Although this view did not gain much momentum during Einstein's lifetime, many schools have come up after Einstein which discuss the so-called hidden variable theories of quantum mechanics. Virendra Singh and Costa de Beauregard have in their articles, reviewed the status of such theories, discussed Bell's inequality, the E.P.R. paradox, etc., whereas R.K. Varma has proposed a deterministic basis for nonrelativistic quantum mechanics and presented the equations of motion which describe a distribution over the deterministic trajectories.

The final section deals with two other important contributions of

Einstein, namely the foundations of statistical mechanics and the photo-electric effect. E.S. Rajagopal has reviewed the theoretical as well as the experimental aspects of fluctuations, critical opalascence and critical point phenomena the basis of which are the works of Einstein and Smoluchowski. He has pointed out that the phenomena near the critical phase transitions offer a situation where the observations of fluctuations yield dramatic results some of which are yet to be properly understood. M.P. Das has described the current state of understanding of the phenomenon of photo emission and how through this phenomenon one can understand the electron structure of solids. He has also indicated how detailed calculation of the work function with the help of many electron theory has provided important check of the photothreshold.

At the end of these articles, we have recorded the abstracts of the contributed papers which were selected for presentation at the meeting. An interesting feature of the Symposium was two public lectures one delivered by D.S. Kothari on "Einstein, Physics and Beyond" and the other by Philip Morrison on "The Great and the Small—Einsteinian Scale". Kothari spoke largely about metaphysics and in particular the philosophical implications of Einstein's work. Morrison touched upon the unity in Nature from the microcosom to the macrocosom and presented very lucidly Einstein's role in making this realisation possible.

We hope the contents of this volume will contribute significantly towards elucidation of Einstein's ideas that have left their indelible mark on modern physics with their richness, variety and profundity.

A.R. Prasanna
J.V. Narlikar
C.V. Vishveshwara

Acknowledgements

It is a great pleasure to place on record our sincere thanks to all the sponsoring institutions and the national bodies who generously financed the symposium. We want to make a particular mention of the Physical Research Laboratory, Ahmedabad, which took the responsibility of hosting the meeting and carried it through successfully with excellent support from the Gujarat University and the Space Applications Centre. We wish to thank all the participants particularly the invited speakers and the chairpersons. Our call for papers received a tremendous response with a large number of contributed papers but as there was a time bound we could not accommodate all of them. We are thankful to many of our colleagues who helped us in refereeing and selecting papers for presentation. Finally we record our gratitude to every institution and every individual who contributed in some way or the other to the success of the Einstein Centenary Symposium at Ahmedabad.

4th February 1979
Ahmedabad

THE ORGANISING COMMITTEE

Contents

GRAVITATION, QUANTA AND THE UNIVERSE

The Foundations and the Mathematical and Observational Aspects of Relativity

Einstein and the Unity of Nature

V.V. NARLIKAR

I

"Who created the world?"

"There was matter somewhere in the cosmos and for a long time it lay there and stank. That stench was the origin of life."

"Where did the matter come from?"

"What is the difference? The main thing is that we have no responsibility—neither to ourselves nor to others. The secret of the universe is apathy. The earth, the sun, the rocks, they're all indifferent, and this is a kind of passive force. Perhaps indifference and gravitation are the same." He spoke and yawned. He ate and smoked.

"Why do you smoke so much?" I asked.

"It keeps me indifferent."

Isaac Bashevis Singer (1978 Nobel prize winner for literature)
in *A Crown of Feathers*, p. 47 (1973)

II

"To raise new questions, new possibilities, to regard old problems from a new angle requires creative imagination and marks a real advance in science."

A. Einstein and L. Infeld,
The Evolution of Physics (1938)

III

"The concept of absolute precision of physical measurements is obviously absurd, a mental abstraction created by mathematicians, to simplify the logic of their systems of thought. It belongs as little to physics as all other systems which are non-verifiable—for example, absolute simultaneity in the theory of relativity.... The principle of elimination of empirically meaningless statements is a heuristic idea, which has proved its worth in all parts of modern physics."

Max Born

It is often remarked that genius is generally stifled by the routine methods of teaching and examination that are adopted at schools and colleges. Einstein's schooling was marked by

(a) a sense of wonder
(b) inquisitiveness
(c) self-study to know

and *a neglect of what was going on* in his class. These qualities were responsible for the growth of his genius and his outstanding original contributions to science. The great contemporary mathematician, Hilbert (1862–1943) had to say: "Every boy in the streets of Göttingen understood more about four-dimensional geometry than Einstein. Yet, in spite of that Einstein did the work and not the mathematicians.... Do you know why Einstein did the most original and profound things about space and time in our generation? Because he learnt nothing (from routine courses) about the philosophy and mathematics of time and space." Einstein was a rebel and a creator and creative genius demands the three qualities that I have referred to at the outset. Born in a small town Ulm in Germany on March 14, 1879 Albert Einstein became famous for the general theory of relativity in 1921 and in the same year was awarded the Nobel prize for physics. Around 1955 physicists had ceased to find much excitement in general relativity. But for the last fifteen years interest has been mounting in this field because of new developments in astrophysics, cosmology, particle physics, etc. The Einstein birth centenary is being celebrated all over the world with enthusiasm wherever science is being studied, and seekers of truth are being gratefully remembered.

In one of his numerous biographies it is mentioned that, during his school days, Einstein did a lot of reading on general science and he was deeply impressed by the reign of law in the universe. This led him to believe in Spinoza's God who reveals Himself in the harmony of all being and not in the common man's God who concerns Himself with the fate and actions of human beings. This important distinction has to be kept in mind when one notes his use of the word 'God' in statements and discussions with others. It was Spinoza's God that enabled Einstein to disclose truths to us hidden from our understanding.

Even at a much earlier age Einstein was struck by the peculiar behaviour of the magnetic needle wherever it was placed. This magical effect in the reality of the world about him was a surprise to him. He was equally surprised around that time to see that, according to some 'law' that can be proved, in any triangle the three perpendiculars from the vertices to the opposite sides must meet at a point. Thus the ground was being prepared for Einstein's later work in natural philosophy.

The Greeks, according to A.N. Whitehead had the genius to be astonished. It were they who gave us mathematics and the methods of proofs and they gave us also the orphic word 'theory'. A theory was supposed to arise out of reflections culminating in an insight into the beauty, the harmony, the magic of nature. This insight could as well be described in a mathematical language. The book of nature seemed to be written in the language of mathematics and from Galileo started the practice of performing experiments and of interpreting the results in an appropriate mathematical language instead of relying on authority, hearsay or tradition. Mathematical truths, otherwise abstract and meaningless to the layman, make sense of our findings of natural phenomena.

H. Weyl said on one occasion, "In the intellectual life of man there can be clearly two domains, the domain of action, of shaping and construction to which active artists, scientists, technicians and statesmen devote themselves, and a domain of reflection of which the fulfilment lies in insight, and which since we struggle in it to find the meaning of our activity, is to be regarded as the proper domain of the philosopher."

In Einstein's life, we find four major periods 1895–1905, 1905–1916, 1927–1938, 1925–1954 during each of which, except the last he developed the insight and the method to come to the solution of his problem. The last problem for which he made many efforts remained unsolved. Meanwhile he had successfully tackled several other problems of theoretical physics (for which, and not for general relativity he was awarded the 1921 Nobel prize for physics, as I have mentioned earlier).

Einstein's 'Autobiographical Notes' was written in 1949, six years before his passing away. The concluding paragraph of the 'Notes' runs as follows: "This exposition has fulfilled its purpose if it shows the reader how the efforts of a life hang together and why they have led to expectations of a definite form."

After establishing the Special Theory of Relativity and a field theory of gravitation providing the equations of gravitatianal motion Einstein was keen on obtaining a unified field theory that would account for the electromagnetic force and the quantum phenomena as well. The recent developments in astrophysics, cosmology, high energy physics, fundamental particles with their strong and weak forces all came later. As far as he knew Einstein *aspired to reveal the unity in the universe through his physical theories.*

With his towering position in the world of science it is good to recall what Einstein once said in all humility: "One thing I have learned in a long life; that all our science, measured against reality, is primitive and childlike—and yet it is the most precious thing we have."

An insight into Einstein's life and personality is gained by the following observations of C. Lanczos (*The Einstein Decade 1905–1915*, p. 39, 1974): "He was not hampered by personal ambition, by the desire to make a career, by the necessity of writing papers for professional advances. He had a burning curiosity to find out what makes the physical universe 'tick'. That was the motivating force of his life, from the early years in Bern, up to his last years in Princeton.

The four periods of investigations illustrate a well-known dictum of

Einstein 'God is subtle but He is not malicious' which he himself paraphrased as 'Nature conceals her secrets by her sublimity and not by trickery'. In 1895, at the age of sixteen, it was a mystery to Einstein as to what the electromagnetic field of a wave of light would look like to an observer keeping pace with the wave. Maxwell's equations did not account for such an oscillatory field at rest, nor was any such field ever observed. This seemed to be another unobservable like the absolute simultaneity of two events at distant places and the absolute velocity of the earth in space. The aether itself in which the light waves were propagated was unobservable.

Lorentz and Poincaré knew the relevant facts. Poincaré himself had used the word 'relativity' in 1904 in a publication. It was also known that the equations of a Galilean transformation

$$x' = x - ut, \quad y' = y, \quad z' = z, \quad t' = t$$

left Newton's laws of mechanics unaffected while the Galilean transformation had to be replaced by the transformation (in the usual notation)

$$x' = \frac{x - ut}{(1 - u^2/c^2)^{1/2}}, \quad y' = y, \quad z' = z, \quad t' = \frac{t - ux/c^2}{(1 - u^2/c^2)^{1/2}}$$

if the Maxwell equations of the electromagnetic field were to remain unchanged in form. The latter transformation was originally called the Larmor-Lorentz transformation and is sometimes referred to as Lorentz transformation or as the Poincaré transformation. J. Larmor found it in 1898 while *W*. Voigt had discovered it separately in 1887.

While all these facts pointed to a chaotic state of observation and theory Einstein could accomplish a grand synthesis out of these in the form of his Special Theory of Relativity, the essence of which is that all inertial frames (x, y, z, t) are connected by the group of Lorentz transformations (of ten parameters) so as to leave:

(i) $ds^2 = dx^2 + dy^2 + dz^2 - c^2dt^2$ invariant
(ii) the velocity of light c invariant
(iii) the velocity of the earth relative to the aether indeterminate
(iv) the space-time homogeneous and isotropic in each inertial frame

In no inertial frame can an inertial particle or an observer attain the limiting velocity which is c and there is no such thing as aether. Neither observations nor theoretical facts support the aether hypothesis. While (iv) accounts for the three principles of conservation of momentum, angular momentum and energy, the symmetry of the Lorentz group satisfied by the inertial frames account for the epoch-making result, $E = Mc^2$. It took time to realise that every form of energy has inertia and that every gram of matter means 9×10^{20} ergs of energy. According to the Special Theory, no particular inertial frame is preferred and all laws and phenomena of physics are expressible in the same form in all inertial frames. In the Minkowski geometry of space-time Maxwell's equations assumed a simple and unified form, bringing out the connection between an electric field, a magnetic field and the relative uniform velocity, as revealed in the equations of transformation of the electromagnetic field. A new synthesis of space and time had been achieved revealing new truths about M, E, c, $\hat{E}$, $\hat{H}$, etc., and unsuspected properties of a moving rod and of a moving clock.

A mathematician or a scientist who is restless until he has the solution to the problem is like a mouse trying hard to get out of the trap. It is a life-and-death struggle and the mouse has to come out within a limited time. Einstein was at his first great problem for ten years before he found his solution in Special Relativity.

The next important period was from 1905 to 1915. This time the problem was gravitation. It was not possible to bring the Newtonian law of gravitation, with a suitable modification, within the framework of Minkowski's geometry. A more sophisticated geometry was needed to advance beyond the principle of equivalence by which the inertial mass m_i is the same as the gravitational mass m_g. No progress could be made towards a relativity of accelerated inertial frames. With the help of Marcel Grossmann, Einstein invented the revolutionary hypothesis that gravitation is a phenomenon of curved Riemannian space-time. Here Einstein's theory benefitted from Gauss's most remarkable discovery that the curvature of a surface is an intrinsic property and that it can be completely determined solely from the distance distortions shown patently in a chart.

In such a space-time there can be no inertial frames and therefore there can be no parallelism of vectors at a distance as in a Minkowski space. The displacement field Γ^i_{lk} gives us, in an invariant measure, the relation between vectors as well as tensors at infinitesimally close points and is therefore a substitute for the inertial fields. Out of the Γ^i_{lk}-field arises the curvature tensor R^i_{lkm}. Hilbert was able to show that Einstein's field equations can be expressed as arising out of a stationary principle, the integral of a world function, taken over space-time being stationary. Interpreting the condition

$$\delta \int R\sqrt{-g}\, d^4x = 0$$

in the usual notation, E.T. Whittaker says, "Gravitation represents a continual tendency of the universe to straighten itself out." The new theory of gravitation was another great revolution in thought and its impact was felt by the world of science when the three crucial tests were found to be satisfied. The general impression gaining ground was that the laws of nature were geometrical laws about geometrical objects. Soon the laws of gravition were expressed in the form

$$R_{ij} - \tfrac{1}{2} R\, g_{ij} = -\, 8\pi \frac{G}{c^2} T_{ij}$$

where $T^j_{i\,;\,j} = 0$ were the laws of conservation, which fitted with the identities $(R_i^j - \frac{1}{2} R g_i^j)_{;j} = 0$. After this revision of 1916* another problem arose. The field equations were not pure field equations because of T_{ij}. Einstein thought that this theory was based on two pillars; one of marble and the other of straw. T_{ij} was the straw pillar and it had no place in a pure field theory. Secondly, following Mach, Einstein believed that the inertia of a particle was due to the presence of all other matter in the universe. An isolated body would have no inertia by itself. The universe as it was then (1917) known was essentially static with stars having some motion though small. To check if the relativity theory was applicable beyond the solar system Einstein considred the presence of a small cosmological constant λ that would counterbalance the gravitational force to give a static finite universe. So the field equations were modified as

$$R_{ij} - \frac{1}{2} R g_{ij} + \lambda g_{ij} = -\frac{8\pi G}{c^2} T_{ij}$$

Later, in 1931, Einstein reconsidered the whole matter in the light of the new knowledge of the universe and put down $\lambda = 0$, as it was an unnecessary addition to the field equations. G. Lemaitre was, however, of the view that "the cosmical constant was the most important contribution of general relativity to physical thought."

There are some cosmologists who consider λ to be necessary to provide a standard of length in general relativity. There are others who find that the cosmic recession of galaxies is being accelerated which cannot be theoretically accounted for without taking $\lambda > 0$. With the constant increase in our knowledge about the galaxies and their distribution, the distance scales adopted in the light of astrophysical researches, cosmic turbulence and other phenomena we may have to wait till the space telescope of the coming years (1983 onwards) provides us the necessary data before we can make up our mind about the value of λ and of our estimates of H_o, the Hubble constant of law of recession and of the red-shifts. No option may then be left in fixing the values of all these parameters so as to be consistent with the known phenomena.

There are critics like V. Fock who do not look upon Einstein's theory of gravitation as *general relativity*, Briefly, the point is that the group of non-singular Gaussian transformations from x, y, z, t to x', y', z', t' does not transform the metric tensor $g_{ij}(x, y, z, t)$ into $g_{ij}(x', y', z', t')$, for the tensor not only defines a co-ordinate system but a set of gravitational potentials of physical significance. Covariance

$$g_{ij}(x, y, z, t) \leftrightarrow g(x', y', z', t')$$

is not what is implied by the term 'general relativity'. E.P. Wigner (*Symmetries and Reflections*, 1967) has also examined this point and agrees with Fock's criticism.**

*Einstein revised the field equations of gravitation several times in 1915.

**"A general principle of relativity, as a physical principle, holding in relation to arbitrary frames of reference, is impossible." Fock, V., (1964), p. 181 *The Theory of Space, Time & Gravitation.*

The new theory was free from the mystery of instantaneous propagation of a force in a strange mysterious medium. Gravitation became a geometrical effect. New geometries attracted the attention of physicists.

The three-body problem was a great challenge to astronomers and mathematicians according to Newton's theory. In Einstein's theory of gravitation the two-body problem was a great challenge. Because of the difficult, non-linear structure of the field equations simple idealized problems involving two particles were being considered. Most such solutions violated reasonable physical or mathematical conditions and had to be rejected. Einstein started seriously on the problem in 1927, in association with J. Grommer. So far, the successes of the theory depended on Schwarzschild's solution according to which the null and timelike geodesics gave all the possible gravitational motions for small bodies like planets, or for effects like the gravitational deflection of stellar light near the sun. But matter and motion cannot be divorced from each other. Matter cannot be given by the field equations and motion by another set of equations of goodesics. Besides, the geodesics could not legitimately be considered as providing the motion of bodies of finite masses and sizes. Thus the problem of a double star, one rotating about the other cannot be solved in this way. But the solution of one body involved a singularity in the representation of an external field. Associated with a singularity is a curvature of space-time. Besides, Mach's principle was violated, no other body being in the field. With two bodies there would be two different sources of curvature in space-time and their co-existence would mean more mathematical singularities and discontinuities. However, if the two sources of curvature continue to move so as to make the necessary adjustments—smoothening—in space-time there may be only two singularities and no undesirable discontinuities may emerge in the field. The process of adjustment could be achieved step by step to various degrees of approximation in series and thus the phenomenon of gravitational motion would emerge, providing equations beyond the first stage of Newtonian approximation. In 1938 this ingenious method with its complicated equations was obtained so that the field equations $R_{ij} = 0$ could give the right correction in the motion of the precession of Mercury. The method was developed by Einstein, Infeld and Hoffmann and the solution for mercury was obtained by H.P. Robertson in a separate paper. Further improvements by Einstein and Infeld followed in 1940. Fock had also developed a similar method in 1938, independently, by using the field equations involving T_{ij}. Although this is a great achievement, much is yet to be done to settle several questions such as those of rapidly changing gravitational field, radiation, etc.

It was while tackling the field equations for the problem of motion that the peculiar structure of Einstein's field equations was noticed, viz., that they are suitably underdetermined and suitably overdetermined. While this is helpful the inherent four-fold arbitrariness of co-ordinates remains at each stage, in a step-by-step approximation in series. The original aim was to seek only the first order departure terms from the Newtonian approximation. At higher approximations, the choice of co-ordinates creates difficulties while identifying terms due to particle motion and gravitational radiation. Einstein's presentation of his theory in 1938 reveals a deeper insight into the phenomenon of gravitation than his theory of 1916.

Thus in three of his four problems where secrets of Nature seemed to be hidden under her sublime appearance, Einstein was able to achieve success after years of concentration and dedication, discovering and using new mathematical techniques. As regards the fourth problem of a unified field theory he could not obtain, up to the end, the necessary insight and the appropriate mathematical technique.

Einstein's first step was to bring into prominence the metric

$$ds^2 = dx^2 + dy^2 + dz^2 - c^2dt^2$$

of Minkowski's geometry, of special relativity. Using curvilinear co-ordinates, this metric can as well be expressed as

$$ds^2 = g_{ij}\, dx^i\, dx^j$$

the four-dimensional metric of Riemannian geometry, provided the curvature tensor $R_{hijk} = 0$. If the metric is to account for gravitation the restriction $R_{hijk} = 0$ goes and the field equations

$$R_{ij} = 0 \quad \text{or} \quad R_{ij} - \tfrac{1}{2}Rg_{ij} = -(8\pi\, G/c^2)\, T_{ij}$$

hold good. The geometry is no longer equivalent to Minkowski's geometry and the group is no longer the Poincaré group but the more general group of continuous transformations of curvilinear co-ordinates. For this reason the theory of gravitation of curved space-time came to be known as general relativity while Special Relativity dealt with flat space-time. Now, in general relativity if we drop T_{ij} mass particles are associated with singularities. But in obtaining the equations of motion of two or more particles we note that the gravitational motion of the particles is the one that does not produce additional (mathematical) singularities in the field. It is pertinent to quote here what Einstein says on this matter in his "Autobiographical Notes".

"Now it would of course be possible to object: If singularities are permitted at the positions of the material points, what justification is there for forbidding the occurrence of singularities in the rest of space? This objection would be justified if the equations of gravitation were to be considered as equations of the total field. (Since this is not the case), however, one will have to say that the field of a material particle may the less be viewed as a *pure gravitional field* the closer one comes to the position of the particle. If one had the field equations of the total field, one would be compelled to demand that the particles themselves would *everywhere* be describable as singularity-free solutions of the completed field equations. Only then would the general theory of relativity be a complete theory."

The creation of the complete theory or the total field theory was the first priority programme of Einstein from 1925 to 1954 before any applications to cosmology or to situations demanding a quantum theory of gravity could be thought of. By total field Einstein originally meant a generalized field including both gravitational and electromagnetic fields and their interactions. Later he imposed a requirement whereby the theory included the Planck's constant h in an unforced manner. During the years when the programme was carried out, putting up one version after another of the unified theory, astronomers and physicists were fast adding to their knowledge of the universe, of thermonuclear reactions, of fundamental particles, etc. New symmetries and new harmonies in nature were coming to light. The principle of complimentarity meant that we are both observers of and participants in the world of physical reality. The field approach seemed to be doomed to fail in the face of all the new facts of observations. In 1951 Einstein confessed: "All these fifty years of conscious brooding have brought me no nearer to the answer to the question 'What are light quanta?' Nowadays every Tom, Dick and Harry thinks he knows it, but he is mistaken." Einstein, in his last paper on the subject, admitted that perhaps the concept of field was inadequate for the unified theory which he was seeking.

During his last years Einstein found a psychological satisfaction in obtaining a complete set of equations for the desired unified field theory. He seemed to agree with what Planck had said in 1913 that physical sciences demanded the acceptance of a real world which functions independently of man. What is observed is what is functioning independently of man and as we learn more and more through observations, using instruments and senses, we gain an ever-deepening insight into the working of that almighty reason which governs nature. In 1925, Einstein used an unsymmetric g_{ij} tensor with ten potentials for gravitation and six for electromagnetism. He had said about his theory, "I now believe I have found the true solution." It did not take long to see flaws in it. He made many attempts during the thirty years that followed, trying different mathematical approaches, in collaboration with several younger colleagues but he could not find in any case the desired solutions free of singularities. His faith was sustained by the belief that "God is subtle but He is not malicious."

Niels Bohr had shown how the principle of complimentarity goes against the classical conception of physical reality in which the observer could maintain his aloofness. The principle of complimentarity had found a place in the physics of quantum theory. P.A.M. Dirac, an authority on quantum mechanics says in his latest book *Directions of Physics* (1978): "We do not try to apply our present quantum theory to particles with very high energies and at very small distances. The foundations have not been correctly established." Dirac further says, "According to the present quantum mechanics, the probability interpretation, the interpretation championed by Bohr, is the correct one. But still Einstein did have a point. He believed that, as he put it, the good God does not play dice, He believed that basically physics should be of a deterministic character.

And, I think it might turn out that ultimately Einstein will prove to be right, because the present form of quantum mechanics should not be considered as the final form."

If and when gravitation is understood as a part of the universal unity of the apparently separate themes in the book of nature such as:

(1) Special and general relativity,
(2) General mechanics including quantum mechanics,
(3) Elementary particle theory,
(4) Irreversible phenomena,
(5) Cosmology, etc.

The word 'gravitation' may assume an unexpectedly new meaning. Einstein, with his faith strengthened in the unity of nature by his earlier researches was justified in devoting so many years of his life at Princeton for a clearer understanding of gravitation, perhaps, knowing fully well that the problem undertaken by him was really meant for succeeding generations of physicists and mathematicians, better equipped with physical data and mathematical techniques.

In his book *The Character of Physical Law* (1965), R. Feynman observes: I must say immediately that one does not, by knowing all the fundamental laws as we know them today immediately obtain an understanding of anything much. It takes a while and even then it is only partial. Nature, as a matter of fact, seems to be so designed that the most important things in the real world appear to be a kind of accidental result of a lot of laws. The thing is that as nuclei begin to be formed from H and He in stellar furnaces a stage soon comes when (a) ^{8}Be is unstable, (b) a resonant level exists in ^{12}C at exactly the right place and (c) a potentially dangerous resonance in ^{16}O happens to be just below the threshold.

All these circumstances favour the existence of life here. That there should be such a level in ^{12}C is "an excellent illustration of the fact that an understanding of the physical laws does not necessarily give you an understanding of things of significance in the world in any direct way. The details of real experience are often very far from the fundamental laws." F. Hoyle suggests that if the dimensionless numbers that arise in physics are responsible for this phenomenon about ^{8}Be, ^{12}C and ^{16}O, it is worth speculating whether the dimensionless numbers differ in another part of the universe, and if so, how they affect the form of life there.

To the seekers of fundamental laws, the principles and their consequences in nature we offer our humble homage and above all to Albert Einstein to whom the world owes so much for his profound researches, and whose birth-centenary we have assembled here to celebrate.

Relativity Since Einstein—A Biased View

E. T. NEWMAN

Let me begin by saying that though there have been some brilliant advances in our understanding of relativity after Einstein's death in 1955, there has been (in my opinion) no major breakthrough in the outstanding physical problems of the wedding of relativity with quantum theory or particle physics. Twenty-four years after his death we still await a new Einstein. They do not come very often.

When I was asked to talk on Relativity Since Einstein, I believed at first it was a simple assignment. I intended to list all the major developments in the subject over the last twenty-five years or so and then organize them into broad topics such as mathematical developments, theoretical ideas and predictions, and finally observations. I thought I would then be able to present them in a unified manner. Unfortunately, it turned out to be more difficult than I had thought. The developments are so varied and rich that I had to make personal selections, and thus the 'biased view'. (If I have left out some topics of importance it is because of some of three reasons: (a) a complete oversight or unawareness of the topic, (b) my lack of understanding of the topic, and (c) my conviction that it was unimportant.) In addition, the three broad topics into which I had hoped to organize the developments are really not adequate—the mathematical and physical ideas are so intertwined that to put items into slots was almost impossible.

In spite of this it appears to me as if many, if not most of the developments arose initially from the introduction or development of new mathematical ideas and techniques into relativity. It is clear that the modern ideas of manifold theory, the co-ordinate free techniques in differential geometry, techniques for global analysis, bundle theory, etc., played crucial roles in many places.

One of the first important accomplishments was the Kruskal extension of the Schwarzschild metric—it finally being understood how the difficulties associated with the Schwarzschild metric, i.e., with the so-called Schwarzschild 'singularity', disappear and that the 'singularity' itself was an effect from a poor choice of co-ordinate system. As the first model of a black hole, this Kruskal manifold became the centrepiece for the rigorous study of black holes. It was from the study of certain topological features of the Kruskal manifold, which remained stable under deformations of the metric, that Penrose was led to the notion of a trapped surface which in turn led to the stability of the blackhole. From this followed the remarkable singularity theorems of Geroch, Hawking, and Penrose—the essential result being that singularities of space-times were not unusual pathological occurrences but were in some sense quite generic.

As an example, consider a cold dilute gas of mass 10^8 solar masses which contracts to the density of the earths atmosphere, i.e., about 10^{-4} g/cc. There is no problem with an equation of state—the density and temperature are in normal range, there is no 'new' physics occurring. Nevertheless, a trapped surface would form and collapse to a black hole and a singularity would be inevitable.

Due to additional mathematical techniques (which we will discuss soon) the entire subject of classical black hole physics developed rapidly and is still developing. Additional (to the Kruskal) black hole solutions of the Einstein equations were found, stable perturbation calculations were investigated, energy extraction mechanisms were studied, final state black hole theorems were proved and perhaps most fascinating, the observational search for black holes began with what appears to be positive results.

Related to the subject of black holes but really much more general in scope is the problem of quantum field theory on a curved space-time. Though the general problem appears to be beset with serious difficulties a great deal of progress has been made. In particular this is true for space-times that have privileged constant time surfaces or timelike Killing vectors. Using these ideas Hawking showed that though classically a Schwarzschild black hole could only grow in size, quantum-machanically it could evaporate. In fact, it radiates with a black body spectrum with temperature proportional to M^{-1}, the total luminosity going as M^{-2}.

I would think that these are among the most important and dramatic developments in recent years. They, however, appear to me (who is a total outsider to the field) to be a first approximation to a theory we do not yet have. Along closely associated lines I would add that a major disappointment must surely be how little we understand what a quantum theory of gravity should be.

Returning to classical questions and the mathematical ideas preceding them, I want to mention that an idea, though perhaps not very profound to start with, did have a profound effect on further research. The idea was that of the Petrov algebraic classification of Weyl curvature tensors. This led first of all (by asking for the different algebraic types) to large families of new solutions of the Einstein equations; the Robinson-Trautman metrics, the plain and plain-fronted waves, the twisting algebraically special metrics which includes in particular the Kerr and charged Kerr metrics. These metrics are the generalizations of the Schwarzschild and charged Schwarzschild metrics to include angular momentum. (Chandrasekar considers the discovery of the Kerr metric to be the most important event in relativity theory since the discovery of the Schwarzschild metric.) A second application of the classification scheme permits a covariant definition of gravitational radiation which in turn was a major impetus for the study of asymptotically flat space-times, the last major topic I wish to discuss.

The study of asymptotically flat space-times is the study of isolated systems. Such systems were needed to understand the structure of gravitational radiation and the definition and role in general relativity of such physical quantities as mass, momentum and angular momentum.

Although the general theory of relativity had been with us for sixty years, it was only in the past 15 years, beginning with the work of Bondi, that rigorous results and a deep understanding of the radiation aspects of the theory have been developed. Bondi's great contribution was the realization that asymptotic flatness and gravitational radiation could best be understood by studying and analyzing the null or characteristic surfaces of a space time. This was in contrast to previous work which concentrated on space-like surfaces and the associated Cauchy problem and which has more recently led to work on the concept of infinity in space-like directions. The null surface approach led immediately to an understanding of the radiation degrees of freedom and their geometric meaning, as well as to an understanding of the related characteristic initial value problem. Although one can relate the local behaviour of gravitational sources to the asymptotic behaviour of the field along null surfaces only through perturbation methods, one can still gain a great understanding of the radiation process. For example, asymptotic multipole moments can be defined and their time evolution studied'. A special case of this is Bondi's beautiful result that gravitational waves carry away positive mass. Bondi's pioneering work was extended by Sachs and Newman and Penrose who systematically introduced complex combinations of real geometric quantities, developing the complex null tetrad formalism, and associated spin-coefficient formalism. This led to great simplifications in the equations of the theory.

The next major advance was made by Penrose who formalized the idea of asymptotic flatness by adding a boundary $\mathscr{I}$ (pronounced 'scri') to space-time. This boundary, attached by the process of conformal compactification, consists essentially of the future ($\mathscr{I}^+$) and past ($\mathscr{I}^-$) end-points of null geodesics. Penrose defined a space-time to be asymptotically flat if this boundary could be attached and then showed that this captured the previous co-ordinate-dependent notions of asymptotic flatness. In this way the difficult and nebulous concept of 'infinity' was made precise and it became possible to speak of relations holding 'on $\mathscr{I}$' rather than just 'asymptotically'. There exists an alternative theory of infinity in space-like directions, 'spi', which has a static, once-for-all character. We will not discuss 'spi'.

The spin-coefficient formalism (while not exclusively adapted to asymptotically flat space-times) has been widely used and two basically diverging lines of work have evolved. In the first, 'practical' or astronomical calculations have used perturbation methods to relate source behavior to asymptotic radiation fields. The second has concentrated on more formal developments, studying the geometrical properties of the field in the neighbourhood of $\mathscr{I}^+$ in the hope of elucidating fundamental questions about the theory and relating it to other branches of physics such as particle physics or quantum theory. Here the sort of questions elucidated were: What relationship does the asymptotically flat region have with (completely flat) Minkowski space? What, if any, asymptotic symmetries are there and how are they related to the Poincaré group? Are there functions analogous to the Poincaré generators, momentum angular momentum? Though some fundamental problems for asymptotically flat space-times still exist, the theory is in quite good shape.

As a final comment I want to raise the question of which one of the recent developments in relativity is most likely to lead to the major advances of future. To answer this would be clearly guesswork on my part mixed with personal tastes and biases. On the other hand, it is in fact fun to speculate and it does no harm if it is not taken too seriously. It seems obvious that the major question in relativity is the marriage between quantum theory and general relativity. Among the more attractive ideas is Hawking's attempt to quantize the gravitational field by means of gravitational instantons in analogy to the quantization of gauge theories. My own taste however, goes towards the Penrose theory of twistors. Twistor theory seems to have both the richest and most natural structure with which to generalize the present laws of nature. Though the absolute probability of twistor theory being the road to 'trush' is probably little, it appears to be large compared to other ideas or theories.

Discussions

A K Pandey: I have a very general query on the consistency of black hole radiation. As more and more radiations come out of the black hole its temperature increases and hence a black hole can never be in stable thermal equilibrium with the conventional heat capacitors. Perhaps this will make zeroth law of thermodynamics non-functioning which requires stable thermal equilibrium between bodies. Have you any comment?

E T Newman: As I understand it an isolated black hole in an asymptotically flat space would be unstable to black body radiation. In order to have equilibrium with the surroundings it must be immersed in a cosmological background.

N Panchapakesan: Could you comment on the new uncertainty principle that Hawking has been talking about?

E T Newman: I belive that the idea is that a black hole emits radiation in an inherently statistical or unpredictable fashion. This is to be thought of as an intrinsic property of black holes having nothing to do with ordinary statistical mechanics or the Heisenberg uncertainty principle.

T Padmanabhan: Has the Bohr-Rosenfeld work on the quantization necessity of a field been generalized for gravitation?

E T Newman: I am quite certain many people have thought and written about it. The few names that come to mind are Feynman, Wigner and James Anderson.

Pankaj Joshi: How seriously should the geodesic incompleteness be considered?

Martin Walker: In regard to the question of geodesic incompletencss, Chris Clark has shown in a sequence of recent papers that if an incomplete geodesic is not locally extensible, then the curvature becomes unbounded along the incomplete geodesic. This can be interpreted as meaning that tidal accelerations become unbounded in the 'neighbourhood of singularity'.

S G Rajeev: Has the technique of expanding the R-R tensor in terms of Rieman co-ordinate been applied to the Yang-Mills field and for quantizing GR?

E T Newman: Not to my knowledge.

Relativity Experiments in Space

RAMANATH COWSIK

1. Introduction

1.1 Parameterised Post-Newtonian Formalism

It is very appropriate that this talk has been scheduled to be deliverad at the Space Applications Centre since my emphasis is going to be on 'space' rather than on the subtle theoretical aspects of 'Relativity'. Also, with the development of interplanetary space flights and other major capabilities in space, I believe that the field of experimental relativity will be predominantly the realm of space engineers and scientists. The General Theory of Relativity (GR) proposed by Einstein in 1915 had remained virtually a pure science for nearly half a century, but during the last decade the attitude towards the subject has been changing because of technological developments and necessity for the precise knowledge of the gravitational effects in many applications such as search for raw materials by means of satellite, meteorology and space-navigation.

There are several excellent discussions available in literature on the subject of 'experimental relativity'; the present talk is mainly based on Ref. 1–5 cited at the end.

The interpretation of experimental results pertaining to various theories of gravitation needs special care since the theories describe not merely the behaviour of the test bodies but also of the reference frames themselves. Therefore, the axioms of every theory of gravity should contain, at least implicitly, a description of the properties of the instruments and the various measurements that can be performed to verify them. Since, in general, a measurement consists in comparing a test body with a standard to ensure reproducibility of results, we need to assume that there exist classes of identical objects and situations which in turn lead to the symmetry group defining the relativity principles of the theory. Invariants of this symmetry group such as energy, momentum, angular momentum, spin, etc., are used in the description of the test objects and experiments.

In GR, inertial motions are geodesics of Riemannian space-time and with the kind of gravitating bodies available within the solar system the deviation between the predictions of GR and that by the simple Newtonian theory of gravity are rather small. Competitive theories of gravity which are developed primarily by changing one or more of the postulates of GR also predict small post-Newtonian effects. Therefore, in order to compare the outcome of the experiments among these competitive theories, one needs a suitable framework. This framework is in the nature of a perturbative expansion; this was developed by Chandrasekhar, Nordtvet, Thorne, Wills and others and is called the 'Parameterised Post-Newtonian Formalism'. Table 1 lists these post-Newtonian parameters and the

Table 1 Some post-Newtonian phenomena and their PPN parameters

Description of phenomenon	Parameters
1. Spatial curvature generated by rest mass	γ
2. Non-linearities in superposition of Newtonian gravitational potentials	β
3. Newtonian-type gravity generated by gravitational energy	$\beta_2 = \frac{1}{2}(\zeta_2 + 3\gamma - 2\beta + 1)$
4. Newtonian-type gravity generated by kinetic energy	$\beta_1 = \frac{1}{4}(\alpha_3 + \zeta_1 + 2\gamma + 2)$
5. Effect of anisotropies in kinetic energy of Newtonian-type gravity	ζ_1
6. Newtonian-type gravity generated by internal energy	$\beta_3 = \zeta_3 + 1$
7. Newtonian-type gravity generated by isotropic part of stresses	$\beta_4 = \zeta_4 + \gamma$
8. Effect of anisotropies in stress on Newtonian-type gravity	η
9. Magnetic-type gravity generated by momentum	$\Delta_1 = \frac{1}{7}(\alpha_1 - \alpha_2 - \zeta_1 + 4\gamma + 3)$
10. The dependence of strength of momentum-generated gravity on direction of momentum (toward observer versus transverse)	$\Delta_2 = \alpha_2 - \zeta_1 + 1$
11. "Preferred-frame effects", i.e., the influence of matter's motion relative to Universe on the gravity the matter generates	α_1, α_2 and α_3
12. "Preferred-orientation effects", i.e. the gravitational influence of the orientation of the experimental apparatus relative to the external universe	α_2, ζ_ω
13. Breakdown in global conservation laws for energy, momentum and/or angular momentum	$\zeta_1 + 2\zeta_\omega$, $\zeta_2 - \zeta_\omega$, ζ_3 $\zeta_4 + (2/3)\zeta_\omega$, α_1, α_2, α_3

physical meaning attached to these.[5] In Table 2 are listed the dependence of some of the well-known tests of GR on the value of these parameters. In this way of parameterisation GR is characterised by $\gamma = 1$ and $\beta = 1$ and the rest of these parameters are zero.

Table 2

Observable effect	Dependence on β, γ
Deflection of light rays and delay of radio signals	$1/2\,(1+\gamma)$
Geodesic precession of gyroscopes	$1/3\,(1+2\gamma)$
Lense-Thirring effect	$1/2\,(1+\gamma)$
Perihelion advance	$1/3\,[2(1+\gamma)-\beta]$

1.2 ADVANTAGES OF SPACE PLATFORMS

As already mentioned the post-Newtonian effects are very small and there are perturbing forces that are large enough to mark these small relativistic effects. Now we shall compare the environment offered by space vehicles with respect to that provided by a terrestrial laboratory.

1. How much room does one have in a space vehicle? Using the space-shuttle as a standard, we can have a space vehicle as large as a Boeing 707, which provides more than enough room for sophisticated computers, hydrogen maser clocks, living room and support equipment for three or more scientists.
2. The total weight of the vehicle could be about 15,000 kg of which at least 5,000 kg would be available for the actual experimental apparatus, if so desired.
3. Weightlessness of space allows large structures $\sim 10^5$ cm $\times$ 10^5 cm to be erected in space without the fear of structural sag or creep which seriously limit the efficacy of earthbound experiments.
4. Spaceships can carry transponders and optical reflectors, land these in orbit about these celestial bodies, thereby increasing tremendously the sensitivity and accuracy of ranging measurements.
5. By going far away from the earth or by approaching the sun very closely, the space vehicles can sample a wide range of gravitational potentials improving the quality and accuracy of gravitational redshift experiments.
6. Spaceships can be made seismically very quiet; the random accelerations can be reduced below 10^{-11} g with long time constants, as compared with levels of $\sim 10^{-6}$ g achievable over short periods on the earth (see point 7).
7. Spaceships can be actively driven along geodesics, the ideal frame for performing the experiments on gravity.

1.3 THE DRAG-FREE DRIVE SYSTEM

A body orbiting the earth is subjected to two forces—the gravitational attraction which tends to drive it along the geodesics, and the drag forces due to solar radiation, solar wind, residual atmosphere, etc. These non-gravitational forces are in general proportional to the surface area so that more massive bodies are better compensated against these perturbations. For example, the drag due to solar radiation at 1 AU is 10^{-8} dynes/cm.^{2}

Thus a spherical object of radius $R = 300$ cm and density $\rho = 1\ g/cm^3$ has a mass $M = 10^8$ g and the ratio of gravitational forces to non-gravitational forces is $\sim 10^{13}$. Bodies with equal values of ρR have the same ratio of gravitational to non-gravitational forces. This advantage of larger bodies in following the geodesic more closely is dramatically brought out by the fact that whereas Eötvos-type experiments performed by Dicke[6] and Braginsky[7] using finite bodies weighing ~ 1 g test equivalence principle to an accuracy of 10^{-12} but the very careful experiments by Fairbanks[8] using an electron could only achieve accuracies of $\sim 10\%$.

Though the amount of drag-compensation on a spherical ball described above is more than adequate for most experiments, the typical ρR for a satellite is much poorer. For example, the spacelab has a mass of $\sim 10^7$ g, $\bar{R} \approx 10^3$ cm so that the drag compensation is only good to $10^{-9} - 10^{-10}$ g even in deep space. Further, in near-earth orbits the residual aerodynamic drag is as large as $\sim 10^{-7}$ g. Thus there is a definite need for actively compensating for these non-gravitational forces. To this end, three different strategies have been evolved and indeed all these are in actual use:

1. Place the artificial satellites in orbit around planets or the moon and use the geodesic of the astronomical objects as reference.
2. Use two space probes with different M but same R and using the difference subtract out the drag effect.
3. Drive the spacecraft actively along the geodesics.

The basic principle[9] of the drag-free drive system is illustrated in Fig. 1.

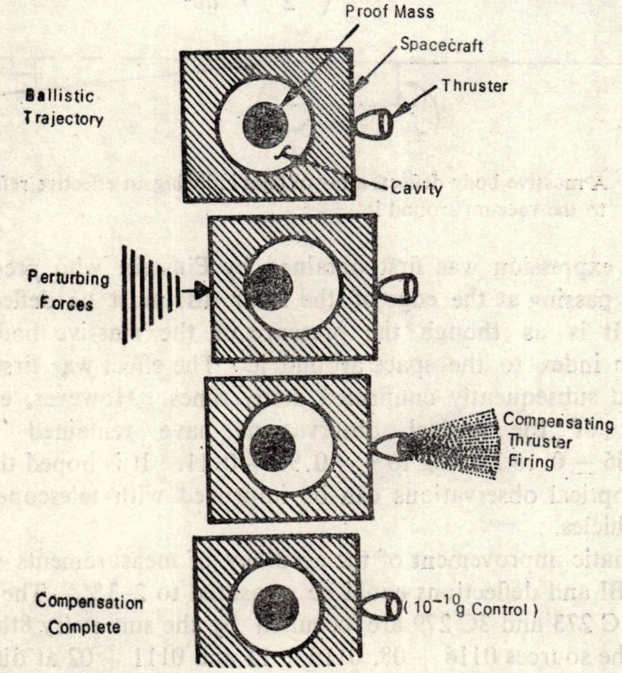

Fig. 1 The basic design concept for the compensation of drag forces.

Consider a spherical test mass, well shielded inside a space vehicle whose position is carefully monitored (say, by means of several beams of lasers reflected off this mass). Any drag force acting on the body of the space vehicle would cause a displacement of the test mass with respect to its fiducial position. Servo rockets are fired to compensate for the drag and the spaceship is driven to bring the test mass to its standard position. Such a system named 'Starlite' was launched by France in February 1975. It was made of ^{238}U weighing about 47 kg. Using about 60 laser beams it could be driven to follow the geodesic around the earth to an accuracy of about 20 cm. The 'La Geos' launched by USA in 1976 had $M = 411$ kg and $R = 60$ cm and could follow the geodesic to about 2 cm, corresponding to a drag compensation to better than 10^{-11} g. These deviations from the geodesic are over long time durations so that further subtraction of the estimated drag forces could be done.

With all these features the experiments in relativity are progressively becoming a challenge and adventure to space engineers and scientists.

2. A review of the experiments

2.1 Deflection of Light Rays and Microwaves

Referring to Fig. 2, the deflection of a light ray which passes a spherical object of mass M at an impact parameter 'd' is given by the expression:

$$\delta\phi = \left(\frac{1+\gamma}{2}\right)\frac{4GM}{dc^2}$$

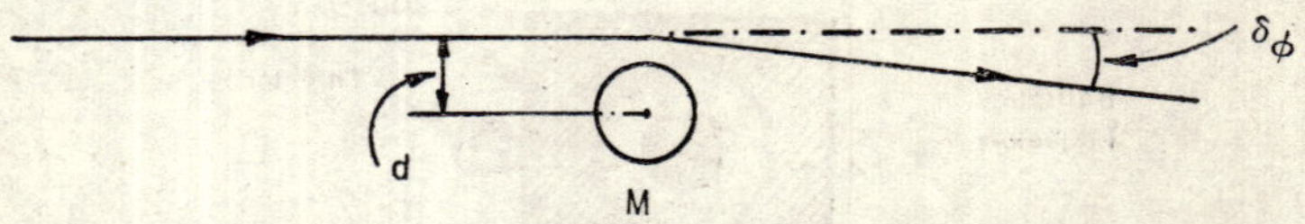

Fig. 2 A massive body deflects a ray of light inducing an effective refractive index to the vaccum around it.

This expression was first obtained by Einstein who predicted that a light ray passing at the edge of the solar disc must be deflected through 1.75″. It is as though the presence of the massive body induces a refractive index to the space around it. The effect was first observed in 1919 and subsequently confirmed many times. However, even now the accuracy of the optical observations have remained rather low: $\delta\phi = 1.66 \pm 0.18$ leading to $\gamma = 0.95 \pm 0.11$. It is hoped that the accuracy of optical observations can be improved with telescopes carried on space vehicles.

Dramatic improvement of the accuracy of measurements was achieved with VLBI and deflections could be measured to 2–3%. The quasi-stellar objects 3C 273 and 3C 279 are occulted by the sun every 8th of October, as also the sources 0116 + 08, 0119 + 11 and 0111 + 02 at different times. Shapiro and Muhleman got $\gamma = 1 \pm 0.12$ in 1967 (the best value for γ now is $\sim 1 + 0.03$). With the latter sources Fomalont and Sramek[10] obtained in 1976 $\gamma = 1.014 \pm 0.018$. In Brans-Dicke theory γ is replaced by $(1+\omega)/(2+\omega)$. Therefore, these observations constrain $\omega > 23$ so that the deviations of the Brans-Dicke theory from GR are less than the experimental errors.

2.2 Time Delay of Radio Signals

The time delay suffered by an electromagnetic signal passing close to a massive object (see Fig. 3) is given by the formula:

$$c\,\Delta t = r + \frac{1+\gamma}{2}\cdot\frac{4GM}{c^2}\cdot\ln\frac{r_0+r_1+r}{r_0+r_1-r}$$

$$= r + \frac{1+\gamma}{2}\cdot\frac{4GM}{c^2}\cdot\ln\left(\operatorname{ctg}\frac{A}{2}\cdot\operatorname{ctg}\frac{B}{2}\right)$$

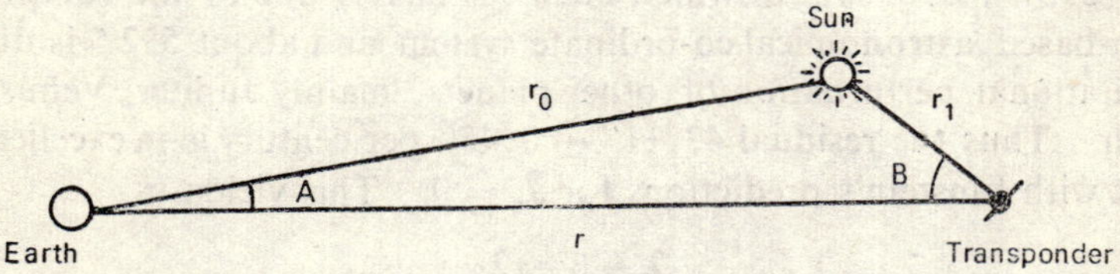

Fig. 3 Radar signals propagate slower near massive bodies; the configuration used in measuring time delays.

Shapiro[11] devised this test in 1964. Just like the refractive index induced by the gravitating mass causes rays to bend, it also slows down the propagation velocity. The effect can be measured by reflecting a radar or laser signal of the surface of a planet as it passes behind the sun. The accuracy off the experiment can be improved by several orders of magnitude by using a transponder whether on a space probe or a lander. For the measurements it is important to know with very high accuracy the distance to the transponder as it passes behind the sun. Therefore, in such experiments very high stability frequency standards are used. Typically the time delay corresponds to a distance of $\sim$60 km which is comparable with the delay introduced by the solar corona when one uses S–band. The best result obtained so far uses a double link using X–band (9500 MHz) making the determination of the distance accurate to $\sim$1 m which in principle should yield a result good to $\sim 10^{-5}$. However, these results obtained using Mariner 6, 7 are still plagued with systematic celestial-mechanical errors and departure of the probes from a geodesic. The results[12] obtained using this technique with transponders aboard Mariner-6 and Mariner-7 is

$$\frac{1+\gamma}{2} = 1.00 \pm 0.03$$

The Viking missions were specially planned to drop landers on to planets thereby reducing geodesic errors (non-gravitational forces) and these are capable of measuring the time delays to very high accuracy. Though these satellites were launched in December 1976, we have not seen any publications on this subject so far.

2.3 Precession of Planetary Perihelia

Conservative metric theories of gravity predict advances of perihelia of bound planetary orbits. This advance per planetary orbit is given by the expression

$$\delta\theta = \frac{2-\beta+2\gamma}{3}\cdot\frac{6\pi G M}{ac^2(1-e^2)} + \text{terms due to solar oblateness } J_2$$

Here a is the semi-major axis and e the eccentricity of the orbit. The general relativistic value for $\delta\theta$ for the planet Mercury is easily estimated (using $a = 5.791 \times 10^{11}$ cm; $e = 0.2056$ and an orbit using frequency 415 per century) to be 43.03″ per century. The actual observations yield a $\delta\theta = 5557 + 0.2''$ out of which 5025″ is merely due to the rotation of the earth-based astronomical co-ordinate system and about 532″ is due to the gravitational perturbation of other planets, mainly Jupiter, Venus and the Earth. Thus the residual $43.11'' \pm 0.45''$ per century is in excellent agreement with Einstein's predictions for $J_2 = 0$. This yields

$$\lambda_p = \frac{2-\beta+2\gamma}{3} = 1.00 \pm 0.01$$

But as pointed out by Dicke and Goldenberg in 1967 this excellent fit could be a mere coincidence if J_2 is not equal to zero. If the sun were oblate with a $J_2 \approx \Delta r/r \neq 0$, then we would expect an additional advance of $1.26''\ J_2 \times 10^5$ per century. If J_2 were $\approx 5 \times 10^{-5}$ and $\omega \approx 6.3$ then the observations can again be made to fit. However, careful observations of the sun yield $J_2 = (0.1 \pm 0.45) \times 10^{-5}$, too small to induce any observable effects, Also, when one combines these results with laser lunar ranging results we get[13]

$$\beta = 1.003 \pm 0.009$$

$$J_2 = (-0.03 \pm 0.41) \times 10^{-5}$$

In fact independent confirmation of the general relativistic value for the advance of perihelia comes from the recent observations on the binary pulsar discovered by Hulse and Taylor. We shall not discuss this fascinating astronomical object too extensively. On the other hand, we will discuss experiments that attempt to measure the solar quadrupole moment $J_{2\odot}$ and solar angular momentum $J_\odot$ directly.

2.4 Precession of a Gyroscope Axis

In 1960, Schiff pointed out that a gyroscope in motion about a massive body such as earth might be expected to undergo relativistic precession in the framework of fixed stars. There are two main effects: the geodesic effect

$$\mathbf{\Omega}_G = \frac{3GM}{2c^2r^3}(\mathbf{r} \times \mathbf{V})$$

and transverse to this the motional effect due to the rotation of the earth

$$\mathbf{\Omega}_M = \frac{6I}{c^2 r^3}\left[\frac{3\mathbf{r}}{r^2}(\boldsymbol{\omega}\cdot\mathbf{r}) - \boldsymbol{\omega}\right]$$

Here m, r and v are respectively the mass, co-ordinate and velocity of the gyroscope, M, I and ω are the mass, moment of inertia and angular velocity of the earth (Fig. 4). In a satellite moving in a 500-mile orbit $\Omega_G = 6.9''$/yr and $\Omega_M = 0.05''$/yr. There are other much smaller effects in the

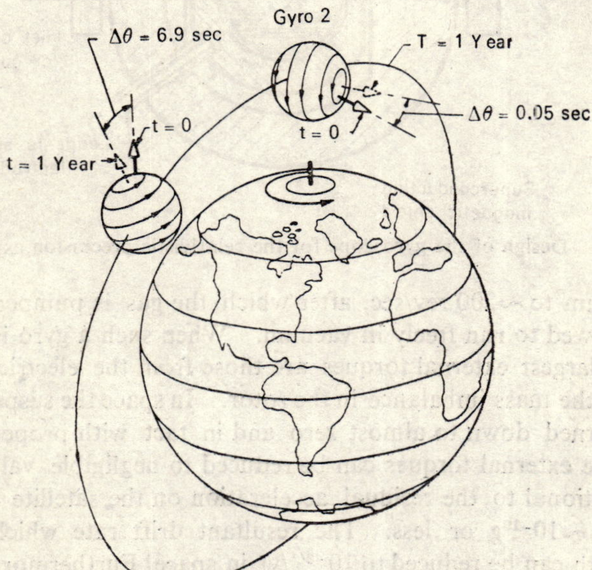

Fig. 4 Relativistic motions of gyroscopes in a $a \sim 500$ mile orbit about the earth.

range $0.01''$ to $0.0001''$ which are due to the quadrupole moment of the earth, rotation of the sun and the orbital motion of the earth about the sun, etc. In the scalar-tensor theory of Brans Dicke Ω_G is multiplied by $(4 + 3\,\omega)/(6 + 3\,\omega)$ and Ω_M by $(3 + 2\,\omega)/(4 + 2\,\omega)$. Thus an experiment to $\sim 1\,\%$ accuracy can check values of ω up to ~ 50. Fairbanks and Everitt and others[14] took upon as a design goal to measure even Ω_M to $\sim 1\,\%$ accuracy, i.e., $0.001''$/yr. This would mean improving the best available gyroscope by 7 orders of magnitude and star-tracking telescope with long term stability about 1000 times better than was available before the space age. Their design which is largely tested in parts makes use of the major developments during the last two decades; (i) space research (ii) large scale low-temperature physics techniques and (iii) certain techniques in electronics instrumentation and control theory.

The gyroscope consists of a ball of quartz 4 cm in diameter coated with a thin film of superconductor, electrically supported by three mutually perpendicular set of condenser plates (Fig. 5). The ball is spun up by a

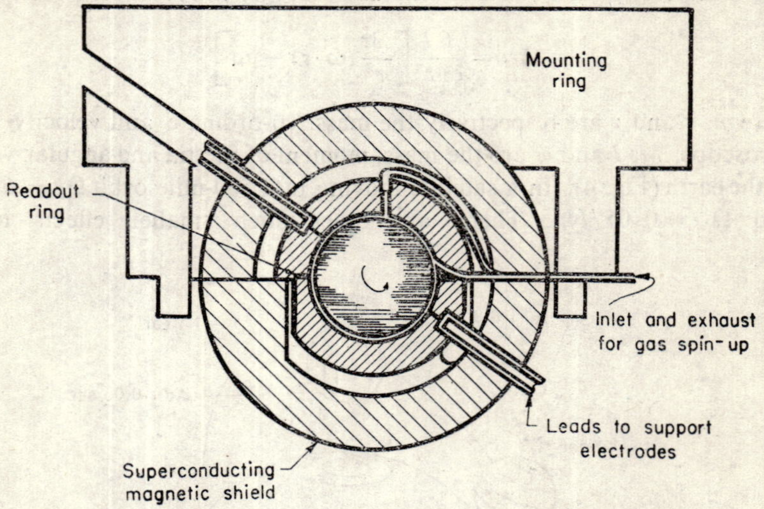

Fig. 5 Design of the gyroscope for the relativistic precession experiment.

jet of helium to ~200 rev/sec, after which the gas is pumped out and the ball is allowed to run freely in vacuum. When such a gyro is operated on earth the largest external torques are those from the electrical suspension and from the mass imbalance in the rotor. In space the suspension voltage may be turned down to almost zero and in fact with proper attention to detail these external torques can be reduced to negligible values since they are proportional to the residual acceleration on the satellite which can be reduced to $\sim 10^{-10}$g or less. The resultant drift rate which is $\sim 10^{5}{}''$/yr on the earth can be reduced to $10^{-3}{}''$/yr in space! Furthermore, going on to a orbiting space platform eliminates errors in telescope reference due to seeing through the atmosphere, and due to creep and sag of the telescope structure in the earths gravity.

However, in ordinary electrically suspended gyroscopes, the interaction of the eddy currents in the spinning rotor with the magnetic fields which have escaped the conventional shielding techniques generate residual drift rates which are several orders of magnitude larger than the desired value. The rotor for this gyroscope being made of non-magnetic quartz coated with superconductor is not susceptible to any magnetic torques provided the superconducting shield is cooled initially in a field below 10^{-6} gauss. The torque due to the interaction of the quadrupole moment of the rotor with the gradient of the earth's field is reduced to negligible levels by keeping sphericity good to $\sim 10^{-6}$ cm and homogeneity of density to $\sim 10^{-6}$.

If the ball is so spherical and homogenous how does one read out the spin axis? The London moment ($H_L \approx 10^{-7}\, \omega_g$ gauss) provides a method

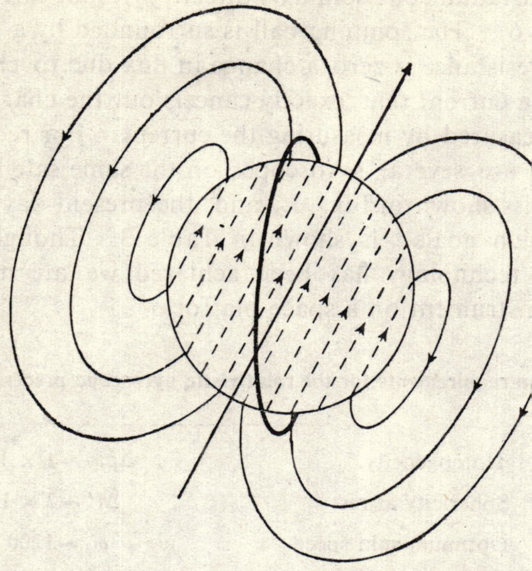

Fig. 6 **Design of the readout of the gyroscope axis using the London moment $H_L = 10^{-7}$ gauss.**

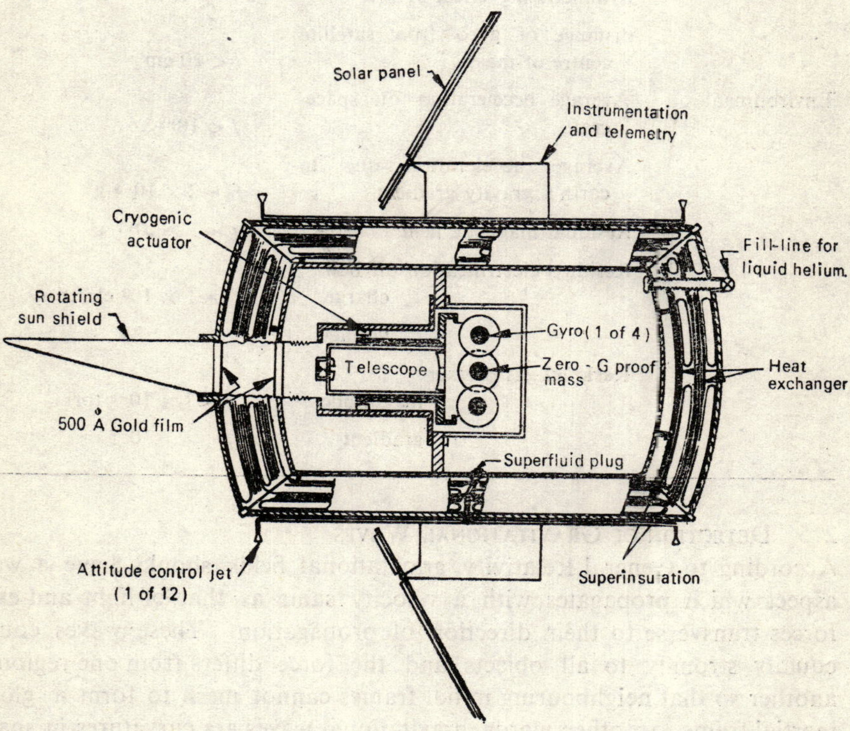

Fig. 7 **A possible satellite configuration for Fairbank's relativistic gyroscope precession experiment.**

of locating the instantaneous spin axis directly without any reference marks on the ball (Fig. 6). The spinning ball is surrounded by a superconducting loop; since the resistance is zero a change in flux due to change in orientation produces a current that exactly cancels out the charge in flux. The orientation is measured by measuring the current. For redundancy it may be advisable to use several gyroscopes on the same satellite. A possible configuration[14] is shown in Fig. 7., and the present-day capabilities of meeting the design goals[15] is shown in Table 3. Though this high level of instrumental technology has been achieved we are not aware of any such operating instrument on a space platform.

Table 3 Design requirements for the relativistic gyroscope precession experiment

Gyro rotor	Homogeneity	$\Delta\rho/\rho \sim 1 \times 10^{-6}$
	Sphericity at rest	$\Delta r' \sim 2 \times 10^{-6}$ cm
	Optimum spin speed	$\omega_s \sim 1200$ rad/sec
Gyro housing & suspension system	Sphericity of cavity	$\Delta r'' \sim 50 \times 10^{-6}$ cm
	Centring accuracy	$t \sim 50 \times 10^{-6}$ cm
	Preload acceleration	$h \sim 3 \times 10^{-6}$ g
	Symmetrizing effect of roll	$\theta_0 \sim 1 \times 10^{-3}$
	distance of gyro from satellite centre of mass	$l < 20$ cm
Environment	Average acceleration on spacecraft	$f < 10^{-9}$ g
	Average acceleration due to earth's gravity gradient	$f_0 \sim 2 \times 10^{-8}$ g
	Residual magnetic field	$H_L \sim 2 \times 10^{-7}$ g
	Residual electrification on ball charge	$Q \sim 1 \times 10^9$ electrons
	voltage	2V
	Residual gas pressure magnitude	$p \sim 1 \times 10^{-9}$ torr
	gradient	$\Delta p/p$ 0.5

2.5 Detection of Gravitational Waves

According to General Relativity, gravitational fields should have a wave aspect which propagates with a velocity same as that of light and exert forces transverse to their direction of propagation. These waves couple equally strongly to all objects and the force differs from one region to another so that neighbouring initial frames cannot mesh to form a global inertial frame. In other words, gravitational waves are curvatures in space-time propagating with the speed of light.

Testing theories of gravity is not the most important pay-off of gravitational wave detection but that it provides a new window into the astrophysical universe. Black holes, close binary systems, galactic nuclei, quasars are some of the objects which might emit gravitational waves copiously. The stresses produced by a gravitational wave can be written in terms of two dimensionless functions representing two possible polarization of the wave. The relative acceleration of two mass points separated by a distance x is given by

$$a = (\ddot{h}_+ + \ddot{h}_\times)\, x$$

and the power carried by the waves is

$$P = \frac{c^5}{16\,\pi G} \langle \dot{h}_+^2/c^2 + \dot{h}_\times^2/c^2 \rangle$$

$$= (7 \times 10^{57}\ \text{erg/sec}) \times \langle \dot{h}_+^2/c^2 + \dot{h}_\times^2/c^2 \rangle$$

A binary star produces an h due to gravitational radiation

$$h = \frac{\Delta l}{l} = \left(\frac{G^2 M_1 M_2}{c^4}\right)\frac{1}{Dr} \approx \frac{10^{-10}\ cm^2}{Dr}$$

where r is the radius of the orbit and D the distance of the binary from the observer.

The search for gravitational waves was pioneered by Joseph Weber; despite early hints of success it now appears that such waves have not been detected with the antennae made of $\sim$1 tonne aluminium bars. Below $\sim$1 Hz where the astrophysical sources like quasars, galactic nuclei and binary stars might radiate dominantly 'Doppler-tracking' of spacecraft is the only promising technique available within the next decade for the observation of these waves. How does the Doppler-tracking gravitational wave detector work? For a good operation[16], we should have

1. A high stability clock on board.
2. A dual frequency tracking so that we have four independent tracking links for redundancy and accuracy (one way up, one way down, round trip up down and round trip down up).
3. Drag-free orbiting capability for the spacecraft.
4. Two spacecrafts for coincidence experiments.

Using precision time keeping a Doppler observation of a spacecraft can be thought of as a series of measurements of the integral phase $\varphi(t)$ or the instantaneous frequency $\nu = \phi/2\pi\ (\approx \Delta\varphi/\tau)$ or a frequency shift $y = (\nu - \nu_0/\nu_0)$ $y(t)$ contains information about clocks, motions of earth and transponder the propagation medium and passing of gravitational wave bursts. If the ground timing speeds up momentarily y will dip momentarily and after the round trip travel time (RTTT) $\approx 2l/c$, it will increase momentarily (Fig. 8). This and other disturbances in propagation effects will have a

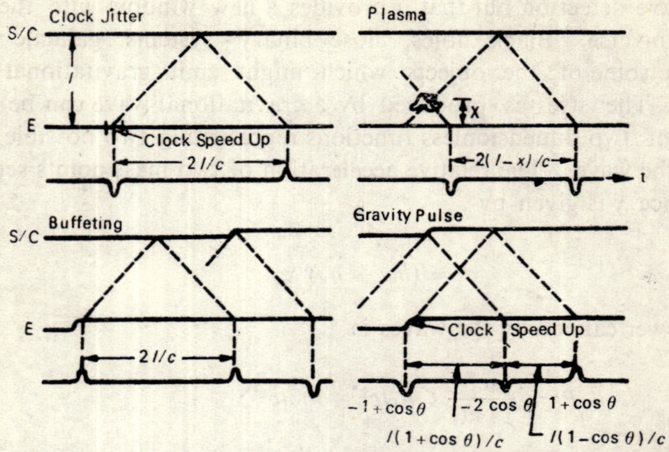

Fig. 8 Doppler signals corresponding to different types of disturbances in the communications link. Notice the unique triple peak due to a gravitational wave-front arriving at angle with respect to the earth-satellite line.

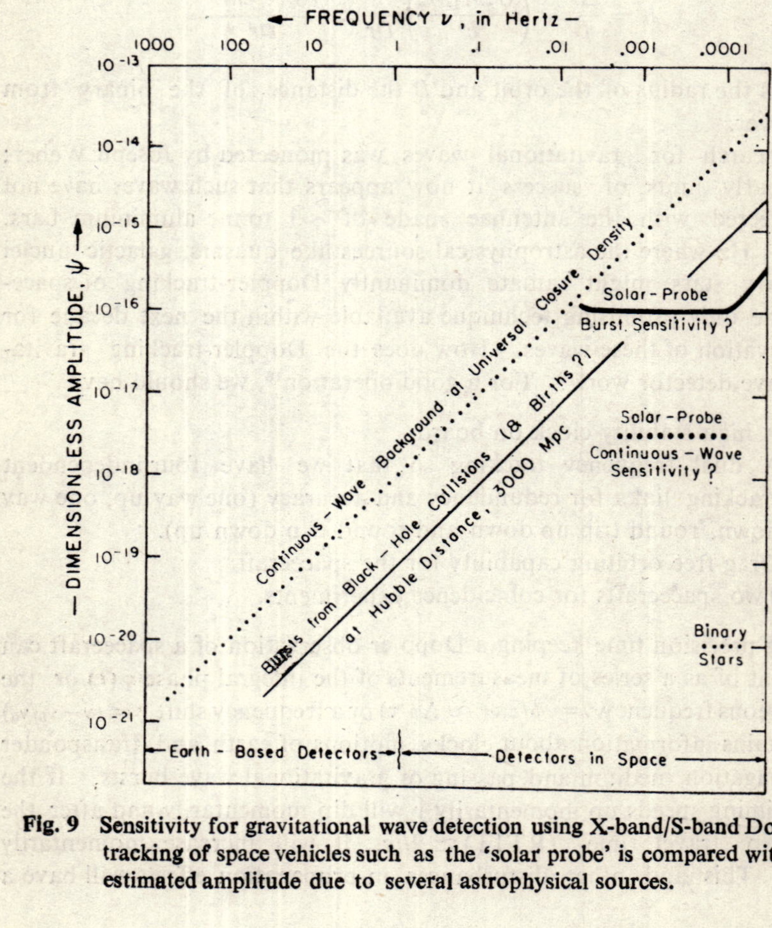

Fig. 9 Sensitivity for gravitational wave detection using X-band/S-band Doppler tracking of space vehicles such as the 'solar probe' is compared with the estimated amplitude due to several astrophysical sources.

double pulse structure. But since gravitational wave burst will produce $(\Delta t/t)$ as well as $(\Delta l/l)$, there will be a three pulse characteristic response because of time delay between the arrival of the wavefront at the ground station and the space vehicle. From tracking of Viking spacecraft, it appears that one can measure stresses of 10^{-16} in pulses and 10^{-18} for continuous gravitational waves. This capability will allow one to see up to the edge of the universe the gravitational waves emitted when massive black holes are formed ($\sim 10^8 M_\odot$) and up to $\sim$1 light year a binary pulsar (like the one discovered by Hulse and Taylor). In Fig. 9 due to Thorne[17] these source strengths and Doppler tracking gravitational wave detector sensitivities are summarised.

Gravitational waves are also detected by using a set up like the Michaelson's interferometer and strains of

$$\frac{\Delta l}{l} = \frac{\lambda}{l\sqrt{n_{ph}}} = \frac{10^{-4}\text{ cm}}{10^5\text{ cm}\sqrt{10^{16}}} \approx 10^{-17}$$

can be measured using a laser power of 0.1 watt, wavelength $\lambda = 1\,\mu$ and a path length of $\sim 10^5$ cm. Only in space stations can one hope to build such large structures free of seismic and other noises. Thus the prospects for gravitational wave astronomy appear quite encouraging in the next decade.

2.6 Tests of the Equivalence Principle

That all bodies irrespective of their composition suffer equal gravitational acceleration or in other words move along geodesics was shown to be true by Eötvos to one part in 10^8 for a variety of materials ranging from Gold to Brazilian snakewood using a torsion balance similar to that used by Cavendish for measuring the gravitational constant. Using a very similar apparatus, Dicke[6], Bragnisky and Panov[7] have checked the equivalence principle to an accuracy of $\sim 10^{-12}$ g. The possibility of testing this principle at a level of $\sim 10^{-17}$ g using satellites has been discussed by many workers and I will review briefly the suggestion of Everitt and Worden Jr.[18] How does going onto an earth's satellite help the efficiency? The vibration environment can be made much quieter and the driving acceleration instead of being derived from the sun, could be the gravitational attraction of the earth which at $\sim$300 nautical miles is $\sim$950 cm/sec^2, roughly 10^3 times the forces Eötvos had to contend with. The apparatus consists of two coaxial cylinders of different materials—gold and aluminium, say, and these orbit the earth in a satellite held oriented in an inertial space (Figs. 10, 11). The masses are supported in superconducting bearings which constrain their lateral motions but leave them free to move along their common axis. Each object is subject to the gravitational pull of the Earth and is balanced by the centrifugal acceleration. Any difference in the ratio of inertial to gravitational mass will result in periodic relative oscillations of an amplitude $\sim$0.7 Å for a violation $\approx 10^{-17}$ g which can be measured by

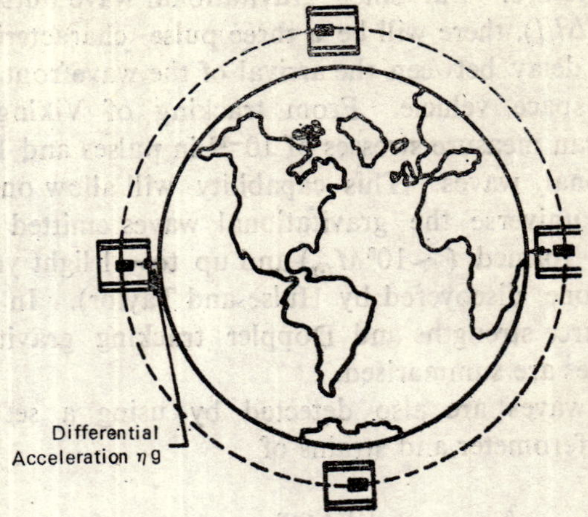

Fig. 10 Concept of an equivalence principle experiment using an earth satellite.

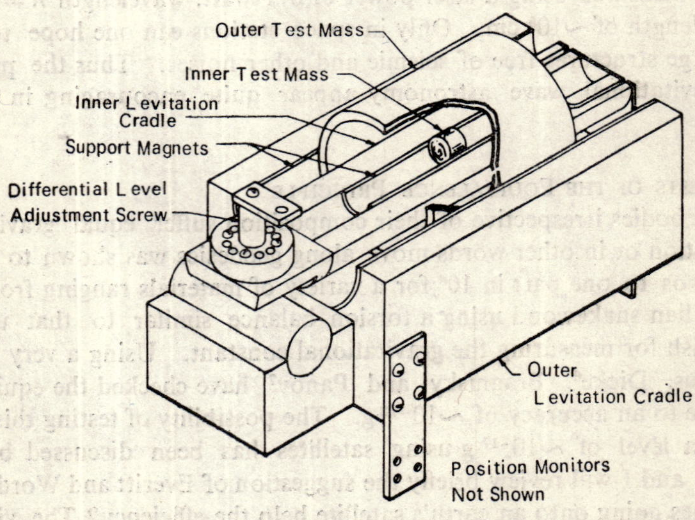

Fig. 11 The accelerometer design for the orbiting equivalenee principle experiment.

means of easily accessible cryogenic techniques. The choice of an experiment based on linear motions as against torsional balance comes from considering effects of gravity gradients which would simulate an apparent violation of the equivalence principle. If the quadrupole moments of the two cylindrical masses are kept same with reasonable accuracy this effect can be reduced to near negligible in this experiment.

2.7 Gravitational Redshift

Experiments to measure the gravitational red-blue shift to optical lines make it possible to establish that universal trajectories of test bodies coincide with geodesics of the metric of the local Lorentz frame of reference and thus make it possible to determine experimentally a class of inertial motions in the Einstein sense. Experiments by Pound, Rebka and Snider using the Mössbauer effect have tested $\Delta\nu/\nu = gh/c^2$ of spectral lines to $\sim 1\%$. With atomic clocks on satellites synchronised with analogous ones on the earth one can measure the expected $(\Delta\nu/\nu) \sim 10^{-9}$ to better than 1 part in 10^4. Indeed such an experiment is in progress conducted by the Smithsonian group using hydrogen maser standards aboard space vehicles but the results are not yet out.

2.8 Nordtvet Effect

A separate class of experiments aims at finding deviations from General Relativity by looking for effects that must be absent in GR but are predicted by other theories of gravitation. In 1968, Nordtvet[19] published a fundamental and general investigation of violations of equivalence principle in metric theories of gravitation. One of the simple questions one wishes to answer is whether an object with a substantial amount of self-gravitation will suffer the same acceleration as other bodies? Moon, for example, has a fractional self energy of

$$\frac{GM_c}{c^2R_c} \approx 3 \times 10^{-11}$$

The breakdown of the equivalence principle would become apparent in the motion of the moon with respect to earth as these objects are perturbed by the sun's gravitational field. In August 1969, a series of very high accuracy measurements was started detailing the orbit of the moon using laser beams reflected off an array of right-angled prisms left there by the Apollo astronauts[20,21]. It was shown with ranging measurements that

$$\frac{M_i}{M_g} = 1 \pm 1.5 \times 10^{-12}$$

Time delays of laser pulses in the same experiment gave even a better confirmation of the equivalance principle

$$\frac{M_i}{M_g} = 1 \pm 7 \times 10^{-11}$$

Even when one uses the more conservative estimates of the errors this measurement restricts the PPN parameter combination

$$(4\beta - \gamma - 3 - \alpha_1 + \tfrac{2}{3}\alpha_2 - \tfrac{2}{3}\zeta_1 - \tfrac{1}{3}\zeta_2 - \tfrac{13}{3}\zeta_\omega) = 0.00 \pm 0.03$$

If one further imposes conservation of momentum in the post-Newtonian limit[22] and that $\alpha_2 = \zeta_\omega = 0$ (see sec. 2.9), one gets

$$(4\beta - \gamma - 3 - \alpha_1) = 0.00 \pm 0.03$$

in terms of Brans-Dicke theory this corresponds to $\omega > 30$.

2.9 Miscellaneous Results

(*a*) *Earth-tide effects*: The PPN parameters α_2 and ζ_ω impose distinctive frequency components of earth tides[23] and the analysis by Warburton and Goodkind[21] of a 18-month record using a superconducting gravity-meter yields the result.[24]

$$\alpha_2 < 2 \times 10^{-3} \quad \text{and} \quad \zeta_\omega < 10^{-3}$$

(*b*) *Preferred-frame effects*: These would be noticeable as yearly variation of earth's rotation[23] and the observations yield

$$|\tfrac{2}{3}\alpha_2 + \alpha_3 - \alpha_1| < 0.2$$

Also the lack of secular terms in the planetary orbits imposes

$$|\alpha_3| < 10^{-5}$$

(*c*) *Isotropy of the gravitational constant*: Using gravimetric data the anisotropies are restricted to

$$\left|\frac{\Delta G}{G}\right| < 10^{-9}$$

(*d*) *Anisotropy of mass*: Because of the anisotropic distribution of mass around us, the vector and tensor forms of Machian theories predict asymmetries $\langle \Delta m/m \rangle$ in the range 10^{-5}-10^{-10}. This effect was put to test experimentally by Hughes, Robinson and Beltran-Lopez[25] who noted that the nuclear levels of the spin 3/2 ground state of Li^7 would split in a magnetic field into four levels and the energy gap between the levels would not be equal if mass were itself anisotropic. Their observations yield

$$\left\langle \frac{\Delta m}{m} \right\rangle < 10^{-20}$$

(*e*) *Time variation of the gravitational constant*: Van Flandern[26] considered the possibility that ephemeris time and the time as measured by atomic clocks did not scale in a time-invariant way. This would lead to a time-variation of the gravitational constant when measured in atomic units. His analysis of lunar eclipses yield

$$\frac{\dot{G}}{G} < (-8 \pm 5) \times 10^{-11}/\text{yr}$$

The errors on this determination is still somewhat too large for an unambiguous comparison with various theories.

3. Binary pulsar

No modern review of experiments on gravitation is complete without at least a brief mention of pulsar in a binary system discovered by Hulse and Taylor[27] in 1974, during their systematic search of the galactic plane for pulsars. This pulsar is called PSR 1913+16 specifying its position in the sky in right ascension and declination. The various parameters[28] of interest here are shown in Table 4. A pulsar for a relativist is an excellent clock, a gyroscope and an object condensed and massive enough to follow to great accuracy the geodesics when in orbit around its companion. Several observational facts indicate that the comparison of PSR 1913+16 is also a condensed object so that the two bodies behave very much like the canonical test bodies in relativity theory. Taylor presented at the Texas symposium, in December 1978, the latest analysis of the radio data on the arrival times of the pulses accumulated over a four year period. It appears that the predictions based on General Relativity including such effects as advance of periastron and damping due to radiatton of gravitational radiation can be observed. It would not be surprising if observations of this pulsar in a binary system in the coming years should yield the most stringent tests of the gravitational theories.

Table 4 Parameters of PSR 1913–16

Right ascention	$\alpha\,(1950.0) = 19\text{ h }13\text{ m }12.484\text{ s} \pm 0.008\text{ s}$
Declination	$\delta\,(1950.0) = +16°\,01'\,08.4'' \pm 0.2''$
Period	$P = 0.05902999527\,(2 \pm 5)$ s
Derivative of P	$\dot{P} = (8.8 \pm 3) \times 10^{-18}$ SS^{-1}
Orbital semi-major axis	$a_i \sin i \equiv \chi = (7.0043 \pm 0.0004) \times 10^{10}$ cm
Orbital eccentricity	$e = 0.61717 \pm 0.00005$
Binary period	$P_b = 27906.980 \pm 0.002$
Rate of periastron advance	$\dot{\omega} = 4.22 \pm 0.04$ deg yr^{-1}
Derivative of P_b	$\dot{P}_b = (2 \pm 6) \times 10^{-10}$
Mass function	$f_1 = \dfrac{(M_2 \sin i)^3}{(M_1 + M_2)^2} = \dfrac{4\pi^2}{G}\,\dfrac{(a_i \sin i)^3}{P_b^2} = 0.13126\,M_\odot$

4. The solar probe

I wish to conclude with a brief description of an idea which is a 'coup de maitre' of space experimentation which would test relativity theories to fantastic precision never imaginable before and at the same time will lead us to the understanding of the sun to a level adequate to yield important clues solving basic astrophysical problems like the solar neutrino emission, its structure and the general problems of the solar system like the zodiacal light, interplanetary dust, solar flares, etc.

This concept of a dedicated space vehicle for experiments on gravitation and for the study of the sun and the solar system was discussed at great depth in a workshop held at the California Institute of Technology in May, 1978, under the sponsorship of NASA[2]. We shall review here briefly some of the design features of the 'Solar Probe' and its scientific capabilities. The planned orbit is shown in Fig. 12; the perihelion is a mere 4 solar radii and aphelion is about one or two astronomical units.

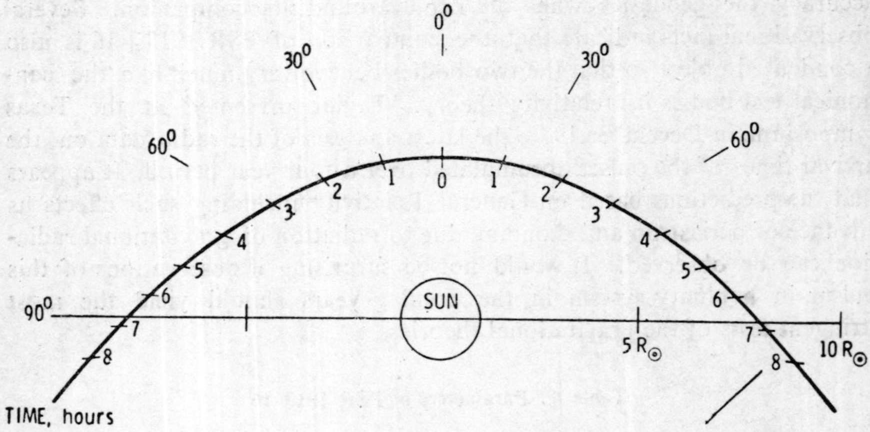

Fig. 12 Planned orbital parameters of the 'solar probe'. Notice the extremely short 4 perihelion of the orbit.

When the spacecraft is passing close to the sun, it is subjected to about 2,500 times as much solar radiation as a satellite in orbit about the earth. Such intense radiation would need a high quality shielding if the spacecraft has to survive and the various instruments should perform properly. Ingenious design consisting of several successive umbrellas (Fig. 13) made of pyrolytic graphite keeps the internal temperatures below 60° despite surface temperatures of ~ 3,000°K during the several hours of closest

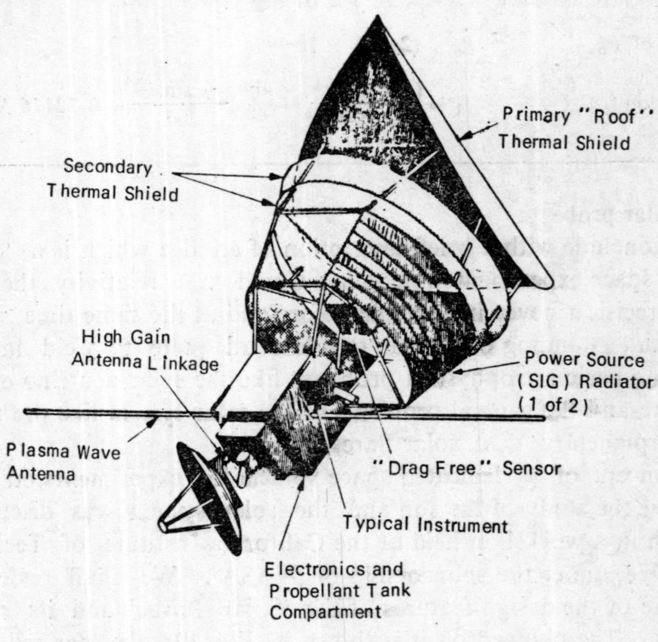

Fig. 13 Design concept for the 'solar probe'.

approach. We list below the capabilities of such a probe for experiments in gravitation:

1. Because of the very high eccentricity of the orbit ($\sim$0.997) there would be a very marked advance in the perihelion and even in a single fly-by the relativistic predictions could be checked to one part in a thousand, if one should use a dual-band tracking system.
2. Gravitational redshifts could be measured to better than one part in a hundred thousand using onboard maser clocks.
3. Using gravity gradiometers even of conventional design (but modified for space applications) the quadrupole term in the Newtonnian gravitational potential of the sun can be measured to an accuracy of $\sim 10^{-8}$. The same experiment is also capable of measuring any time varying quadrupole terms induced by oscillations of the star.
4. The solar angular momentum can be measured using the Lense-Thirring precession of the gyroscope axis.
5. As described earlier with dual band radar ranging capabilities the probe can be used as an antenna to detect gravitational waves with very high sensitivity.
6. The PPN parameters γ and β would be measured to an accuracy of $\sim 10^{-4}$.

5. Conclusions

In Table 5, we list the various results cited in this paper. It is remarkable that inspite of several decades of ingenious experimentation, not a single contradiction to the predictions of Einstein's general theory of relativity has been observed nor have any of the predictions of other theories of gravity been substantiated.

Table 5

Test	Experimental result
Deflection of light rays	$\gamma = 1.014 \pm 0.018$
	($\omega > 23$ in Brans-Dicke theory)
Time delay of radar signals	$\frac{1+\gamma}{2} = 1.00 \pm 0.003$
Precession of the perihelion of mercury	$\lambda_p = \frac{2-\beta+2\gamma}{3} = 1.00 \pm 0.01$
	$\beta = 1.003 \pm 0.009$
	$J_{2\odot} = (-0.03 \pm 0.41) \times 10^{-5}$
Geodetic precession of gyroscopes and the Lense-Thirring effect	[1% capability]
Gravitational waves	[$\sim 10^{-16} - 10^{-17}$ capability]
	Radiation effects probably observed in the binary pulsar.

Test	Experimental result
Equivalence principle	10^{-12}; [10^{-17} capability]
Gravitational redshift	1%
	[0.01% capability]
Nordtvet effect	$(4\beta - \gamma - 3 - \alpha_1) = 0.00 \pm 0.03$
	($\omega > 30$ in Brans-Dicke theory)
Earth-tide effects	$\alpha_2 < 2 \times 10^{-3}$
	$\zeta_\omega < 10^{-3}$
Preferred frame effects	$\lvert \tfrac{2}{3}\alpha_2 + \alpha_3 - \alpha_1 \rvert < 0.2$
	$\lvert \alpha_3 \rvert < 10^{-5}$
Isotropy of the gravitational constant	$\left\lvert \frac{\Delta G}{G} \right\rvert < 10^{-9}$
Anisotropy of mass	$\left\langle \frac{\Delta m}{m} \right\rangle \leqslant 10^{-20}$
Time-variation of the gravitational constant	$\left\lvert \frac{\dot{G}}{G} \right\rvert < (-8 \pm 5) . 10^{-11}$/yr
Solar quadrupole moment	[$\sim 10^{-8}$ capability]
Solar angular momentum	[$\sim 1\%$ capability]

References

1. Konopleva N.P., (1977) *Usp. Fiz. Nauk.* **123,** 537.
2. Neugebauer, M.N. and Davies, R.W. (1978) Ed. *Close-up of the Sun*, JPL-Publication 78–70.
3. Proceedings of the International School of General Relativistic Effects in Physics and Astrophysics: Experiments and Theory, 1977 (Max-Planck Institut preprint MPI-PAE/Astro 138).
4. Weinberg, S. (1972), *Gravitation and Cosmology*, Wiley.
5. Braginsky, V.B., Caves, C.M and Thorne, K.S. (1976), Orange Aid preprint 478.
6. Roll, P.G., Krotkov, R. and Dicke, R.H., (1964), *Ann. Phys.* (USA), **26,** 442.
7. Braginsky, V.B. and Panov, V.I., (1971), *Zh. Eksp. Teor. Fiz.* **61,** 873.
8. Fairbank, W.M., Witteborn, F.C., Madey, J.M.J., and Lockhart, J.M., (1974), in *Experimental Gravitation* Bertotti, B. (Ed.), Academic Press, p. 310.
9. Randolph, J.E. (1978), in *Close-up of the Sun*, JPL pub. 78–70, M.N. Neugebauer and R.W. Davies (Eds.) p. 521.
10. Fomalont, E.B., and Sramek, R.A., (1976), *Phys. Rev. Lett.*, **36,** 1475.
11. Shapiro, I.I., (1964), *Phys. Rev. Lett.* **13,** 789.
12. Shapiro, I.I., Ash, M.E., Ingalls, R.P., Smith, W.B., Campbell, D.B., Dyce, R.B., Jurgens, R.F., and Pettengill, G.H., (1971), *Phys. Rev. Lett.* **26,** 1132.
13. Anderson, J.D., (1977), Proc. Inter. School of General Relativistic Effects in Physics and Astrophysics: Experiments and Theory, (Max-Planck Institut preprint MPI-PAE/Astro 138, München):

14. Everitt, C.W., Fairbank, W.M., and Hamilton, W.O., (1972), in *Magic without Magic*, Freeman, San Francisco.
15. Everitt, C.W.F., (1977), Proc. Inter. School of General Relativistic Effects in Physics and Astrophysics: Experiments and Theory, (Max-Planck Institut preprint MPI-PAE/Astro 138, München), p. 104.
16. Vessot, R.F.C. and Levine, M.W., (1978) in *Close-up of the Sun*, JPL Pub. 78–70, M.N. Neugebauer and R.W. Davies, (Ed.) p, 457.
17. Thorne, K.S., Ibid.
18. Everitt, C.W.F., (1977), Proc. Inter. School of General Relativistic Effects in Physics and Astrophysics: Experiments and Theory, (Max-Planck Institut preprint MPI-PAE/Astro 138, München), p. 104.
19. Nordvett, K., Jr., (1968), *Phys. Rev.* **169,** 1014.
20. Shapiro, I.I., Councelman, C.C. and King, R.W., (1976), *Phys. Rev. Lett.* **36,** 555.
21. Warburton, R.J. and Goodkind, J.M., (1976), *Ap. J.* **208.** 881.
22. Lee, D.L., Lightman, A.P. and Ni, W.T., (1974), *Phys. Rev.* **D10,** 1685.
23. Will, C.M., (1974), in *Experimental Gravitation,* B. Bertotti (Ed.), Academic Press.
24. Anderson, J.D., (1977), Proc. Inter. School of General Relativistic Effects in Physics and Astrophysics; Experiments and Theory, (Max-Planck Institut preprint MPI-PAE/Astro 138, München), p. 156.
25. Hughes, V.W., Robinsson, H.G. and Beltran-Lopez, V., (1960), *Phys. Rev. Lett.* 4,342.
26. Van Flandern, T.C., (1975), *MNRAS* **170,** 333.
27. Hulse, R.A. and Taylor, J.H., (1974), *Ap. J. Lett.* **191,** L-59.
28. Taylor, J.H., Hulse, R.A., Fowler, L.A., Gullahorn, G.E. and Rankin, J.M. (1976), *Ap. J. Lett.* **206,** 153.

Discussion

C Sivaram: (1) Is not the solar radiation pressure larger than the solar wind pressure? You said that the main non-gravitational force was the solar wind. (2) The system of K-mesons places a limit of 1 part in 10^{14} to the accuracy of the equivalence principle.

R Cowsik: (1) You are right. It was a slip. (2) I am not familiar with this result but probably it is model dependent.

Phil Morrison: (1) What is the expected all-up mass of such a solar probe? (2) Does the recoil momentum of a CR burst show up in the various motion detectors (at levels near 10^{-17})?

R Cowsik: (1) I should guess a mass $= 10^4$ kg. (2) I guess not.

Nitin Nitsure: The method of ensuring the free fall of a spaceship by using a small freely falling object inside it and observing it with laser rays has the following drawback:–when the spaceship undergoes some non-gravitational force, this will cause an asymmetric doppler shift in the light falling on the object which would disturb it. To avoid this, won't the amplitude of the laser rays have to be kept very small?

R Cowsik: Yes. The limiting compensations to the drag force are $10^{-16} - 10^{-17}$ g.

S D Verma: How is the sensitivity of gravitational wave cryo-genic experiment of Fairbanks on ground compared with the space experiment?

R Cowsik: The expected rate of precession on the geodesic of the earth in the Sun's field is 10^{-3} of the effect on gyroscopes in satellites in geodesics of the earth's field. The experimental sensitivity is just barely sufficient to detect this.

S Ramamurty: In measuring gravitational radiation, e.g., by solar probes, will there not be a background gravitational radiation due to remote objects in the universe?

R Cowsik: Yes, but at this stage we are still trying to see the brightest of the sources.

A K Pandey: I have one comment on the existence of gravitational waves. I feel that for experimental detection of gravitational radiation, the solution to the field must be purely retarded. If we recall the work of Hogarths and subsequently by Hoyle and Narlikar in lines of absorber theory in case of electromagnetism we require the existence of perfect absorber along the future light cone for gravitation as well. The same is required if gravitational waves have to be emitted from a massive particle. The problem is whether for such a weak interaction universe can act as an absorber along future light cone.

R Cowsik: I do not know.

The Asymptotic Structure of Space-time and the Relativistic Kepler Problem

MARTIN WALKER

1. Introduction and philosophical background

The notion of an 'isolated system', which is used very often in physics, is an idealization which only approximates the truth since all localized concentrations of matter—be it a star with planets revolving round it, a binary star system, or a galaxy—are constantly bathed in various kinds of radiation coming from the distant matter surrounding them. For example, all systems in nature are immersed in the 3°K microwave electromagnetic background radiation, discovered by Penzias and Wilson, and thought to be left over from the creation of the universe. Experience has taught us, however, that in most circumstances localized system evolve to a very high degree of accuracy, as if this incoming radiation were not there, and as if the rest of the matter in the universe were not there. It has, therefore, become customary for the sake of mathematical economy and convenience, to construct idealized, theoretical models which treat localized concentrations of matter as if they were completely isolated from the rest of the universe. This is as true in Einstein's theory of gravitation as it is in Newton's, and has been highly successful in the description of quasi-stationary systems such as the solar system.

There are, however, complications when we wish to study radiating systems. That is to say, systems whose evolution is expected to be strongly influenced by the emission of radiation, be it electromagnetic or gravitational. Throughout the discussion to follow, Maxwell's classical electrodynamics will be considered in parallel with Einstein's (or Newton's) theory of gravitation. Our greater familiarity with electrodynamics is a great help in understanding the greater complexity of gravitation. By 'field' we shall understand the following: for Newton's theory of gravitation, the scalar potential, V, satisfying Poisson's equation; in Maxwell's electrodynamics, the antisymmetric electromagnetic field tensor, F^{ab}; and in Einstein's theory of gravitation, the curvature tensor in vacuum, C^{a}_{bcd}.

Perhaps the main difference between Newton's and Einstein's theories of gravitation is that changes in the gravitational field are felt instantaneously in the former, while in the latter they propagate through space with a finite speed. Mathematically, this difference is reflected in the fact that Newton's field equation is elliptic, while Einstein's is hyperbolic in character. A consequence of this difference is that Einstein's equation admits wave-like solutions, which describe 'gravitational radiation'. This radiation propagates mainly along the light cones of space-time (which, since the field equation is non-linear, are themselves determined by gravitation. By contrast, in electrodynamics, the light cones, along which e m radiation propagates, are fixed a priori, once and for all). Gravitational radiation can carry away energy, momentum, angular momentum, and higher multipole structure from the material sources of the gravitational field, thus damping their evolution.

A second difference between Newtonian gravitation on the one hand and Maxwellian electrodynamics and Einsteinian gravitation on the other is that while the field in the former is described by a single scalar function in space, in the latter theories the fields are described by tensor-fields on space-time; the fields having several different components. This difference has important consequences for the problem of specifying the uniqueness of the solution to the corresponding field equation. Uniqueness is achieved by imposing asymptotic conditions on the field. (We shall assume throughout that the material sources of the field are localized spatially at any finite time, in accordance with the remarks made above. The cosmological problem is deliberately ignored.)

In Newtonian gravitation, the field V satisfies Laplace's equation outside the matter. Far away from the matter, therefore, there are only two possibilities: either V falls off like $1/r$, or it blows up like r. For an isolated system, one takes that solutions which vanish at infinity in space, and since this single requirement uniquely specifies the solution to the problem, that is the end of the story.

The situation in electrodynamics and relativistic gravitation is not so simple. First of all, the fields have several different components, and these can fall off at different rates. Secondly, in space-time, it is not clear where 'infinity' is located. If one is interested in outgoing radiation, for example, one would follow the behaviour of the field along future-directed, outgoing, light rays. If one were interested in incoming radiation, as we shall be for the most part, one would follow the behaviour of the field along light rays receding from the system into the past (since this is where the incoming radiation would have come from). So you see that the phrase, 'the field vanishes asymptotically at infinity', embodies a good deal of ambiguity. Not only can different components of the field fall off at different rates, but as we shall see, the same component can fall off differently depending on whether one recedes from the system into the future or into the past. The issue is further complicated in Einstein's gravitation since the disappearance of the field (the curvature of space-time) reflects the approach of the curved space-time metric to flatness far from the material sources.

The remainder of this article will be devoted to a description of the recent report at the Ninth Texas Symposium on Relativistic Astrophysics in Munich of Joe Taylor and his colleagues on the observed effect of the emission of gravitational radiation on the binary pulsar PSR 1913+16, and to some new results on the problem of incoming radiation and the asymptotic structure of space-times in the past, due to myself and Clifford

Will. Sec. 2 briefly reviews the description of gravitational radiation according to the linearized approximation to Einstein's theory. Sec. 3 will review the exciting new measurements of the binary pulsar. Sec. 4 discusses what must be achieved theoretically in order to provide a consistent description of the binary pulsar within Einstein's theory. Sec. 5 will review what is known about the asymptotic boundary conditions referred to above. Finally, Sec. 6 will present some new results on this question.

2. Description of gravitational radiation according to the linearized approximation

Uptil now, no one has been able to provide a description of the generation of gravitational waves or of the effect of their emission on the evolution of gravitationally interacting systems, which goes beyond the linearized approximation to Einstein's theory. What is known from the exact theory is how to identify outgoing gravitational radiation in the asymptotic field in future, and how radiation relates to the change in total mass-energy of the system. This understanding, which is primarily due to Bondi et al. (1962) and to Sachs (1964), may be (very) briefly described as follows. The effect of gravitational waves on orbiting test particles far from the source is to produce a quadrupole distortion in the spatial configuration of the particles. Thus, an initially circular distribution of particles would become elliptical when a wave passes. This distortion is measured quantitatively in terms of the asymptotic shear of an outgoing null hypersurface. The shear measures two quantities, the ratio of major to minor axes of the ellipse, and the orientation of the major axis. These are conveniently gathered into a single complex scalar, denoted by σ (see Sachs, 1964, Penrose, 1964, or Penrose, 1968. for details). The complex function, σ, has the asymptotic behaviour, $\sigma = \sigma^0 r^{-2} + O(r^{-3})$ in an asymptotically flat space-time, σ^0 depending on a retarded time parameter, u, and on angles on a sphere. The derivative of the complex-conjugate of σ^0 with respect to u is called the (Bondi-Sachs) news function, and the rate of loss of total mass-energy of the system described by the asymptotically flat space-time at hand is given by an integral over a spherical section of the outgoing null hypersurface, $u =$ constant, which has been slid out along the hypersurface to infinity:

$$\dot{M}(u) = -(1/4\pi)\int |\,\dot{\sigma}^0\,|^2\, dS \tag{1}$$

In addition, the r^{-1}, and r^{-2}, parts of the asymptotic curvature along $u =$ constant are given by angular derivatives of the news function (see Exton et al., 1969). If there is no news (i.e. no gravitational waves are present), then the total mass-energy is constant, and the curvature tensor falls off asymptotically as r^{-3} or faster in future null directions. This fact will be important for the discussion of the behaviour in the past, which will follow in Sec. 6.

What is so far lacking in the exact theory is a relation between the asymptotic news function and the behaviour of the material sources. In the linearized approximation to Einstein's theory, the asymptotic shear can be shown to be equal to minus the second derivative with respect to retarded time of a certain spatial component of the trace-free retarded mass-quadrupole moment of the material sources. This identification leads to the so-called 'quadrupole formula' for energy-loss due to gravitational radiation first derived by Einstein in 1918: the rate of loss of total mass-energy of the system is negatively proportional to the square of the third time-derivative of the retarded quadrupole moment. It is to be emphasized here that all of the quantities mentioned in the above discussion are evaluated along future-directed, outgoing null rays (i.e., one follows the outgoing radiation).

3. New timing results on the binary pulsar PSR 1913+16

One of the reasons why progress on gravitational radiation theory has been so slow in the past has been the absence of observational data. One cannot help feeling that, had the data been there, the mathematical difficulties would have been overcome. Therefore, there was great excitement at the Ninth Texas Symposium on Relativistic Astrophysics in Munich in December, 1978, when Joseph Taylor announced the latest results of his collaboration with Lee Fowler and Peter McCulloch on timing the pulsar in a binary system (Taylor et al., 1978). In their own words: "Measurements of second- and third-order relativistic effects in the orbit of binary pulsar PSR 1913+16 have yielded self-consistent estimates of the masses of the pulsar and its companion, quantitative confirmation of the existence of gravitational radiation at the level predicted by general relativity, and detection of geodetic precession of the pulsar spin axis." I will not go into technical detail on this beautiful result, but will confine myself to some, more or less, general remarks on the binary pulsar itself, and on the data-reduction and interpretation. For details, see Blandford and Teukolsky (1976), Smarr and Blandford (1976), Epstein (1977), and references quoted there.

Pulsars are thought to be rapidly rotating neutron stars of mass one to two solar masses, and radii of the order of 10 km. They are, therefore, very compact objects. They apparently have a frozen-in dipole magnetic field, whose axis is not aligned with the rotation axis of the star itself. By various mechanisms, electromagnetic radiation of radio frequency is produced near one of the magnetic poles, and beamed like a search-light out into space. A distant observer, located in the beam path, therefore receives a pulse of radio signal each time the beam sweeps around. The pulsar in PSR 1913+16 rotates about 16 times per second, and serves as a very stable, accurate clock.

What was exciting about this pulsar from the beginning (Hulse and Taylor, 1975) was that an alternate red- and blue-shifting of this 0.059s pulsar period indicated that the pulsar was a member of a binary system. (In fact, PSR 1913+16 is still the only known pulsar in a binary system).

Since the projected semi-major axis of the elliptical orbit was about one solar radius, while the pulsar itself was not eclipsed, the unseen companion object must also be rather compact and certainly not a normal star. Moreover, the eccentricity of the orbit was very large, 0.617, so general relativistic effects could easily dominate the evolution of the system. The problem with the earlier timing measurements was that, without taking relativistic effects into account, only a certain function (the mass function) of the masses of the two objects and the angle of inclination of the orbit to the plane of the sky could be measured, but neither the individual masses themselves, not the actual size of the orbit. This meant that there were several possible candidates for the companion star, some of which could affect the future evolution by means of non-gravitational effects such as tidal friction tidally induced quadrupole moments on the companion.

By substantially improving the accuracy of the timing measurements (from 1 ms in 1975 to 50 μs in 1978), Taylor and his colleagues can now fit a theoretical model for the arrival time of the pulses which includes the relativistic periastron advance, the gravitational redshift and transverse doppler shifts of the pulsar clock, and the rate of decrease of the binary period itself. It turns out that the model can be fit perfectly by the data, with the residual errors showing no periodic behaviour, and being apparently randomly distributed over each binary period of $7\frac{3}{4}$ hrs. This suggests that the companion object behaves like a point mass, and is most probably also a neutron star. What is more, one of the relativistic parameters is redundant in the model. This means that the most accurately determined of the relativistic parameters can be used to determine the angle of inclination of the orbit, and the individual masses of the objects, thus specifying the system completely. With this information, the final redundant, parameter can be computed theoretically. This predicted value can then be compared with the value for the parameter obtained by fitting the experimental data to the model, as a consistency check.

Now, as remarked in the previous section, a prediction of a decrease in binary period due to the emission of gravitational radiation can only be made using the linearized approximation to Einstein's theory, more particularly, by using the quadrupole formula (see Landau and Lifschitz, 1975). Let us denote this rate of decrease by $\dot{p}_b$ (Q pole), and the fitted value of the decrease by $\dot{p}_b$. Then, as Taylor announced at Munich,

$$\dot{p}_b = \dot{p}_b(Q \text{ pole})\,(1.33 \pm 0.25) \qquad (2)$$

If this excess is a real effect (the error is still rather large), then it would appear that the quadrupole formula predicts a rate of decrease of binary period due to gravitational radiation-damping, which is too small by about 30%. The interpretation would be that non-linear effects in Einstein's theory have to be consistently taken into account in order to obtain a correct description of the binary pulsar.

4. The relativistic Kepler problem

In view of the discussion of the latest observational results given above, it behooves us (theorists) to take a closer look at Einstein's theory of gravitation, to take the non-linearities in the theory seriously, and to work out as rigorously as possible the consequences for a binary system. A hint in this direction comes from a computation of the energy loss in small-angle gravitational scattering by Arnold Rosenblum (1978), in which Einstein's equation was iterated three times, starting from no interaction in flat space-time. He finds a result which is somewhat larger than the value predicted by the quadrupole formula. Unfortunately, Rosenblum works with point particles which are, strictly speaking, not compatible with Einstein's theory. The computational difficulties involved in working with extended bodies will be vastly greater than Rosenblum's already very complex calculations.

How should we proceed to work out the consequences of Einstein's theory for the evolution of a binary system? The first step is to make an idealization which can be handled mathematically, This idealization is called an 'isolated system'. The rationale behind it was discussed in Introduction. It should be pointed out, however, why the assumption made here, that one has two bodies interacting gravitationally, which are alone in an otherwise empty universe, so necessary in gravitational theory, is unnecessary in electrodynamics. The reason is simply that electromagnetic effects can be shielded, that an electromagnetically interacting system can be prepared, put into a Faraday cage, and allowed to run. It is of the nature of gravitation that this cannot be done for gravitationally interacting bodies. One might expect, on intuitive grounds, that a system should evolve independent of how it arose. One way to ensure this is to build a model in which the system has been isolated for all time. This train of thought leads to the following formulation of what Cliff Will and I call "the relativistic Kepler problem": Find a solution, (M, I_{ab}) defined on M, of Einstein's gravitational field equation, satisfying the following conditions:

1. The stress energy tensor T_{ab} of g_{ab} has support consisting of a pair of timelike world-tubes, which are initially disjoint.
2. T_{ab} describes physically reasonable matter.
3. g_{ab} is asymptotically flat.
4. There is no incoming radiation in M.

The present state of knowledge on this problem is easily surveyed; no precisely rigorous results are known. The best-understood aspects of the problem are discussed now.

5. Asymptotic behaviour in the future

The best understood aspect of the relativistic Kepler problem that concerns the asymptotic structure of space-time is: how fast the gravitational field should fall off as one recedes from the material sources. In electrodynamics, the situation is simpler than in gravitational theory because the flat (Minkowski) space-time is specified a priori; one merely investigates the behaviour of fields 'painted onto' it. In gravitational theory one has, in addition, to supply a definition of 'asymptotic fiatness' (for details see Penrose, 1964, 1968) I but will restrict myself to a description of how the fields fall off. Since most previous work has been devoted to outgoing radiation, discussed here is only the asymptotic behaviour along outgoing light rays in the future.

Let us begin with electrodynamics in flat space-time, E. Since the e m field has six components, it is convenient to choose a certain collection of vectors, four at each event, with respect to which components can be taken. Choose one particular time like straight line in F, which may be thought of as the unaccelerated centre of mass of the physical system being investigated. Consider the future-directed light cones with vertices on this line. Denote by $\mathbf{l}^a$ the null vector tangent to the null straight lines generating these cones. Similarly, denote by $\mathbf{n}^a$ the null vector tangent to radially ingoing light rays, generating past light cones with vertices on our timelike straight line. If $\mathbf{t}^a$ is the unit timelike vector tangent to the line, $\mathbf{l}^a$ and $\mathbf{n}^a$ may be normalized by $\mathbf{l}^a\mathbf{t}_a = \mathbf{l}^a\mathbf{n}_a = 1$. (The signature of space-time is $(+ - - -)$.) The null vectors $\mathbf{l}^a$ and $\mathbf{n}^a$ are orthogonal to a spacelike 2-plane at each event. In this plane, choose an orthonormal pair of vectors, and then take $2^{-1/2}$ times one plus i times the other. Denoting the result by $\mathbf{m}^a$, $\mathbf{m}^a$ is a complex vector satisfying $\mathbf{m}^a\mathbf{m}_a = 0$, $\mathbf{m}^a\bar{\mathbf{m}}_a = -1$, $\mathbf{m}^a\mathbf{l}_a = \mathbf{m}^a\mathbf{n}_a = 0$. Then the Minkowski metric is given by $\eta_{ab} = 2\mathbf{l}_{(a}\mathbf{n}_{b)} - 2\mathbf{m}_{(a}\bar{\mathbf{m}}_{b)}$ (round brackets denote symmetrization).

With respect to this 'null tetrad' at each event, define the following components of the e m field tensor, F_{ab}:

$$\phi_0 = F_{ab}\,\mathbf{l}^a\mathbf{m}^b,$$
$$\phi_1 = \tfrac{1}{2}F_{ab}\,(\mathbf{l}^a\mathbf{n}^b + \mathbf{m}^a\bar{\mathbf{m}}^b),$$
$$\phi_2 = F_{ab}\,\bar{\mathbf{m}}^a\mathbf{n}^b \qquad (3)$$

Asymptotically, along outgoing null rays to the future (in the direction of $\mathbf{l}^a$), these three complex scalars have the following physical interpretation: ϕ_0 describes the strength of the e m dipole field, ϕ_1 measures the Coulomb-part of the field (the charge is given by the integral of r^4 times $\phi_1\bar{\phi}_1$ over a sphere at infinity), and ϕ_2 measures the outgoing e m radiation (the rate of loss of energy due to e m radiation is given by the integral of r^2 times $\phi_2\bar{\phi}_2$, over a sphere at future null infinity).

What is more, there is a theorem about how these components fall off, for retarded e m fields, along outgoing null straight lines (Goldberg and Kerr, 1964): ϕ_0 falls off like r^{-3}, ϕ_1, like r^{-2}, and ϕ_2 like r^{-1}. This, so-called 'e m peeling theorem' (actually, there is more to the peeling theorem than this, but we do not require the additional information here. It has to do with the 'principal null directions' of the field) is the precise statement of the idea that the e m field should vanish at future null infinity.

The gravitational case is similar, though more complicated. In an asymptotically flat space-time, vectors similar to $\mathbf{l}^a$, $\mathbf{n}^a$ and $\mathbf{m}^a$ can be introduced, with $\mathbf{l}^a$ tangent to the generators of a family of outgoing null hypersurface in curved space-time (see Newman and Penrose, 1962). Since the curvature tensor in vacuum has 10 independent components, five complex components are defined. They are called ψ_0, ψ_1, ψ_2, ψ_3, and ψ_4 (their definition is not important for descriptive purposes, and may be found in the quoted literature). In the linearized approximation, in which the asymptotic field can be related to the material sources, the ψ's have the following physical interpretation: ψ_0 measures the strength of the retarded guadrupole moment, ψ_1 measures the angular momentum, ψ_2 measures the mass, and ψ_3 and ψ_4 measure outgoing gravitational radiation. (It is the leading terms in ψ_3 and ψ_4 which are given as angular derivatives of the news function, as mentioned is Sec. 2.) The peeling theorem of gravitation follows from a sufficiently strong definition of asymptotic flatness at future null infinity (see Penrose, 1968), and states that ψ_0 falls off like r^{-5}, ψ_1 like r^{-4}, ψ_2 like r^{-3}, ψ_3 like r^{-2}, and ψ_4 like r^{-1}, asymptotically along future-directed outgoing null geodesics. The qualification 'sufficiently strong' is made here because it will turn out that this peeling behaviour does not hold in the past, so that the notion of asymptotic flatness at past null infinity has to be weakened somewhat from that in the future.

6. New results on asymptotic behaviour in the past

The fourth criterion for a solution of the relativistic Kepler problem enunciated above was that there should be no incoming radiation. A physically reasonable way of obtaining such a no-incoming-radiation condition is to consider the expression for the flux of incoming energy at infinity (Leipold and Walker, 1977), but before doing this, let us ask what the fall-off near past null infinity would be if the time-reversal of the future bahaviour held. From the description of the null tetrad given above, it is evident that the vectors $\mathbf{l}^a$ and $\mathbf{n}^a$ reverse their roles when one goes from the future to the past. Glancing at the definition of the complex components ϕ_0, ϕ_1, and ϕ_2 in eqn. (3) then reveals that $-\phi_2$ plays the role of ϕ_0 in the past, $-\bar{\phi}_1$ that of ϕ_1, and $-\bar{\phi}_0$ that of ϕ_2. Therefore, if the time-reverse of the future asymptotic behaviour of the field were to hold in the past, we would expect $\phi_0 = O(r^{-1})$, $\phi_1 = O(r^{-2})$, and $\phi_2 = O(r^{-3})$ along null straight lines receding from the system into the past. In particular, just as ϕ_2 measured outgoing radiation in the future, ϕ_0 measures incoming radiation in the past. Indeed, the rate of gain of energy of the system due to incoming e m radiation would be

$$\dot{E} = \lim_{r\to\infty} (1/4\pi) \int r^2 \, |\phi_0|^2 \, dS \tag{4}$$

where the integral is over a spherical section of a past light cone, and the limit is taken holding advanced time, $v = t + r$ constant, i.e., along such a past cone. The physical requirement that there be no gain of energy due to incoming radiation leads to the following chain of implications. (Strictly speaking, $\dot{E} = 0$ only implies $\phi_0 = 0\,(r^{-1})$, but we shall assume $\phi_0 = O\,(r^{-2})$ for simplicity.)

$$\dot{E} = 0 \Rightarrow \lim_{r\to\infty} r^2\,|\phi_0|^2 = 0$$
$$\Rightarrow \lim_{r\to\infty} r\,\phi_0 = 0$$
$$\Rightarrow \phi_0 = O\,(r^{-2}) \tag{5}$$

Another way of saying this is that the 'r^{-1}-part' of the e m field should vanish at past null infinity.

The gravitational case is only slightly more complicated. The time-reverse of the future peeling behaviour would be

$$\psi_0 = O\,(r^{-1}),\ \psi_1 = O\,(r^{-2}),\ \psi_2 = O\,(r^{-3}),\ \psi_3 = O\,(r^{-4}),\ \psi_4 = O\,(r^{-5})$$

along past-directed, outgoing null geodesics, i.e., near past null infinity. The condition that there should be no gain of total mass-energy due to incoming gravitational radiation requires the vanishing of the past Bondi-Sachs news function at past null infinity. Recalling that the amplitudes of the r^{-1}— and r^{-2}— parts of the curvature tensor were given by angular derivatives of the news function in the future, we see that the vanishing of the past news function requires,

$$\psi_0 = O\,(r^{-3}),\ \psi_1 = O\,(r^{-3}) \tag{6}$$

along null geodesics receding from the material sources into the past.

Equations (5) and (6) for the behaviour of the e m and gravitational fields along outgoing, past-directed null rays are our conditions for the absence of incoming radiation.

Uptil now, we have proceeded as if the behaviour of fields near past null infinity were, apart from the question of incoming radiation, just the time reverse of the behaviour of retarded fields near future null infinity. This, however, is not so. Advanced fields might behave in this way, but causality considerations lead one to take retarded solutions of Maxwell's equations (retarded solutions of Einstein's equation are, strictly speaking, not defined. We shall come to this problem shortly). Now the behaviour of the retarded solution of Maxwell's equations depends on what the sources were doing at the corresponding retarded time. Therefore, the asymptotic behaviour of the solution along light rays receding from the sources into the past, as $r \to \infty$ depends on the behaviour of the sources themselves, as $t \to -\infty$. But this latter behaviour is by no means freely specifiable. It is determined by the equations of motion of the sources, and the retarded nature of the interaction. One is not free to impose asymptotic conditions on the e m fiield in the past; whether the retarded e m field satisfies the condition (5) or not depends on the equations of motion of charged bodies.

I have investigated this question for small angle scattering and for the bound state (it turns out to be the same thing in the past), using two charged pointparticles obeying the Lorentz-Dirac equation with retarded e m fields. I assume that the particles are sufficiently far apart, that the interaction is weak and the motion of the charges is slow, then integrate the equation of motion for retarded interactions between the particles into the past. In this case, the radiation-reaction can be treated as a small perturbation of the Coulomb interaction. The result is that an initially elliptical orbit becomes more and more eccentric in the past, the eccentricity eventually going to a limiting value greater than or equal to unity as $t \to -\infty$. Therefore, the electromagnetically bound particles were captured from an initial scattering state in the infinite past. Computing the behaviour of the corresponding retarded e m field along light rays going out into the past leads to the following asymptotic behaviour:

$$\phi_0 = O(r^{-2}),\ \phi_1 = O(r^{-2}),\ \phi_2 = O(r^{-2}) \tag{7}$$

Therefore, the equation of motion for retarded-interacting charged particles is such that the retarded field indeed obeys the condition for absence of incoming radiation. On the other hand, the asymptotic behaviour is not what would have been anticipated from time reversal of the future behaviour for the component ϕ_2, which falls off more slowly ($O(r^{-2})$ instead of $O(r^{-3})$) in the past.

Together with Clifford Will, I (Walker and Will, 1979) have carried out a similar analysis for the gravitational case, which I describe here. The first difficulty in the gravitational problem is the following: In electrodynamics, Maxwell's field equations are logically independent of any equations of motion for the sources of the e m field. This means that Maxwell's equations can be solved for an arbitrary charge-current distribution, independently of whether the charge-current distribution satisfies any equations of motion, or not (for example, by taking the retarded solution). This is, however, not the case for Einstein's gravitational equation. Einstein's equation determines both the behaviour of the field, and the evolution of the material sources. The first step, therefore, is to separate Einstein's equation into a truncated field equation, which can be solved independently of the source motion, and a separate equation for the motion of the sources (see Ehlers et al., 1976). In doing this, I shall follow Epstein and Wagoner (1975). The separation can be achieved as follows. The unaccelerated centre of mass world line (Schattner, 1978) of the matter, with its proper time parameter, can be considered as a time-like straight line in an auxiliary, flat, Minkowski space-time, F. An inertial co-ordinate chart, x^a, in F, whose spatial origins lie on this line, can be defined (see below). With respect to this chart, the space-time metric is split into a flat part and a deviation, h_{ab}, from flatness, as follows:

$$g_{ab} = \eta_{ab} + h_{ab} \tag{8}$$

Instead of h_{ab}, the 'trace-reversed' objects,

$$\bar{h}^{ab} := h^{ab} - \tfrac{1}{2}(trh)\,\eta^{ab}, \tag{9}$$

are taken as the independent variables in the analysis. In terms of these new variables, a four-index object, with the symmetries of the Weyl tensor, is defined:

$$H^{abcd} := 2\eta^{a[d}\bar{h}^{c]b} + 2\eta^{b[c}\bar{h}^{d]a} \qquad (10)$$

the square brackets denoting anti-symmetrization. Finally, in terms of H^{abcd}, a symmetric object, t^{ab}, is defined:

$$16\pi t^{ab} := \partial_c\partial_d H^{abcd} - 2G^{ab} \qquad (11)$$

∂_c denoting partial differentiation with respect to the chart x^c, and G^{ab} denoting the Einstein tensor of the metric g_{ab}. Because of (8) and (9), the Einstein tensor is a functional of the variables $\bar{h}^{ab}$. Einstein's equation, $G^{ab} = 8\pi T^{ab}$, with T^{ab} denoting the energy of matter, can be rewritten as

$$\partial_c\partial_d H^{abcd} = 16\pi (T^{ab} + t^{ab}) \qquad (12)$$

Observe that t^{ab} is a complicated non-linear functional of $\bar{h}^{ab}$ and its first and second derivatives, so that although (12) is linear in $\bar{h}^{ab}$ on the left hand side, this is only because all of the non-linearities have been shuffled over onto the right hand side. A solution of (12) evidently defines also a solution of $G^{ab} = 8\pi T^{ab}$.

Equation (12) is now split into two parts, one of which governs the evolution of the sources, and the other of which is a reduced field equation, which can be solved independently of the source behaviour. The first part is

$$\partial_a \bar{h}^{ab} = 0 \qquad (13)$$

This is analogous to the Lorentz gauge in electrodynamics, but its role here is completely different. With the help of eqn. (13), eqn. (12) reduces to

$$\Box_\eta \bar{h}^{ab} = 16\pi (T^{ab} + t^{ab}) \qquad (14)$$

$\Box_\eta$ denoting the flat-space scalar-wave operator. Since the Green function for $\Box_\eta$ is known, and provided the variables $\bar{h}^{ab}$ satisfy certain fall-off conditions in the past (see J. Ehlers, 1977), (14) is equivalent to the integral equation

$$\bar{h}^{ab}(t, \vec{x}) = 4\int (T^{ab} + t^{ab})(t - |\vec{x} - \vec{x}'|, \vec{x}')\,|\vec{x} - \vec{x}'|^{-1}\, d^3\vec{x}' \qquad (15)$$

This idea is to iterate eqn. (15), having specified a particular source model (e.q., a perfect fluid) for T^{ab}, to obtain $\bar{h}^{ab}$ as a functional of the source variables (pressure, density, matter flow-vector). This functional is then inserted into eqn. (13), which determines the source variables as function of the co-ordinates x^a. One has then solved the problem. (In a non-linear, circular fashion, the condition (13) also determines the co-ordinate chart, x^a.) Due to extreme mathematical complexity, and the difficulty of carrying out the retarded integration in eqn. (15), this programme has never been carried out beyond the first iteration

$$\left(\bar{h}^{ab}_{(0)} = 0,\ \bar{h}^{ab}_{(1)} = 4\int T^{ab}_{\text{ret}}\right)$$

What Cliff Will and I actually did was the following: We first used the retarded, quadrupole, radiation-reaction potential proposed, e.g., by Misner et al. (1973) together with the method of perturbation of osculating orbital elements from celestial mechanics (see Smart, 1953), to estimate the behaviour of the orbit of the relativistic gravitational two-body problem in the infinite past. We then computed the linearized, retarded gravitational field due to this motion, and investigated its asymptotic behaviour along past-directed null rays in the flat, background space-time. We found that the orbit went over into a scattering state in the past, with the limit of the eccentricity of the osculating orbit in the infinite past being generically greater than unity. Therefore, in the idealization employed here (isolated, gravitationally interacting point particles), all bound systems were formed by capture from an initial scattering state by emission of gravitational radiation. (Of course, other possibilities can arise if the idealization is changed slightly. For example, one could have an infinitely thin cloud of dust in the past, filling all space, but with density initially zero in the infinite past, with initial motion such that two bodies contract out of the dust and go into orbit about one-another. Such a description would, however, be much more complicated than the one we have made.)

The behaviour of the field along past-directed light rays in the flat background space-time was found to be as:

$$\psi_0 = O(r^{-3}),\ \psi_1 = O(r^{-3}),\ \psi_2 = O(r^{-3}),$$
$$\psi_3 = O(r^{-4} \ln r),\ \psi_4 = O(r^{-4}) \tag{16}$$

As in the e m case, the gravitational equation of motion seems to be such that retarded fields indeed satisfy the condition for absence of incoming radiation (eqn. 6). But the remaining components fall off slower than would have been expected from a time reversal of the future behaviour (discussion preceding eqn. (6). This means that the strong form of asymptotic flatness, which implied the peeling result in the future, cannot in general be expected to hold in the past. Rather, a somewhat weaker form of asymptotic flatness in the past must be used. We expect, but have not proved, that the behaviour (16) will hold also along the true, curved space light rays.

To conclude, a remark on the significance of this result. If the condition for absence of incoming radiation had not been satisfied, this would have called into question the whole notion of an isolated system, as far as describing radiation was concerned. The very idea of an isolated system could have become incompatible with the theory. The fact that this is not so is very comforting, and should encourage us to really carry out the programme of iterating the reduced field equation for realistic source models in order to be better able to describe such systems as the binary pulsar.

References

1. Blandford, Roger and Teukolosky, Saul, (1976) *Ap. J.* **205**, 580.
2. Bondi, H, van der Burg, M.G.J. and Metzner, A.W.K., (1962), *Proc. Roy. Soc.* (London) **A269** 21.
3. Ehlers, J., in MPI-PAE/Astro 138, Max-Planck Institute, München (1977).
4. Ehlers, J., Rosenblum, A., Goldberg, J.N. and Havas, P., (1976), *Ap. J. Lett.* **208**, L 77.
5. Epstein, Reuben and Wagoner, Robert V., (1975), *Ap. J.* **197**, 717.
6. Epstein, Reuben, (1977), *Ap. J.* **216**, 92.
7. Exton, A, Newman, E.T. and Penrose, R., (1969) *J. Math. Phys.* **10** 1566.
8. Goldberg, J.A. and Kerr, R.P., (1964), *J. Math. Phys.* **5**, 172.
9. Hulse, R.A. and Taylor, J.H., (1975), *Ap. J. Lett.* **195**, L51.
10. Landau, L. and Lifschitz, E.M., (1975), *The Classical Theory of Fields* Pergamon, London.
11. Leipold, G. and Walker, Martin, (1977), *Ann. Inst. H. Poincare*, **37**, 61.
12. Misner, C.W., Thorne, K.S. and Wheeler, J.A., (1973), *Gravitation*, Freeman, San Francisco.
13. Newman, E.T. and Penrose, R., (1962), *J. Math. Phys.*, **3**, 566.
14. Penrose, R. 1964, in *Relativity, Groups and Topology*, C. De Witt and B. De Witt (Eds.) Blackie and Son, London-Glasgow, p. 570.
15. Penrose, R., (New York-Amsterdam) in C.M. De Witt and J.A. Wheeler (Eds.), *Battelle Rencontres*, W.A. Benjamin, (1968) p. 184.
16. Rosenblum, A., (1978) *Phys. Rev. Lett.* **41**, 1003.
17. Sachs, R, (1964), in *Relativity*, Groups and Topology, op. cit.
18. Schattner, R., 1978, *Gen. Rel. Grav.*, in press.
19. Smart, Larry, L., and Blandford, R., (1976) *Ap. J.* **207**, 574.
20. Smart, W.M., (1953), *Celestial Mechanics* Longmans, Green & Co.
21. Taylor, J.H., Fowler, L.A. and McCulloch, P.M., (1978) preprint, Deptt. of Physics and Astronomy, University of Massachusetts, Amherst, MA, U.S.A.
22. Walker, Martin and Will, Clifford M., (1979), to appear in *Phys. Rev.* D.

Discussions

E T Newman: Is there a means of introducing a past null infinity in your case?

M Walker: Yes, the hypersurface at past null infinity exists. The rate at which the physical curvature tensor falls off asymptotically can be described in terms of the degree of differentiability of the conformally rescaled metric tensor on $\mathscr{I}^-$. What we have found is that this degree of differentiability is too low to ensure the usual scaling behaviour for the curvature, but the hypersurface itself still exists. The part of the rescaled curvature tensor which is regular on $\mathscr{I}^-$ can be identified in an invariant way.

D Brill: You showed two ways to go to infinity, in the null direction and in the spacelike direction. Does this exhaust the possibilities? What is the relation between the asymptotic behaviour in the spacelike and in the null direction?

M Walker: A third possibility is to go to timelike infinity, following the source into the infinite past or infinite future. This has been studied by Eardley and Sachs a few years ago, using a projective structure of space-time. The relation between null and spacelike infinity is still unclear. One can impose asymptotic conditions on a space-time which are sufficiently strong to guarantee that, for example, the past limit of the future Bondi-Sachs energy-momentum vector is equal to the ADM energy-momentum measured at spacelike infinity. The difficulty is that one has no idea whether sufficiently many solutions of Einstein's equations to cover all cases of physical interest exist, which satisfy these conditions. One peculiar feature in the spacelike infinity case is that many physical questions seem to depend very sensitively on the degree of differentiability of the rescaled fields at spatial infinity. This has been discussed in recent papers by Ashtekar and Hansen, and Ashtekar.

Nitin Nitsure: Going to infinity in the null or spacelike directions involves going to infinity in the Euclidean sense. Going to infinity in the timelike direction is not going to 'infinity' at all in the Euclidean sense (Remark).

M Walker: It may not be in the Euclidean sense, but it is in the Lorentzian sense. Any region which is 'infinitely far away' from a finite event, be it infinite in time, or in space, or in both, is 'at infinity' as far as the finite event is concerned.

C Sivaram: Do you get the usual well-known result that for emission of gravitational radiation the eccentricities of the orbits decrease at future ∞?

M Walker: Yes, of course, but I would like to emphasize two points in this regard. First, we do not use a radiation-reaction potential which has been averaged over several orbits, as is usually done (by Peters, and by Peters and Mathews, for example, in the gravitational case). Rather, we use an instantaneous potential since one cannot average when the bound orbit goes over into an unbound one as it is followed into the past. Secondly, it does not seem reasonable to follow the predicted evolution very far into the future since, as the orbit becomes smaller, the approximations of weak fields and slow motion will eventually break down. In the past, however, they are fulfilled even better.

S Ramamurthy: (1) If there is no unique time-direction in curved space-time as we have heard Prof. Newman say, how can we speak of time reversed fields and their asymptotic fields? (2) Taking a general 2^n pole field, will its asymptotic behaviour be $1/r^{2n+1}$?

M Walker: This is only to be understood in a qualitative way, not as a precise mapping. The rate of fall-off of the field components is a qualitative behaviour, so any discrete map which sends future into past would do. This map need not preserve any specific structure.
(2) Yes.

S B Khadkikar: (1) I would like to know the effect on the asymptotic behaviour of e m radiation and fields when quantum nature of orbits is taken into account. Does it remain the same? (2) In contrast to time reversal symmetry, the condition on no excess radiation seems to be artificial. Is there deeper significance for this in relation to the structure of the theory?

M Walker: (1) I have not the foggiest idea, but if the system does not radiate, I would expect $\phi_0 = O\,(r^{-3})$, $\phi_1 = O\,(r^{-2})$ $\phi_2 = O\,(r^{-2})$ for e m fields near future null infinity.
(2) Time reversal symmetry is somethtng which may be expected of microphysics, but probably does not hold for gravitation. I am not sure that I understand what you mean by 'excess radiation'. What I meant was that one should exclude incoming radiation, since otherwise one could hardly call a system 'isolated'.

Positivity of Total Energy in General Relativity*

DIETER R. BRILL

The importance of questions about energy in general relativity is pointed out in several contributions to this symposium. For well-known physical and mathematical reasons there is no covariant notion of energy-momentum density in general relativity (unless one introduces preferred frames[1]); however, it has been known since Einstein and Klein[2] that in asymptotically flat space-time one can define a unique *total* energy-momentum. It is essential for the physical interpretation that the total mass-energy be positive, and that the total energy-momentum vector be timelike. For example, if solutions with negative mass energy existed, a simple scaling would yield arbitrarily negative mass-energy. This would have bizarre consequences, such as the possibility of extracting an infinite amount of energy from the vacuum. In this paper is reviewed to-date progress in proving that the total mass-energy is positive.

1. The '3+1' decomposition and the positive mass conjecture

The total mass is a quantity defined by the asymptotic behaviour of the geometry of asymptotically flat space-times. However, such space-times have several different asymptotic regions, viz., spacelike infinity and future and past null infinity. We confine our attention to the conserved, total 'spacelike' mass, rather than the 'Bondi mass' at future null infinity, about which a positivity conjecture is not so easily formulated.

In order to define this total mass it is advantageous to think of space-time as a family of space-like surfaces, and write the Einstein equations in '3+1' decomposition in terms of geometrical quantities associated with these spacelike surfaces. These quantities are the intrinsic 3-dimensional geometry g_{ij} on the surface, and its extrinsic curvature (second fundamental form) K_{ij}. For asymptotic flatness we demand fall-off of these quantities (with e_{ij} = flat metric)

$$g_{ij} = e_{ij} + O(1/r), \quad \partial_k g_{ij} = O(1/r^2), \quad \partial_k \partial_s g_{ij} = O(1/r^3)$$
$$K_{ij} = O(1/r^2), \quad \partial_s K_{ij} = O(1/r^3)$$

Although more general and elegant fall-off conditions have been formulated[3], but those mentioned above would serve our purpose.

The total mass** and momentum are then defined by integrals over a

*Work supported in part by the National Science Foundation. An expanded version, with Pong Soo Jang, is to be published in the G R G Centenary Volume.

**It has become customary to denote by 'mass' the total energy P^0, rather than the 'invariant mass, $(P^0)^2 - (P^i)^2$.

'sphere' S of co-ordinate radius r

$$m = P^0 = \lim_{r \to \infty} \frac{1}{16\pi} \int_S (\partial_j g_{ij} - \partial_i g_{jj})\, dS_i$$

$$P^i = \lim_{r \to \infty} \frac{1}{8\pi} \int_S (K^i{}_j - \delta^i{}_j K)\, dS_j$$

This surface integral definition does not explicate the relation between the total mass and the mass density of the sources of the gravitational field (if any). The situation is similar to electromagnetism, where the surface integral definition of the total charge involves cnly the E-field and makes no reference to charge density. The latter only enters when the 'constraint' $\operatorname{div} E = \rho$ is used to convert $Q = \int E ds$ into a volume integral. Similarly, in general relativity the value of the total mass in terms of the freely specifiable matter and gravitational variables is determined by the constraint equations,

$$R + K^{ij}K_{ij} + K^2 - 2\rho = 0$$
$$(K^{ij} - Kg^{ij})_{|j} - J^i = 0$$

Here ρ, J^i are the matter energy and current density, | denotes 3-dimensional covariant differentiation, R is the 3-dimensional scalar curvature, and we use the abbreviations

$$K = K^l_l, \ \pi^{ij} = \sqrt{g}\,(K^{ij} - Kg^{ij})$$

Unfortunately, there is no simple way of solving the constraints and cxpressing m in terms of independent variables in a manifestly positive fashion. The problem therefore is to prove in a less direct way the *positive mass conjecture*: *The mass m of any asymptotically flat initial data set is non-negative, and vanishes, if and only if the initial data correspond to flat space-time.*

By 'initial data set' we mean a regular solution g_{ij}, K^{ij}, ρ, J^i of the constraint equation satisfying the local energy condition $\rho \geqslant (J_i J^i)^{1/2}$.

If the positive mass conjecture is true, the energy-momentum four-vector $\mathbf{P}^\mu$ cannot be spacelike. This is so because $\mathbf{P}^\mu$ transforms as a four-veetor under asymptotic Lorentz transformations. If $\mathbf{P}^\mu$ were spacelike, one could then find a spacelike surface on which $P^0 = 0$. But this cannot happen if the positive mass conjecture holds.

2. Weak field limit

If the initial data are a small perturbation of flat space data, and the total mass is evaluated to second order, the positive mass problem has the most satisfactory solution possible[4]: We write

$$g_{ij} = e_{ij} + \delta_{ij} + \ldots, \quad \pi^{ij} = \delta\pi^{ij} + \ldots$$

and decompose δg_{ij}, $\delta\pi^{ij}$ into transverse traceless and longitudinal parts. The former (denoted by 'TT') can be freely specified, and the latter are determined by the constraints. To second order, the total mass then takes the manifestly positive form

$$8\pi\delta^2 m = \int [\tfrac{1}{4}(\partial_i \delta g_{jk}^{TT})^2 + (\delta\pi^{ijTT})^2 + \delta^2\rho]\, d^3x \geqslant 0$$

(here $(\delta\pi^{ij})^2 = \delta\pi^{ij}\,\delta\pi_{ij}$, etc.). For the momentum one finds

$$8\pi\delta^2 P^i = \int [(\partial^i \delta g_{jk}^{TT})\,\delta\pi^{jkTT} + \delta^2 J^i]\, d^3x$$

Since we have at each point $|(\partial\delta g)\,\delta\pi| \leqslant \frac{1}{4}(\partial\delta g)^2 + (\delta\pi)^2$ and $|\delta^2 J^i| \leqslant \delta^2\rho$, we can verify directly that $|\delta^2 P^i| \leqslant \delta^2 m$. Finally, if $\delta^2 m = 0$, then $\partial_i \delta g_{jk}^{TT} = 0 = \delta\pi^{ijTT}$, from which one can show the first order form of the Gauss-Codazzi equation for flat space-time,

$$\delta R_{ij} = 0 \quad \partial_{[i}\delta K_{j]k} = 0$$

3. Special cases

The formulae given above for $\delta^2 m$ in the weak field theory hold on any spacelike surface which is a small perturbation of a plane in the 'comparison Minkowski space'; thus for example, $\delta\pi$ can be arbitrary. However, little further progress in the strong field case has been made without assuming (in addition to the constraints) that $\pi = 0$. It is reasonable to impose this condition, since physically it means that we discuss m not on an arbitrary spacelike surface, but on a 'maximal' spacelike surface*—a natural generalization of the planes in Minkowski space. Hence the condition $\pi = 0$ may be considered a co ordinate condition on the time co-ordinate. Nonetheless, one should show that it can always be satisfied, i.e., asymptotically flat space-times always contain at least one maximal spacelike surface. This proof has not yet been given in general; but it has been shown[3] for all space-times in some neighbourhood of Minkowski space that can be deformed to Minkowski space while maintaining the strong energy condition.

If our space-time contains a maximal surface the metric on that surface can always be related to another, simpler metric which has a smaller mass. Hence, the positive mass conjecture only needs to be proved for these simple metrics. The exact statement is the Theorem (O'Murchadha and York[5]): *Every asymptotically flat metric with $R \geqslant 0$, $R \not\equiv 0$, has greater mass than the unique conformally equivalent metric with $R = 0$.*

This theorem applies to the maximal ($K = 0$) case, since in this case the constraint equation implies $R = K^{ij}K_{ij} + 2\rho \geqslant 0$. The conformally related metric is obtained as $g^1{}_{ij} = \varphi^4 g_{ij}$ by solving

$$O = \varphi^5 R' = R\varphi - 8\nabla^2\varphi$$

*A surface whose 3-volume is an extremum compared to that of nearby spacelike surfaces.

for the conformal factor φ. The metric g'_{ij} can be considered as initial data where $K_{ij} = 0$, $J^i = 0$, $\rho = 0$, the so-called time symmetric case. Hence the above theorem is sometimes stated by saying: time symmetric data sets are minima of mass. Its significance for the positive mass question is that it allows us to confine attention to these time-symmetric data sets.

It follows immediately from this theorem that conformally flat initial data on a maximal spacelike surface have positive mass. By various other special computations one can show that the mass is positive for axially symmetric initial data on a maximal spacelike surface[6], and for spherically symmetric initial data on an arbitrary asymptotically flat spacelike surface.[7]

4. Approaches to the general proof

A number of methods have been used to attack the general positive mass conjecture, without assumptions of special symmetries. Each of these approaches gives some particular insights, therefore, I discuss several of them. The existence of a maximal spacelike surface is always assumed (so that $K = 0 = \pi$), hence one can always restrict the discussion to the 'time symmetric' case, $R = 0$, due to the O'Murchadha and York theorem.

Local solution

Suppose the the initial data g_{ij} can be connected to the flat data by a 'curve' of metrics $g_{ij}(t)$ such that $g_{ij} = g_{ij}(1)$, $e_{ij} = g_{ij}(0)$. This curve lies in the infinite dimensional space $\mathfrak{m}$ of all metrics which the initial surface could possess. For a tangent vector to this space at g, $h_{ij} \in T_g\mathfrak{m}$, define the norm and hence a metric on $\mathfrak{m}$ by

$$\|h\|^2 = \langle h, h\rangle \equiv \int (\nabla_i h_{jk})^2 \, dV$$

and as the curve $g_{ij}(t)$ choose a geodesic in this metric. Choquet-Bruhat, Fischer and Marsden[8] have shown that along such a geodesic the inequality $m(g) \geqslant C\|g - e\|$ holds for some constraint $C > 0$, provided $\|g - e\|$ is sufficiently small. Hence, in some small but finite 'local' region near the flat metric the mass is always positive.

Solutions by 2 + 1 decomposition

One of the difficulties in proving the positive mass conjecture appears to be that there is no general theorem about the sign of integrals over the scalar curvature in 3-dimensional spaces. For example, suppose that some time-symmetric initial data g_{ij} are conformally flat to $O(1/r)$, we can then write $g_{ij} = \psi^4 \bar{g}_{ij}$ with g_{ij} flat to $O(1/r^2)$, and ψ behaving asymptotically as $1 + m/2r + O(1/r^2)$. By writing $R = 0$ in terms of $\bar{g}$ it is easily seen that

$$2\pi m = \int [(\nabla\psi/\psi)^2 - \bar{R}/8] \, dV$$

Whereas the first term on the right is positive definite, the second term has no definite sign, The situation is different in two dimensions, where the Gauss-Bonnet theorem determines the value of $\int {}^2R\,dA$, This suggests a '2 + 1' decomposition of the 3-dimensional spacelike surface into a family of 2-dimensional surfaces on which one can use the Gauss-Bonnet theorem.

For example, Geroch[9], and Jang and Wald[10] use a sequence of topological spheres parameterized by $s =$ constant and with extrinsic mean curvature ${}^2K$, such that $|\nabla s| = {}^2K$. If such surfaces can be constructed one can write

$$m = \lim_{s\to\infty} A^{1/2} f / 64\pi^{3/2}$$

when A is the area of the surface, $s =$ constant, and f is defined by

$$f(s) = \int_{s=\text{const.}} (2\,{}^2R - ({}^2K)^2)\, dA$$

These authors then show that $df/ds \geqslant -\frac{1}{2}f$ by using the Gauss-Bonnet theorem, and the time-symmetric constraint $R = 0$ (or, more generally, the energy condition $R \geqslant 0$). Typically the sequence of nested spheres will start at a point with $s = -\infty$, hence $f(-\infty) = 0$. The inequality therefore implies that f, and hence the mass, is positive at $s \to \infty$.

If there is an apparent horizon, with ${}^2K = 0$, we may choose it as the surface $s = 0$. From the above definition of f and the Gauss-Bonnet theorem we find $f(0) = 16\pi$. Since $A(s) = A(0)\, e^s$ we can evaluate the limit for m as

$$m \geqslant A(0)^{1/2}/4\pi^{1/2} > 0$$

This proof of the positivity of m does not use any information, such as energy condition, from the interior of the horizon. This is appropriate since the energy measured at infinity should be independent of anything that occurs inside a horizon. Jang and Wald have also noted that the relationship $16\pi m^2 \geqslant A(0)$ is also obtained from the cosmic censorship hypothesis.

Another proof by 2 + 1 decomposition is due to Schoen and Yau.[11] The idea of the proof is to derive a contradiction from $m < 0$ by showing that an extremal geometrical object must exist on topological grounds, but cannot be the required extremum due to restrictions on its second variation by the Einstein equations. This general method applied to geodesics is frequently used in proving the 'singularity theorems'; here it is applied to 2-dimensional extremal objects, namely, minimal surfaces.

Let x, y, z be the co-ordinates (of rectangular type) on the spacelike-surface. For a minimal surface with boundary the Euclidean circle $x^2 + y^2 = R^2$, $z = 0$. If the mass associated with the metric on the spacelike surface is negative, according to Schoen and Yau one obtains a regular, complete minimal surface S by letting $r \to \infty$. But a computation of the area's second variation shows the contradiction, that such a surface cannot have minimum area; the second variation of the area due to a motion of the surface along the normal n^i is

$$d^2A/ds^2 = -\int [R_{ij}n^in^j + {}^2K_{ij}\,{}^2K^{ij} - ({}^2K)^2]\,dA$$

the 2+1 decomposition of the scalar curvature gives us

$$R = {}^2R + {}^2K^{ij}\,K_{ij} - ({}^2K)^2 + 2R_{ij}\,n^in^j = 0$$

Use this together with the Gauss-Bonnet equation for this unbounded surface, $\int {}^2R\,dA > 0$ to find

$$d^2A/ds^2 = \frac{1}{2}\int ({}^2R - {}^2K_{ij}\,{}^2K^{ij})\,dA < 0$$

Hence, the surface's area cannot be a minimum, and the assumption $m < 0$ must be false. We thus conclude $m \geqslant 0$. Schoen and Yau further proved the second part of the positive mass conjecture, that $m = 0$ implies flat space-time.

5. Conclusion

In summary, one can say that the positive mass conjecture has been proved for most physically important situations. The main restriction is that the existence of a maximal spacelike surface must be assumed. (A minor restriction of the present Schoen-Yau proof is that the asymptotic dependence is not quite as general as is allowed by asymptotic flatness.) A number of interesting details, such as the necessity of energy conditions inside horizons and a simplification of the many steps in the present proof remain to be explored.

References

1. Moller, C., *Mat, Fys. Medd. Dan. Vid. Selsk*, **35** no. 3 (1966).
2. Einstein, A., Sb. Preuss, *Akad. Wiss.*, 448 (1918); Klein, F. *Göttinger Nachr.*, 394.
3. Cantor, Fischer, Marsden, O'Murchadha, York *Comm. Math. Phys.* **49**, 187 (1976); O'Murchadha, N., reprint (1978).
4. Brill, D., Deser, S. *Ann. Phys.* **50**, 548 (1968).
5. O'Murchadha, N. and York, J.W., *Phys Rev.*, **D10**, 2345 (1974).
6. Brill D.R., *Ann. Phys.*, **7**, 466 (1959).
7. Leibovitz, C. and Israel, W., *Phys. Rev.* **D1**, 3226, (1970); Jang, P.S., *J. Math.* **17**, 141 (1976).
8. Choquet-Bruhat, Fischer and Marsden, Proc. of the 1976 Summer School Varenna (Italy).
9. Geroch, R., *Ann. N. Y. Acad. Sci.* **224**, 108 (1973).
10. Jang, P.S. and Wald, R, *J. Math. Phys.* **18**, 41 (1977).
11. Schoen, R. and Yau, S.T., preprint, (1978).

Discussions

Martin Walker: You have spoken about the total mass measured at spacelike infinity. Are there any results on the positivity of the Bondi-mass measured at null infinity, which decreases when the system radiates?

D Brill: There seem to be two separate issues: (i) Must Bondi mass be positive at "the start" (retarded time- ∞)? (ii) Can more mass than the starting mass be radiated away? The first issue probably depends on the relation between Bondi mass and spacelike mass, which is still terra incognita. The second issue probably depends on cosmic censorship, because there is no restriction on what may be emitted by a naked singularity.

I Adamut: Did you calculate the size order of this local energy for some natural configuration?

D Brill: I can write down solutions with any value of m between 0 and ∞. These solutions can for example be obtained from each other by simple scaling.

T Padmanabhan: (1) Consider an arbitrary closed region of space-time which is finite. Can we talk meaningfully about the mass inside the region in some way? (2) Can we suppose that the mass calculated in a particular topology is same for two arbitrary surfaces, or for two regions of integration with same surface?

D Brill: (1) This may be possible in some cases, e.g. when the space-time is static, so that one can use the killing vector to define an energy density. (2) In the definition of mass I gave, the surface must go to infinity in the limit. Therefore, there could be two distinct surfaces only if there are two (or several) aymptotically flat regions. There seems to be no particular relation between the possible values of m in these different regions, except that if $m=0$ in any one region, the whole space-time is flat (i.e. there cannot be any distinct asymptotically flat regions).

M Sohnius: Could eventually 'physics', i.e. the actual matter coupled to gravity provide simpler means to render m positive like it does in supergravity?

D Brill: Of course, the discussion I gave allows classical matter via $\mathcal{J}$ and J. The matter in supergravity is very special—a kind of quantum 'square root' of gravity—and indeed allows a manifestly positive expression for m. It would be interesting to find viable, classical matter sources which guarantee positivity of m.

E Gunzig: You mention that your definition agrees with the momentum four-vector $\mathbf{P}_\mu$ associated to the pseudo-tensor $t_{\mu\nu}$; what is exactly the present status of $t_{\mu\nu}$ in the definition of the gravitational total energy?

D Brill: The total energy can be defined as a surface integral without ever using $t_{\mu\nu}$. However, to show that it transforms as a four-vector under asymptotic Lorentz transformations it seems simplest to re-express it in terms of $t_{\mu\nu}$.

B R Iyer: In the axially symmetric case what can you say about the angular momentum density integrated over volume of space?

D Brill: Total angular momentum can be defined—whether the system is axially symmetric or not—provided K_{ij} falls off sufficiently rapidly at infinity. It should be limited by some expression like (mass times suitably defined linear extent of the system). In the axially symmetric case this could probably be spelt out in detail, but I am not aware of work on this.

C Sivaram: (1) Is time-symmetry built into the proof of the positivity of the energy? The solution need not be time-symmetric. (2) Is the procedure valid for asymptotically non-flat spaces? (3) Does the positivity of the energy depend on gauge invariance like in electrodynamics?

D Brill: (1) The usual meaning of 'time symmetry' is $K_{ij}=0$ (i.e. the whole extrinsic curvature tensor vanishes; the spacelike surface is totally geodesic). This is *not* necessary to prove positivity of mass. However, all proofs known to me, except the weak-field case have to assume $K=g^{ji}K_{ij}=0$ (the mean curvature vanishes; the spacelike surface is one of maximal volume).

(2) I know of no useful definition of 'mass' in spaces that are not asymptotically flat. The procedures used here may still be valid, but prove other things. For example, in closed spaces the procedure used for weak fields may yield second order conditions on first order perturbations (linearization instability).

(3) No invariance is assumed except the asymptotic Lorentz invariance of asymptotically flat spaces.

Gravitational Radiation—The Theoretical Problems

A. PAPAPETROU

The field equations of general relativity being differential equations of the hyperbolic type they predict the existence of gravitational radiation. This is seen most clearly when one discusses the problem of gravitational *shock waves,* one of the few problems which have been solved exactly in general relativity.[1–4] Indeed, the results of this discussion are completely analogous to those obtained for electromagnetic shock waves in special relativity. More intuitively, though indirectly, the existence of gravitational radiation follows from the non-existence of periodic solutions of the Einstein eqns.[5,6]; if there is no gravitational radiation one should expect the possibility of strictly periodic motions of binary stars, with the two components moving on circular orbits. However, there is no *exact* method known for determining the energy of the gravitational radiation emitted by an insular distribution of matter. The problem is very difficult because of the complicated non-linear structure of the Einstein equations.

In the linear approximation valid for weak gravitational fields, the problem of gravitational radiation is treated in complete analogy with the corresponding problem of the Maxwell theory. This has been shown by Einstein who in 1918 derived the formula giving the energy of the radiation emitted by the gravitational 4-pole.[7] This formula was obtained with the help of the pseudotensorial conservation law of energy and momentum by an integration on a 2-dimensional sphere at $r \to \infty$. The energy of the radiation can also be calculated by an integral over the sources of the field according to a method developed by Eddington[8]. Fddington's method has been completed by Peres[9] and it was then proved that the total energy of the gravitational radiation calculated by this method is exactly the same as the one obtained by the use of the pseudotensorial conservation law.[10]

There is no doubt that the total energy of the gravitational radiation calculated thus is physically meaningful, at least in the case of a radiating material system which starts from a stationary initial state and also tends to a stationary final state. However, it is not difficult to see that these methods of calculation are applicable only to material systems whose internal motions are due mainly to non-gravitational forces. Indeed, in the linear approximation the first-order term $h_{\mu\nu}$ of $g_{\mu\nu}$

$$g_{\mu\nu} = \eta_{\mu\nu} + h_{\mu\nu} + \dots \tag{1}$$

has to be calculated from the field equation (2)

$$\Box \gamma_{\mu\nu} = -2k\, T_{\mu\nu},\ \gamma_{\mu\nu} \equiv h_{\mu\nu} - \tfrac{1}{2}\eta_{\mu\nu}\, \eta^{\rho\sigma}\, h_{\rho\sigma} \quad (2)$$

the $\gamma_{\mu\nu}$ satisfying the coordinate condition

$$(\eta^{\nu\rho}\, \gamma_{\mu\nu}),_{\rho} \equiv \gamma^{\rho}_{\nu},_{\rho} = 0 \quad (2a)$$

The source term $T_{\mu\nu}$ has to satisfy the energy-momentum conservation law of special relativity,

$$T^{\nu}_{\mu,\nu} \equiv (\eta^{\nu\rho}\, T_{\mu\rho}),_{\nu} = 0 \quad (3)$$

and thus independent of the condition (2a).

The fact that the matter tensor satisfies eqn. (3) instead of the exact equation

$$T^{\nu}_{\mu;\nu} = 0 \quad (4)$$

means that the linear approximation is applicable only to material systems in which the gravitational forces are negligibly small in comparison to non-gravitational forces. Therefore the Einstein formula for the energy of the gravitational 4-pole radiation is not directly applicable to the case of a binary star, where there are only gravitational forces present. It seems of course reasonable to assume that the 4-pole formula should be applicable to the case of the binary star *qualitatively* as well. However, it must be remembered that this is an assumption which has to be checked by direct calculation. This calculation will be needed in the near future, as it will become clear from the following remarks concerning the possibilities of observing and measuring gravitational radiation.

Estimates of the energy of gravitational radiation by means of the 4-pole formula clearly show that radiation produced in the laboratory, i.e. emitted from terrestrial sources, will be too weak for experimental detection, at least in the near future. Therefore, experiments aiming at the detection of gravitational radiation have to concentrate on radiation emitted by astrophysical sources.

There are two different types of such sources of radiation considered to be observable in the near future: supernovae explosions and binary stars. In the first type the forces acting inside the system are mainly non-gravitational, and consequently the results obtained from the linear approximation are applicable. But it has to be remembered that the detection of the radiation emitted in a supernova explosion will constitute only a proof of the existence of gravitational radiation and not a quantitative proof of general relativity; the reason being that we do not know in sufficient detail what happens in the explosion.

In the case of a binary star the intensity of the emitted gravitational radiation increases with decreasing period of rotation of the system. But the estimate is that even for the most rapidly rotating binary stars the emitted radiation will be too weak to be observed directly by any type of receiver in the near future. There is, however, a cumulative effect which offers a possibility for determining indirectly the energy of the gravitational radiation emitted by the system: the loss of energy of the system leads to a contraction and, therefore, to a decrease of the period of rotation, and this change of the period may become observable, especially in the case of rapidly rotating binaries. Evidently, observations of this type will not only constitute a proof of the existence of gravitational radiation, but also allow a quantitative comparison with the predictions of general relativity. This shows clearly the need for deriving the exact formula giving the energy of the radiation emitted by a binary star.

In principle this formula should be obtained from the solution of the so-called problem of motion in general relativity. This is the problem of determining the equations of motion of bodies interacting through their gravitational field from the field equations of general relativity. The case of spherically symmetric bodies moving with velocities $v \ll c$ has been treated in detail by two different weak-field approximation methods up to the first post-Newtonian approximation.[11–16] Radiation effects are not contained in these calculations and it is doubtful whether this could be obtained by continuing these methods to higher approximations. Indeed, a basic feature of these methods is the following. Using the weak field assumption one reduces the Riemannian d'Alembert operator.

$$_{(g)}\Box \equiv g^{\mu\nu}\,\partial_\mu\partial_\nu, \tag{5}$$

appearing in the main part of the field equations, to the corresponding operator in Minkowski space,

$$\Box \equiv \eta^{\mu\nu}\,\partial_\mu\partial_\nu = \Delta - \partial_t\partial_t;\ \Delta = \partial_x\partial_x + \partial_y\partial_y + \partial_z\partial_z \tag{6}$$

And then again, using the slow motion assumption $v \ll c$, one retains only the operator Δ treating the term $-\partial_t\partial_t$ as a perturbation. It is almost certain that in order to determine radiation effects one has to use an approximation method retaining the operator $\Box$. In the different steps of approximation the field equations will then be inhomogeneous wave equations in Minkowski space, of which one has to determine the retarded solutions. This is not an easy task, as the number of source terms increases rapidly with the order of approximation. Finally, besides the question of convergence, since the solution obtained in this manner is retarded with respect to the Minkowski metric one is confronted by an important question: Will it also be retarded with respect to the actual Riemannian metric, i.e. will it be free of any incoming gravitational radiation?

Because of these difficulties there has been until now no systematic discussion up to radiation effects of the two-body problem. Two simpler problems containing purely gravitational motions have been discussed recently leading to interesting results. These shall be briefly described now.

The first of these problems has been considered by Cooperstock and is the following.[17, 18] Two bodies of equal masses are initially kept at rest by a support balancing their mutual gravitational attraction. The support is broken at $t = 0$; more exactly, it is assumed that the stresses keeping the two bodies at rest for $t < 0$ are removed gradually in the interval $0 < t < t^*$. Therefore, the two bodies will start moving at $t = 0$ and their motion will be a free fall for $t > t^*$. The problem is simpler than that of the binary star because the field is stationary for $t < 0$: it is sufficient to determine a formal solution of the field equations for $t \geqslant 0$ which joins satisfactorily with the initial static solution in order to be sure that there is no incoming radiation. Further simplification results from the fact that the field is axially symmetric. Moreover, if we assume that the initial positions of the two bodies are at the points $z = \pm a$ of the symmetry axis Oz, then the field is symmetric also with respect to the plane $z = 0$.

The results of the detailed discussion of this problem are as follows. During the stress-breaking time interval $0 < t < t^*$, the linear approximation can be considered sufficient. However, for $t > t^*$ it is indispensable to solve the field equations up to the third order of approximation. Moreover, it was also necessary to consider the interior of the two bodies. The authors assumed that each one of the bodies consisted of perfect fluid enclosed in an envelope of radius r_0. Discussing carefully each one of the numerous source terms appearing in the inhomogeneous wave equations, to which reduce the field equations, they arrived at the following result: During the time interval $t^* < t < t^* + \delta t^*$, where δt^* is so small that the corresponding variation δa of the half distance a of the bodies satisfies $\delta a \ll a$, the energy of the emitted gravitational radiation is given by the Einstein expression for the gravitational 4-pole multiplied by the number $(a/r_0)^4$. The radius r_0 cannot be arbitrarily small, because it has been assumed that the field is weak everywhere. Nevertheless, we can have $a/r_0 \gg 1$. It is highly desirable to have this physically surprising result checked by an independent calculation, and also to see how does the result depend on the internal structure of the bodies, if it does. Moreover, it would be interesting to know what happens for $t > t^* + \delta t^*$.

The second problem taken up by Rosenblum[19] is as follows: Consider two bodies of equal masses approaching from infinity to initial orbits which are parallel lines at a distance $2a$. Their initial velocities are assumed to be sufficiently large in order to have only a small scattering angle. Again the simplification is due to the fact that for $t \to -\infty$ we have the two bodies at infinite distance moving with constant velocities, so that the initial field contains no gravitational radiation. The calculalions have been based on the assumption of point particles. The author hopes that certain divergences resulting from this assumption will be transformed away by renormalization of the active gravitational mass. The final result is the following; the total energy of the gravitational radiation emitted during the scattering exceeds the energy calculated from the formula for the gravitational 4-pole by a factor nearly equal to 2.35.

The preceding results underline the necessity of solving the problem of motion for the binary star up to radiation effects. The experience obtained by the discussion of these two special problems will certainly be useful for the solution of the problem of the binary star. But it will be indispensable to learn how to solve the radiation problem without the assumption of the field being initially time-independent.

References

1. Stellmacher, F.K., *Math. Ann.* **115**, 740 (1938).
2. Treder, H., *Ann. Physik* **7** (2), 225 (1958).
3. Papapetrou, A., and Treder, H., *Math. Nachr.* **20** 53 (1959); **23**, 371 (1961).
4. Lichnerowicz, A., *C.R. Acad. Sc. Paris* **A273**, 528 (1971); **A276** 1385 (1973).
5. Papapetrou, A., *Ann. Physik* **6** (20), 399 (1957).
6. Papapetrou, A., *Ann. Physik* **7** (1), 186 (1958).
7. Einstein, A., *Berl. Berichte*, p. 154 (1918).
8. Eddington, A.S., *The Mathematical Theory of Relativity*, Cambridge, p. 248 (1954).
9. Peres, A., *Phys. Rev.*, **128**, 2471 (1962).
10. Papapetrou, A., *Ann. Inst. H. Poincare* **1**, 117 (1964).
11. Einstein, A., and Grommer, J., *Sitzber. Berl. Akad.* **2** (1927).
12. Einstein, A., Infeld, L., and Hoffmann, B., *Ann. Math.* **39**, 65 (1938).
13. Einstein, A. and Infeld, L., *Ann. Math.* **41**, 455 (1940).
14. Einstein, A., and Infeld, L., *Can. J. Math.* **1**, 209 (1949).
15. Fock, V.A., *J. Phys.* USSR **1**, 81 (1939).
16. Papapetrou, A., *Proc. Roy. Soc.* (London) **A64**, 57 (1951).
17. Cooperstock, F.I., *Phys. Rev.* **D10** 3171 (1974).
18. Cooperstock, F.I., and Lim, P.H., *Phys. Rev.* **D10**, 2105 (1977).
19. Rosenblum, A., *Phys. Rev. Lett.* **41**, 1003 (1978).

Discussions

M Sohnius: How valid do you think were the theoretical values with which Joseph Taylor compared his measurements in the binary pulsar system?

A Papapetrou: Evidently, the comparison has been made using the Einstein 4-pole formula, as we have at present no other theoretical result for the binary star.

S D Verma: What is the present experimental situation as regards the sensitivity of detection of gravitational radiation by a few of the best experiments (e.g., cooled experiments of Fairbanks of Stanford)?

A Papapetrou: It is hoped that improved receivers of gravitational radiation will have the sensitivity needed for observing the radiation emitted in a supernova explosion.

E Gunzig: Does the solution for a two bodies free-fall depends on the initial preparation ($0 \leqslant t \leqslant t^*$), i.e. on the initial conditions for this problem?

A Papapetrou: The Einstein's 4-pole formula is sufficient for the interval $0 < t < t^*$, independently of the detailed manner in which the stresses have been brought to the value zero. But one has to keep $t^* > 0$, since the limit $t^* \to 0$ would lead to some divergent terms.

P C Vaidya: The two bodies you took were not point particles but of finite radius r_0. Is this true? Why not point particles?

A Papapetrou: The field has to be weak everywhere. Consequently, if m is the mass of each one of the bodies, we must have $r_0 \gg m$.

A K Pandey: The observations and measurement of electromagnetic radiation by photoelectric effect requires an 'absorber' at least 10^{-8} cm thick. For observing gravitational radiation correspondingly we require it to be 10^{40} times (e^2/m^2G) thicker? It is the size of universe. Any comment of such observability?

A Papapetrou: For comparison we have to put the mass of the receiver (which is a macroscopic apparatus) and not the mass of the electron in place of m in the expression $e^2/m^2 G$.

S Ramadurai: How will the radiation affect the results from binary pulsars?

A Papapetrou: It is difficult to make an estimate of this effect.

Stationary, Axially Symmetric Solutions of Einstein's Equations

D.M. CHITRE

1. Introduction

Soon after Einstein gave formulation of General Theory of Relativity, Weyl[1] (1917) showed that the static, axially symmetric class of vacuum gravitational field equations can be solved completely. The stationary, axially symmetric fields are the ones which should logically be considered next, both from a mathematical and a physical point of view. It took about forty-five years before a physically realistic solution to these equations could be found. This is the celebrated Kerr[2] metric representing gravitational field of a spinning mass, and was found as an example of algebraically special metrics. In 1968, Ernst[3] gave a formulation of the field equations in which the Kerr metric took on a particularly simple form. Using the Ernst formulation, Tomimatsu and Sato[4] found a few more solutions by generalizing the Kerr metric. Thus, in spite of the fact that the field equations are remarkably simple and should certainly possess a large class of physically realistic solutions, only a handful of such solutions were known till a year ago.

A systematic study of the field equations show a remarkable amount of symmetry and internal structure. Geroch[5] in a couple of papers in 1971 and 1972 showed that Einstein's equations for stationary, axisymmetric systems give rise to an infinite hierarchy of potentials. Furthermore, generalising Ehlers'[6] transformation, he showed the existence of an infinite-dimensional symmetry group of transformations which leave Einstein's equations invariant. Geroch conjectured that this group freedom should be sufficient to generate all stationary solutions. There were mainly two major difficulties in trying to prove that conjecture. First, it was not clear how Geroch's formalism related to the known stationary, axisymmetric solutions. Secondly, the transformations were written only in their infinitesimal forms and they typically involved an infinite number of new potentials. Thus, it was not clear how one could get a finite transformation out of it and whether it could lead to a physically realistic metric.

In the next section, we will rewrite and generalize Geroch's formalism as started by Kinnersley[7] and given by Kinnersley and me.[8] This will enable us to write down the infinitesimal generators of the entire group, *K*—the symmetry group for the vacuum field equations. In the third section, we will give an algebraic classification scheme[9] to characterize certain stationary solutions of Einstein's equations. It categorizes an infinite series of solutions, the lowest (non-flat) members of which are the

Schwarzschild solution for the static case and the Kerr solution for the stationary case. The characterization is based on certain algebraic relationships among the hierarchy of potentials; and it gives the Tominatsu-Sato solutions as the higher members. In the last section, we discuss various interesting subgroups of K_0. First, we identify[10] that subgroup which generates all static metrics starting from the flat space-time. Next, we study[11] a certain subgroup $B \subset K$. This subgroup preserves flat space-time. Furthermore, when applied to the static metrics of the above classification scheme, the transformations can even be written down in a finite form. For example, when it is applied to the Schwarzschild metric, it gives rise to the Kerr metric.The subgroup maps the second static member into a new five-parameter stationary solutions which contoin $\delta = 2$ Tomimatsu-Sato solution as a special case. Furthermore, it is shown[12] how to construcs an infinite parameter, stationary solution when the subgroup B is applied on any Weyl metric. Lastly, we briefly discuss a certain subgroup A^P, recently found by Hoenselaers, Kinnersley and Xanthopoulos[13]. It generates asymptotically flat space-times with arbitrary multipole moments.

The lack of space restricts us to talk only about the structure of the internal group, K and its application to generate asymptotically flat solutions. Interested readers should refer to Maison[14], and Belinski and Zakharov[15] for applications of inverse acettering method to Einstein's equations, and to Harrison's[16] work for the application of Bäcklund transformation to generate new solutions.

2. Formalism

Stationary, axially symmetric gravitational fields are space-times which contain two commuting Killing vectors. Let us then choose the metric to be independent of two of the co-ordinates, say $x^1 = t$ and $x^2 = \phi$. We also assume that the space-time is orthogonally transitive.[17] We can then write the metric in the following form

$$ds^2 = f_{AB}dx^A dx^B - \exp(2\Gamma)\,\delta_{MN}dx^M dx^N$$

$$A, B = 1, 2; \quad M, N = 3, 4 \tag{1}$$

where f_{AB}, Γ are functions of x^3, x^4.

We have,

$$f^{Ax} f_{xB} = -\rho^2 \delta^A{}_B \tag{2}$$

where indices are raised and lowered using $\varepsilon_{AB} = \pm 1$.

We introduce the two-dimensional gradient operators

$$\nabla = (\partial_3, \partial_4), \ \tilde{\nabla} = (\partial_4, -\partial_3) \tag{3}$$

The Einstein equations fall into three groups: R_{AB}, R_{AM} and R_{MN} of which R_{AM} vanish identically for our block diagonal metric. The $R_{MN} = 0$ equations can be shown to lead to integrable equations for Γ, which can be solved by quadrature, once f_{AB} are known. The $R_{AB} = 0$ equations involve only f_{AB} and can be cast in the form of a vanishing divergence

$$\nabla.[\rho^{-1} f^{Ax} \nabla f_{xB}] = 0 \tag{4}$$

The above field equations imply the existence of potentials ψ_{AB} such that

$$\nabla \psi_{AB} = -\rho^{-1} f_A^x \tilde{\nabla} f_{xB} \tag{5}$$

Analogous to the Ernst formulation, the complex combination

$$H_{AB} = f_{AB} + i\psi_{AB} \tag{6}$$

satisfies
$$\nabla H_{AB} = -i\rho^{-1} f_A^x \tilde{\nabla} H_{xB} \tag{7}$$

One can show that these field equations imply the existence of an infinite hierarchy of potentials $\overset{n+1}{H}_{AB}$, which obey the following field equations:

$$\nabla \overset{n+1}{H}_{AB} = -i\rho^{-1} f_A^x \tilde{\nabla} \overset{n+1}{H}_{xB} \tag{8}$$

where $\overset{n+1}{H}_{AB}$ is constructed from the lower order potentials as

$$\overset{n+1}{H}_{AB} = i(\overset{1n}{N}_{AB} + H_{Ax} \overset{n}{H}{}_B^x) \tag{9}$$

where
$$\nabla \overset{1n}{N}_{AB} = H_{xA}^* \nabla \overset{n}{H}{}_B^x \tag{10}$$

The hierarchy is constructed starting with $n = 1$, with $\overset{1}{H}_{AB} \equiv H_{AB}$. The field equations for $\overset{n}{H}_{AB}$ also guarantee the existence of the potentials $\overset{mn}{N}_{AB}$ which satisfy

$$\nabla \overset{mn}{N}_{AB} = \overset{m}{H}{}_{xA}^* \nabla \overset{n}{H}{}_B^x \tag{11}$$

where m and n take on all non-negative integral values with the provisio that

$$\overset{0}{H}_{AB} = i\varepsilon_{AB} \tag{12}$$

and
$$\overset{0n}{N}_{AB} = -i \overset{n}{H}_{AB} \tag{13}$$

Notice that the hierarchy of potentials is defined only up to a constant. We would use this gauge freedom whenever necessary.

Now we turn to the question of how to calculate these potentials for a given space-time, The approach is to use generating functions $F_{AB}(t)$ and $G_{AB}(s, t)$ defined by,

$$F_{AB} = \sum_{n=0}^{\infty} t^n \overset{n}{H}_{AB}, \tag{14}$$

and
$$G_{AB} = \sum_{n=0}^{\infty} \sum_{m=0}^{\infty} s^m t^n \overset{mn}{N}_{AB} \tag{15}$$

In this section, we will give a method of computing F_{AB} and relegate the calculation of G_{AB} to the fourth section, where we will see that in a perticular gauge, G_{AB} can be expressed as an algebraic combination of F_{AB}'s.

From the field equations (8) and the recursion relations (9) and (10), one can show that F_{AB} obeys

$$\nabla F_{AB} = - i\rho^{-1} f_A^x \tilde{\nabla} F_{xB} \tag{16}$$

$$\nabla F_{AB} = it\,[(H_{Ax} + H_{xA}^*)\,\nabla F_B^x + F_B^x\,\nabla H_{Ax}] \tag{17}$$

Writing out the $A = 1$ component of eqn. (16) and with suitable manipulation, one gets

$$f_{11}\nabla^2 F_{1B} = \nabla H_{11}\cdot F_{1B} \tag{18}$$

The $A = 2$ component similarly yields

$$f_{22}\nabla^2 F_{2B} = \nabla H_{22}\cdot \nabla F_{28} \tag{19}$$

The eqn. (17) give rise to the following first order differential equations for F_{1B} and F_{2B}:

$$(1 - 2tz)\,\tilde{\nabla} H_{11}\cdot\nabla F_{1B} + 2t\rho\nabla H_{11}\cdot\nabla F_{1B} = it\,(\nabla H_{11}\cdot\tilde{\nabla} H_{12})\,F_{1B} \tag{20}$$

$$(1 - 2tz)\,\tilde{\nabla} H_{22}\cdot\nabla F_{2B} + 2t\rho\,\nabla H_{22}\cdot\nabla F_{2B} = it\,(\nabla H_{21}\cdot\tilde{\nabla} H_{22})\;F_{28} \tag{21}$$

where we have now taken the co-ordinates X^3 and X^4 as ρ and z respectively.

Equation (20) can be solved by method of characteristics. Its solution will contain one arbitrary function, say g. When this is inserted in eqn. (18), we get a linear second-order ordinary differential equation for g. The two linearly independent solutions correspond to F_{11} and F_{12}. F_{2B} may be obtained algebraically, once F_{1B} is known.

3. Classification

Appropriate field equations given in eqn. (8) can be combined to give

$$\nabla(i\overset{n+1}{H}_{11} - \overset{n}{H}_{12}) = - i\rho^{-1}\,[f_{12}\,\tilde{\nabla}(i\overset{n+1}{H}_{11} - \overset{n}{H}_{12}) - f_{11}\,\tilde{\nabla}(i\overset{n+1}{H}_{21} - \overset{n}{H}_{22})] \tag{22}$$

and

$$\nabla(i\overset{n+1}{H}_{21} - \overset{n}{H}_{22}) = - i\rho^{-1}\,[f_{22}\,\tilde{\nabla}(i\overset{n+1}{H}_{11} - \overset{n}{H}_{12}) - f_{21}\,\tilde{\nabla}(i\overset{n+1}{H}_{21} - \overset{n}{H}_{22})] \tag{23}$$

which imply
$$i\overset{n+1}{H}_{11} = \overset{n}{H}_{12} \tag{24}$$

if and only if
$$i\overset{n+1}{H}_{21} = \overset{n}{H}_{22} \tag{25}$$

Taking the gradient of eqn. (9) and using eqn. (10), one obtains

$$\nabla(i\overset{n+2}{H}_{11} - \overset{n+1}{H}_{12}) = i\,[(H_{11} + H_{11}^*)\,\nabla(i\overset{n+1}{H}_{21} - \overset{n}{H}_{22}) - (H_{21}^* + H_{12})$$
$$\nabla(i\overset{n+1}{H}_{11} - \overset{n}{H}_{12}) + (i\overset{n+1}{H}_{21} - \overset{n}{H}_{22})\,\nabla H_{11} - (i\overset{n+1}{H}_{11} - \overset{n}{H}_{12})\,\nabla H_{12}] \tag{26}$$

It follows from eqns. (24), (25) and (26) that if

$$i \overset{n+1}{H}_{11} = \overset{n}{H}_{12} \tag{27}$$

for some value of n, then it would hold good for all higher values also.

We classify our solutions by the above single (complex) algebraic condition. Thus, the lth order solutions are characterized by the fact that l is the lowest integral value of n for which eqn. (27) is satisfied.

We find that for the static case, the Weyl solution of Zipoy-Voorhees class given by

$$f_{11} = \left(\frac{x-1}{x+1}\right)^l \tag{28}$$

is the lth order solution in our classification. x and y are the usual prolate spheroidal co-ordinates related to cylindrical co-ordinates ρ and z by

$$\rho^2 = (x^2 - 1)(1 - y^2)$$

$$z = xy \tag{29}$$

For example, for $l = 1$, the metric is the Schwarzschild solution, where the unit of length is chosen to be mass, m. Simple computation gives

$$i\overset{2}{H}_{11} = 2i(x-1)y = \overset{1}{H}_{12} \tag{30}$$

For general l, we compute the potentials by the method given in the last section. For the metric given by eqn. (28)

$$\left.\begin{aligned} H_{11} &= F_{11} = \left(\frac{x-1}{x+1}\right)^l \\ H_{12} &= 2iy(x-l) \end{aligned}\right\} \tag{31}$$

Equation (20) then becomes

$$(1 - 2txy)\, F_{1B,\,y} - 2t(x^2 - 1)\, F_{1B,\,x}$$

$$= 2t(x - l)\, F_{1B} \tag{32}$$

The corresponding characteristic equations are

$$\frac{dy}{1 - 2txy} = \frac{-dx}{2t(x^2 - 1)} = \frac{dF_{1B}}{2t(x - l)\, F_{1B}} \tag{33}$$

with solutions

$$x - 2ty = a(x^2 - 1)^{1/2} \tag{34}$$

and

$$F_{1B} = \frac{b}{(x^2 - 1)^{1/2}} \left(\frac{x-1}{x+1}\right)^{l/2} \tag{35}$$

Eqn. (34) determines a family of characteristic curves in the x, y plane, and eqn. (35) tells how F_{1B} varies along the curves. Here a, b are constant along each curve, and we may regard b as an arbitrary function of a. To determine $b(a)$, we substitute eqns. (34), (35) into eqn. (18). We find

$$(a^2 + 4t^2 - 1)\, b'' + 3ab' + (1 - l^2)\, b = 0 \tag{36}$$

with solutions $R^{-1}(a \pm R)^l$, where

$$R = (a^2 + 4t^2 - 1)^{1/2}$$
$$= S(x^2 - 1)^{-1/2} \quad (37)$$

with $$S^2 = (1 - 2tz)^2 + (2t\rho)^2 \quad (38)$$

Thus, we obtain the generating functions

$$F_{11} = \frac{c(t)\,t}{S}\left(\frac{x - 2ty - s}{x + 1}\right)^l \quad (39)$$

$$F_{12} = \frac{id(t)}{s}\left(\frac{x - 2ty + s}{x + 1}\right)^l \quad (40)$$

where $c(t)$, $d(t)$ are essentially undetermined. Notice, that the choice of $c(t)$ and $d(t)$ would restrict the gauge. The simplest choice (and hence the simplest gauge) would be $c = d = 1$. Then $F_{11} + itF_{12}$ will be a polynomial in t of degree l. Hence

$$i\overset{n+1}{H}_{11} = \overset{n}{H}_{12} \text{ for } n \geqslant l \quad (41)$$

For the stationary (non-static) case, one finds that the Kerr solution satisfies (27) for $n = 1$, i.e. one can choose a suitable gauge in which the appropriate potentials are

$$\left.\begin{aligned}
f_{11} &= \frac{p^2x^2 + q^2y^2 - 1}{(px + 1)^2 + q^2y^2}, \quad v\psi_{11} = \frac{2qy}{(px + 1)^2 + q^2y^2} \\
f_{12} &= \frac{2q\,[(1 - y^2)\,x + (x^2 - y^2)\,p]}{(px + 1)^2 + q^2y^2} \\
\psi_{21} &= -\frac{2y[(px + 1)^2 - q^2]}{p[(px + 1)^2 + q^2y^2]} \\
\text{and} \qquad i\overset{2}{H}_{11} &= \overset{1}{H}_{12}\,(= f_{12} + i\psi_{21} + 2ixy)
\end{aligned}\right\} \quad (42)$$

The higher-order stationary solutions in our classification scheme would contain the TS solutions; this has been checked only for $\delta = 2r$, TS solution.

It is not clear, how one would obtain all solutions belonging to a particular class. For example, if we restrict ourselves to the simplest possible case, i.e. the static $n = 1$ case; the condition

$$i\overset{2}{H}_{11} = \overset{1}{H}_{12}$$

reduces to the following two equations:

$$(\xi^2 - 1)\left(x\frac{\partial\xi}{\partial x} - \xi\right) = \xi^2\left(\frac{1}{y}\frac{\partial\xi}{\partial y} - y\frac{\partial\xi}{\partial y}\right) \quad (43)$$

and $$(\xi^2 - 1)\left(y\frac{\partial\xi}{\partial y} - \xi\right) = \xi^2\left(\frac{1}{x}\frac{\partial\xi}{\partial x} - x\frac{\partial\xi}{\partial x}\right) \quad (44)$$

where $$\xi = \frac{f_{11} + 1}{1 - f_{11}} \quad (45)$$

$\xi = x$ or $\xi = y$ satisfy eqns. (43) and (44). The most general solutions of these equations are given by

$$F\left[\frac{\xi}{x}, \frac{\xi^2-1}{y^2-1}\right] = 0 \tag{46}$$

and

$$G\left[\frac{\xi}{y}, \frac{\xi^2-1}{x^2-1}\right] = 0 \tag{47}$$

where F and G are arbitrary functions. It is not clear whether one can find suitable forms for F and G so as to produce a solution for ξ which is neither x nor y.

4. The group K

Ehlers[6] found out a remarkable internal symmetry in the field equations. His transformation, the gravitational duality rotation, leads to a one-parameter family of physically different solutions, when applied to any stationary vacuum solution. Geroch[5] generalized Ehlers's work to show that there actually exists an infinite-dimensional symmetry group K for the stationary vacuum Einstein equations. The infinitesimal part of K can be built up by induction. We construct the group K by taking all possible commutators, until closure is obtained. It turns out that the transformations can be written down as $\overset{k}{\gamma}_{AB}$, where k is any integer $-\infty < k < \infty$. The infinitesimal parameters $\overset{k}{\gamma}_{AB}$ are real and symmetric. To write down the corresponding effect they produce on the hierarchy of potentials, we use a gauge in which the potentials satisfy the following relations:

$$\overset{mn}{N}_{AB} - \overset{nm}{N}{}^{*}_{BA} = \overset{m}{H}{}^{*}_{xA}\overset{n}{H}{}^{x}_{B} \tag{48}$$

and

$$\overset{m,n+1}{N}_{AB} - \overset{m+1,n}{N}_{AB} = i\overset{m1}{N}_{Ax}\overset{n}{H}{}^{x}_{B} \tag{49}$$

Eqn. (48) is obtained by integrating (11), and (49) is arrived at by inserting (8) in (11). The choice of gauge consists in setting certain constants to be zero.

The $\overset{k}{\gamma}_{AB}$ transformations are then given for infinitesimal values of the parameter by

$$\begin{aligned}\overset{k}{\gamma}_{AB}\colon \overset{mn}{N}_{AB} \to \overset{mn}{N}_{AB} &+ \overset{k}{\gamma}_{Ax}\overset{m+k,n}{N}{}^{x}_{B} + \overset{k}{\gamma}_{xB}\overset{m,n+k}{N}{}^{x}_{A}\\ &+ \overset{k}{\gamma}{}^{xY}\Big(\sum_{s=1}^{k}\overset{ms}{N}_{Ax}\overset{k-s,n}{N}_{YB}\Big)\end{aligned} \tag{50}$$

First, we notice that $\overset{k}{\gamma}_{12}$ can give rise to all static metrics starting with flat space-time. The proof[10] depends on two observations: (i) $\overset{k}{\gamma}_{12}$ leave static metrics static, (ii) the action of $\overset{k}{\gamma}_{12}$ can be written simply as

$$\overset{k}{\gamma}_{12}\colon \overset{01}{N}_{11} \to \overset{01}{N}_{11}\,[1 - 4\overset{k}{\gamma}_{12}\,(2\gamma)\,P_k\,(\cos\theta)] \tag{51}$$

Since it is no longer non-linear, thefin etriansforma tion can be written very easily as

$$\overset{01}{N}_{11} \rightarrow \overset{01}{N}_{11} \exp\left[-4\overset{k}{\gamma}_{12}\,(2\gamma)^k\, P_k\,(\cos\theta)\right] \tag{52}$$

Thus, if we start with flat space-time, any Weyl solution may be generated, provided only that we choose the origin to be a point where the ln (F_{11}) is non-singular.

Next, we consider the commuting group of transformations

$$\overset{k}{\beta} = \overset{k}{\gamma}_{11} + \overset{k+2}{\gamma}_{22},\ k = 0, 1, 2, \ldots \tag{53}$$

Using eqn. (50), we can calculate the effect of $\overset{k}{\beta}$ on $\overset{mn}{N}_{AB}$:

$$\begin{aligned}
\overset{k}{\beta} : \overset{mn}{N}_{11} \rightarrow \overset{mn}{N}_{11} + \overset{k}{\beta}\,[i\,(&\overset{m+k+1,n}{N_{11}} - i\ \overset{m+k,n}{N_{21}}) \\
&- i\,(\overset{m,n+k+1}{N_{11}} + i\ \overset{m,n+k}{N_{12}}) + \overset{m1}{N}_{11}\,(\overset{k+1,n}{N_{11}} - i\overset{kn}{N}_{21}) \\
&+ (\overset{m,k+2}{N_{11}} + i\ \overset{m,k+1}{N_{12}})\ \overset{0n}{N_{11}} \\
&+ \sum_s\ \overset{m,s+1}{N_{11}} + i\,\overset{ms}{N}_{12})\quad (\overset{k+1-s,n}{N_{11}} - i\ \overset{k-s,n}{N_{21}})]
\end{aligned} \tag{54}$$

and also similar expressions for $\overset{mn}{N}_{12}$, $\overset{mn}{N}_{21}$, and $\overset{mn}{N}_{22}$. However, the form of Eqn. (54) strongly suggests that a simplification would follow if we used instead certain linear combinations of the N_{AB}. We therefore define

$$\begin{aligned}
P_{on} &= \overset{on}{N}_{11} + i\,\overset{o,n-1}{N}_{12}, \\
P_{mn} &= \overset{mn}{N}_{11} - i\ \overset{m-1,n}{N_{21}} + i\ \overset{m,n-1}{N_{12}} + \overset{m-1,\,n-1}{N_{22}},\quad m > 0
\end{aligned} \tag{55}$$

In particular, since $\overset{00}{N}_{AB} = \varepsilon_{AB}$, we have

$$P_{01} = \overset{01}{N}_{11} + i = -\,i\,(H_{11} - 1) \tag{56}$$

The transformation $\overset{k}{\beta}$ can now be written in terms of what it does to P_{mn}:

$$\begin{aligned}
\overset{k}{\beta}\ :\ P_{on} &\rightarrow P_{on} + \overset{k}{\beta}\,(-\,2i\ P_{o,n+k+1} \sum_s P_{os}\,P_{k+2-s,n}) \\
P_{mn} &\rightarrow P_{mn} + \overset{k}{\beta}\,(2i\ P_{m+k+1,n} - 2i\ P_{m,n+k+1} \\
&\qquad + \sum_s P_{ms}\,P_{k+2-s,n})
\end{aligned} \tag{57}$$

where the summations run from $s = 1$ to $s = k + 2$.

For flat space-time all P_{mn}'s vanish. We see from eqn. (57) that after a transformation of $\overset{k}{\beta}$ (even a finite one), they must continue to vanish. Thus, the flat space-time will remain flat in the same gauge and in the same co-ordinates. If we now start with a space-time which is asymptotically flat, we can pick a gauge for the potentials such that $P_{mn} \rightarrow 0$ at spatial infinity. Since, $\overset{k}{\beta}$ preserves this condition, it automatically preserves asymptotic flatness, in the simplest sense that the Riemann tensor tends to zero.

The difficulty in getting a new asymptotically flat metric by the above method rests squarely in the problem of summing an infinite series whose first two terms are given by eqn. (57). The other equivalent way of looking at the problem is that we would in general be facing an infinite set of first order differential equations; the solution of which would give us a new space-time. We will see in turn two different ways of tackling the above problem.

(i) We apply the above subgroup to the metrics classified in the third section. It turns out that since the higher order potentials are related to the lower order ones in these cases, the infinite set of differential equations truncates into a finite one, and one can devise a method to solve these finite first order differential equations to get new metrics. Before one proceeds to do that, one should be careful about the choice of the gauge. In the third section, we chose the simplest gauge in which we set $c(t) = d(t) = 1$ in eqn. (39) and (40). But, notice that when we wrote the transformations $\overset{k}{\gamma}_{AB}$, we used the restricted gauge in which eqns. (48) and (49) held good. This choice of the gauge corresponds to choosing

$$c(t) = d(t) = (1 - 4t^2)^{-l/2} \tag{58}$$

for the metric

$$F_{11} = \left(\frac{x-1}{x+1}\right)^l \tag{59}$$

Let us now apply the transformations $\overset{k}{\beta}$ to the Schwarzschild metric, given by eqn. (59), with $l = 1$. One can show that there are only four independent P_{mn}'s and all of the higher ones are fixed linear combinations of them. Thus,

$$\begin{aligned} P_{01} &= \frac{2i}{x+1}, \quad P_{11} = \frac{4iy}{x+1} \\ P_{02} &= 0, \qquad P_{12} = 4i \\ P_{03} &= 2P_{01}, \quad P_{13} = P_{31} = 2P_{11} \\ P_{21} &= 2P_{01} - 4i \end{aligned} \tag{60}$$

and

and so forth. Furthermore, these relationships are preserved under $\overset{k}{\beta}$. Thus, for $\overset{0}{\beta}$ there are only two distinct equations:

$$\left.\begin{aligned} P_{01} &\to P_{01} + \overset{0}{\beta} P_{01} P_{11} \\ P_{11} &\to P_{11} + \overset{0}{\beta} (P_{11}P_{11} + 8i P_{01} + 16) \end{aligned}\right\} \tag{61}$$

while for $\overset{1}{\beta}$ there is only one:

$$P_{01} \to P_{01} + 4\overset{1}{\beta} (P_{01} P_{01} - 2i P_{01}) \tag{62}$$

The same recursion relations also cause the $\overset{k}{\beta}$ to repeat. For example, we find that

$$\overset{2}{\beta} = 4\overset{0}{\beta} \tag{63}$$

To obtain a finite transformation, one can look upon eqn. (61) as two coupled first order differential equations, the solutions of which are

$$P_{01} = \frac{2i}{a \cos 4\beta - ib \sin 4\beta + 1}$$
$$P_{11} = 4i\left(\frac{b \cos 4\beta - ia \sin 4\beta}{a \cos 4\beta - ib \sin 4\beta + 1}\right) \tag{64}$$

where a and b are integrating constants. From the initial conditions, we obtain $a = x$, $b = y$. Substituting in eqn. (64) we thus obtain the Kerr metric in the Ernst form.

A similar analysis has been carried through for the next static metric in the classification, i.e. the Zipoy-Voorhees metric with $l = 2$. Here one can show that there exist six independent P_{mn}'s, namely, P_{01}, P_{02}, P_{11}, P_{12}, P_{21} and P_{22}. The higher ones are given by the recursion relations

$$\left.\begin{aligned} P_{0,2l+n} &= 4^l P_{on}, & n &= 1, 2 \\ P_{2k+m,\, 2l+n} &= 4^{k+l} P_{mn}, & m, n &= 1, 2 \end{aligned}\right\} \tag{65}$$

Once more, one can show that the transformations $\overset{k}{\beta}$ preserve the above recursion relations. The $\overset{k}{\beta}$'s repeat themselves for $l = 2$ also, starting this time with

$$\overset{4}{\beta} = 8\overset{2}{\beta} - 16\overset{0}{\beta} \tag{66}$$

for each $\overset{k}{\beta}$, six transformation equations will be necessary. The six non-linear coupled first order differential equations turn out to be of the Riccati type and thus can be solved to generate a five parameter family of stationary, asymptotically flat space-time.[11]

(ii) We develop a technique here to get a stationary metric by applying $\overset{k}{\beta}$ to any static space-time. Thus, in general one would obtain an infinite parameter stationary metric starting with any Weyl solution. We start with the Riccati equation for the following infinite matrix:

$$\mathbf{P} = \begin{pmatrix} 0 & P_{01} & P_{02} & \cdots \\ 0 & P_{11} & P_{12} & \cdots \\ 0 & P_{21} & P_{22} & \cdots \\ \vdots & \vdots & \vdots & \end{pmatrix} \tag{67}$$

To handle all of the $\overset{k}{\beta}$'s at once, we let $\overset{k}{\beta}$ denote the final constant values. The instantaneous values will be $\lambda\overset{k}{\beta}$, where $0 \leqslant \lambda \leqslant 1$ is the variable of integration. Then, the original set of transformation eqns. (57) can be written as

$$\frac{d\mathbf{P}}{d\lambda} = \mathbf{PAP} + \mathbf{BP} + \mathbf{PB}^{\dagger} \tag{68}$$

where
$$\mathbf{A} = \begin{pmatrix} 0 & 0 & \overset{0}{\beta} & \overset{1}{\beta} & \cdots \\ 0 & \overset{0}{\beta} & \overset{1}{\beta} & \overset{2}{\beta} & \cdots \\ \overset{0}{\beta} & \overset{1}{\beta} & \overset{2}{\beta} & \overset{3}{\beta} & \cdots \\ \vdots & \vdots & \vdots & \vdots & \end{pmatrix} \tag{69}$$

and
$$\mathbf{B} = 2i \begin{pmatrix} 0 & 0 & 0 & 0 & 0 & \cdots \\ 0 & 0 & \overset{0}{\beta} & \overset{1}{\beta} & \overset{2}{\beta} & \cdots \\ 0 & 0 & 0 & \overset{0}{\beta} & \overset{1}{\beta} & \cdots \\ \vdots & \vdots & \vdots & \vdots & \vdots & \end{pmatrix} \tag{70}$$

where $\mathbf{B}^{\dagger}$ is the Hermitian conjugate of $\mathbf{B}$.

It was shown[10] that the solution of eqn. (68) can be written as

$$\mathbf{P}(1) = e^{\mathbf{B}} P(0) [I - C(1) P(0)]^{-1} e^{\mathbf{B}^{\dagger}} \tag{71}$$

where
$$C(1) = \int_0^1 e^{\lambda \mathbf{B}^{\dagger}} \mathbf{A} e^{\lambda \mathbf{B}} \, d\lambda \tag{72}$$

Eqn. (71) is only a formal solution, as it involves inverting an infinite matrix.

To get an explicit solution, we use the method of generating functions. The eqn. (71) can be re-expressed as

$$e^{-\mathbf{B}} \mathbf{P}(1) e^{-\mathbf{B}^{\dagger}} = \mathbf{P}(0) + e^{-\mathbf{B}} \mathbf{P}(1) e^{-\mathbf{B}^{\dagger}} C(1) \mathbf{P}(0) \tag{73}$$

Let us define

$$P'(1) = e^{-\mathbf{B}} P(1) e^{-\mathbf{B}^{\dagger}} \tag{74}$$

To illustrate the method, let us take only $\overset{0}{\beta} \neq 0$ for simplicity. Then, it is easy to show that

$$P_{01}(1) = \sum_{n=1}^{\infty} P'_{on}(1) \frac{(-2i\overset{0}{\beta})^{n-1}}{(n-1)!} \tag{75}$$

Writing eqn. (73) as

$$P'_{on}(1) = P_{on}(0) + \sum_{ij} P'_{oi}(1) C_{ij}(1) P_{jn}(0) \tag{76}$$

and computing $C_{ij}(1)$ explicitly from (72), one can show[12], that one obtains an integral equation of the Volterra type for

$$E(t) = \sum_{n=1}^{\infty} \frac{P'_{on}(1) \, t^{n-1}}{(n-1)!} \tag{77}$$

where the Kernel in the integral equation is of the form

$$K(t,t') = \sum_{i,j=1}^{\infty} \frac{t^{(i-1)} P_{ij}(0)\, t'^{(j-1)}}{(i-1)!\,(j-1)!} \tag{78}$$

The Kernel can be computed by doing Laplace integrals over

$$Q(t,t') = \sum_{m,n=0}^{\infty} t^m t'^n P_{mn}(0) \tag{79}$$

Thus, starting with any static metric, one can compute the Kernal $K(t, t')$, then the new stationary space-time is given as a solution of an integral equation. [$Q(t, t')$ is an algebraic combination of $G_{AB}(t, t')$ which can be obtained algebraically from $F_{AB}(t)$ in this particular gauge.]

Recently, Hoenselaers, Kinnersley, and Xanthopoulos[13] have found a new set of transformations, called $\mathbf{A}^{\mathrm{P}}$. They are built from infinite linear combinations of the $\overset{k}{\gamma}_{22}$. These transformations do not leave flat space-time invariant, but they map it into asymptotically flat space-time. Furthermore, they can be described very simply in terms of the generating functions and hence their applications require only algebraic manipulations. Their method can be used to produce exact, stationary solutions with any prescribed set of multipole moments.

References

1. Weyl, H., 'Zur Gravitationstheorie', *Ann. Physik* **54**, 117 (1917).
2. Kerr, R., 'Gravitatioal field of a spinning mass as an example of algebraically special metrics', *Phys. Rev. Lett.* **11**, 237 (1963).
3. Ernst, F., 'New formulation of the axially symmetric gravitational field problem I', *Phys. Rev.* **167**, 1175 (1968).
4. Tomimatsu, A. and Sato, H., 'New Exact Solution for the Gravitational Field of a Spinning Mass', *Phys. Rev. Lett.* **29**, 1344 (1972); 'New series of exact solutions for gravitational fields of spinning masses', *Prog. Theor. Phys.* **50**, 95 (1973).
5. Geroch, R., 'A method for generating new solutions of Einstein's equations I', *J. Math. Phys.* **12**, 918 (1971); II, *J. Math. Phys.* **13** 394 (1972).
6. Ehlers, J., in *Les Theories Relativistes de la Gravitation*, CNRS, Paris (1959).
7. Kinnersley, W., 'Symmetries of the stationary Einstein-Maxwell field equations I', *J. Math. Phys.* **18**, 1529 (1977).
8. Kinnersley, W. and Chitre, D.M., 'Symmetries of stationary Einstein-Maxwell field equations I', *J. Math. Phys.* **18**, 1538 (1977).
9. D.M. Chitre, 'Characterization of certain stationary solutions of Einstein's equations', *J. Math. Phys.* **19**, 1625 (1978).
10. Kinnersley, W. and Chitre, D.M., 'Symmetries of stationary Einstein-Maxwell field equations III', *J. Math. Phys.* **19**, 1926 (1978).
11. Kinnersley, W. and Chitre, D.M., 'A group transformation that generates the Kerr and Tomimatsu-Sato metrics', *Phys. Rev. Lett.* **40**, 1608 (1978); 'Symmetries of the stationary Einstein-Maxwell equations IV, transformations which preserve asymptotic flatness', *J. Math. Phys.* **19**, 2037 (1978).

12. Chitre, D.M., 'Asymptotically flat, stationary axially symmetric metrics as solutions of an integral equation', a paper in preparation.
13. Hoenselaers, C., Kinnersley, W. and Xanthopoulos, B.C., 'Symmetries of the stationary Einstein-Maxwell equations, VI. Transformations which generate Asymptotically Flat Space-times with arbitrary multipole moments', Preprint, Montana State University.
14. Maison, D., 'Stationary, axially symmetric Einstein spaces, a completely integrable Hamiltonian system?' *Phys. Rev, Lett.* **41**, 521 (1978).
15. Belinsky, V.A. and Zakharov, V.A., in Proceedings of the Meeting on Sources of Gravitational Radiation, Battelle Seattle Research Center, Seattle, July 1978.
16. Harrison, B.K. 'Backlund transformation for the Ernst equation of general relativity". *Phy. Rev. Lett.* **41,** 1197 (1978)
17. Kundu, P., 'Projective tensor formalism for stationary axisymmetric gravitational fields', *Phys. Rev.* **D18,** 4471 (1978).

Discussions

J Pachner: The stationary axially symmetric exterior fields under discussion must be generated by some stationary source (a body). Do you think that some of your solutions may be eliminated by the requirement that the exterior solution must be a continuation of an interior solution?

D M Chitre: No. The problem is usually the other way round. Given an exterior solution, how to find a matching interior solution; the answer for which is not known even for the Kerr case.

C Sivaram: Can you generate Tomimatsu-Sato solutions with any arbitrary δ?

D M Chitre: Yes. The algebraic computation, however, gets rather lengthy for $\delta \geqslant 3$.

The Theory of $\mathcal{H}$-Space

E.T. NEWMAN

The $\mathcal{H}$-Space theory concerns the complex vacuum Einstein equations approached from very specific point of view. Though there are three large groups (Plebanski's group in Mexico City, Penrose's in Oxford and ours) I will be discussing mainly the work of our group in Pittsburgh. It should, however, be mentioned that our work would have been stillborn without the very active collaboration of Penrose and his group.

Before getting into the main topic of $\mathcal{H}$-space I would like to make a few preliminary remarks—mainly about notation. We are going to consider 4-complex dimensional manifolds and work in a particular patch with the co-ordinates z^a. We will introduce 10 holomorphic (complex analytic) functions on that patch, calling them g_{ab} and talk about a complex holomorphic Riemannian metric

$$ds^2 = g_{ab}(z^c)\, dz^a\, dz^b \neq g_{ab}\, dz^a\, \overline{dz^b}$$

(It may be emphasized that there is no idea here of hermiticity; these are not hermetian metrics—they are complex holomorphic metrics.) The very simplest example is to take the complex extension of Minkowski space and talk about complex Minkowski space with metric

$$ds^2 = \eta_{ab}\, dz^a\, dz^b;\ \eta = \mathrm{diag}(1, -1, -1, -1),\ z^a = x^a + iy^a$$

In the more general situation, if there is a complex holomorphic metric one can formally define on that space a complex holomorphic connection

$$\Gamma^a_{bc} = \tfrac{1}{2} g^{ad}(g_{bd,\, c} + g_{cd,\, k} - g_{bc,\, d})$$

and then formally write out the Riemann tensor as one would normally do:

$$R^a_{bcd} = \Gamma.:,\ . - \Gamma.:,\ . + \Gamma.:\ \Gamma.:$$

One can then attempt to find the vacuum solution to the Einstein equations,

$$R_{ab} = 0$$

Because we are dealing with complex metrics and curvature tensors rather than real ones there is a further condition one can impose, namely, the Weyl tensor be either self-dual or anti-self-dual, i.e.

$$\overset{*}{C}_{abcd} = iC_{abcd}$$

or

$$\overset{*}{C}_{abcd} = -iC_{abcd}$$

These then are the basic ideas we will be talking about.

Why were we interested in this? In the beginning I really did not know. We were studying the asymptotic behaviour of *real* vacuum solutions of the Einstein equations and these complex ideas appeared in a most natural fashion. This idea forms the Principal message of this paper: From the real Einstein theory of gravitation, if certain natural questions are asked, one is led directly into $\mathcal{H}$-space theory with no ad hoc assumptions imposed. (Actually, $\mathcal{H}$-space theory is intimately connected with the Penrose theory of twistors, so that if one is interested in twistors one should be interested in $\mathcal{H}$-space.) Towards the end of the paper are given some possible connections to physical ideas.

One of the many distinctions between Minkowski space and a curved space is in the behaviour of the light cones. On the upper left hand side of Fig. 1 we have a Minkowski space light cone and on the upper right side

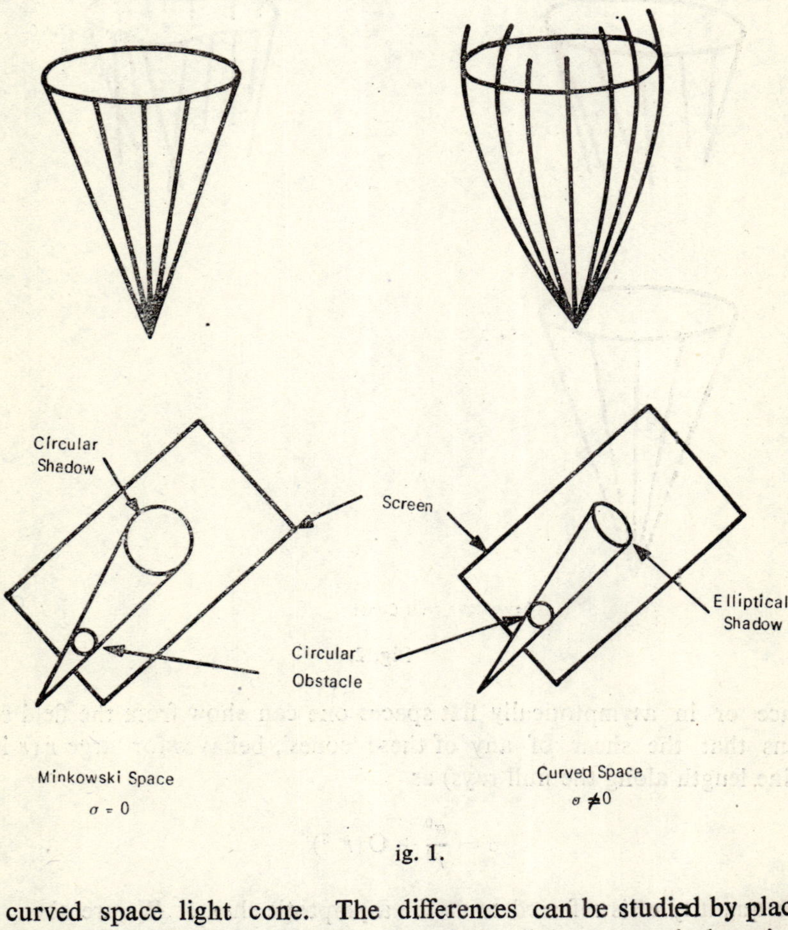

ig. 1.

a curved space light cone. The differences can be studied by placing a small circular obstacle in the path of the light rays and observing the distortion of the shadow on a screen a short distance back. For the Minkowski space light cone (Fig. 1, lower left) the disc expands but has no distortion while for the curved space case (lower right) the circular disc forms an elliptical shadow. This is technically stated thas for the Minkowski space cone the shear vanishes but for the curved space cone the shear is non-vanishing. In other words, the shear symbolized by the letter σ, measures the distortion. This idea of shear plays a dominant role in our development of $\mathcal{H}$-space.

We can now talk about more general null (or light-like surfaces), surfaces that do not necessarily emanate from a point. In particular, we can discuss asymptotic null cones, that is null surfaces with S^2 cross-sections which look like null cones asymptotically but if followed backwards do not focus at a point (see Fig. 2). In general, either in Minkowski

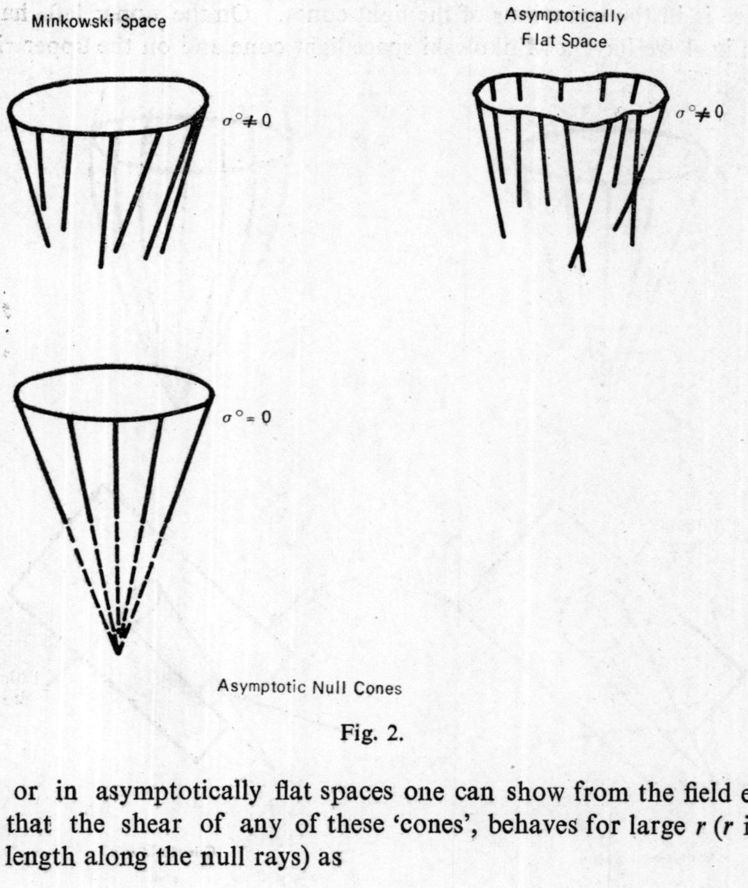

Fig. 2.

space or in asymptotically flat spaces one can show from the field equations that the shear of any of these 'cones', behaves for large r (r is the affine length along the null rays) as

$$\sigma = \frac{\sigma^0}{r^2} + O(r^{-3})$$

The quantity σ^0 is referred to as the asymptotic shear. We are then confronted with a theorem (not very difficult to prove) which states: In Minkowski space if an asymptotic null cone has vanishing asymptotic shear σ^0 then it is the ordinary null cone from a point.

One can thus stay out close to $r = \infty$ and reconstruct the points in Minkowski space by simply looking at these asymptotically shear free null cones. These special asymptotic cones are referred to as 'good' cones and we see that they are in one-to-one correspondence with the points of Minkowski space, i.e. we can speak of the manifold of 'good' cones as being equivalent to Minkowski space. (We will eventually show how this idea of making a manifold from the good cones can be generalized to curved space.) In order to simplify our picture of what is taking place we can add a boundary (called null infinity) to the Minkowski space (which is technically called conformal compactification) by adding to the space the end points of all the null rays that get out to infinity. We can then talk about the intersection of the good cones with null infinity (see Fig. 3). The intersections, which are spheres are called 'good' cuts. For each good cut, there is a good cone and hence a corresponding point of Minkowski space. A question then arises: Can you determine or reconstruct the Minkowski metric without going into the interior (simply knowing the positioning of the good cuts)? The answer is yes.

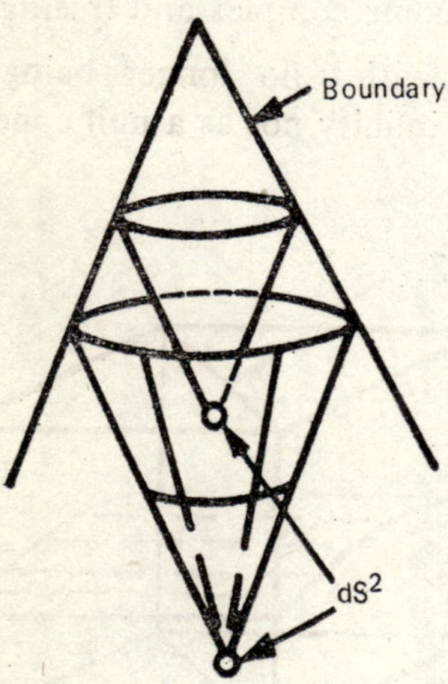

Two Neighboring Null Cones

Fig. 3

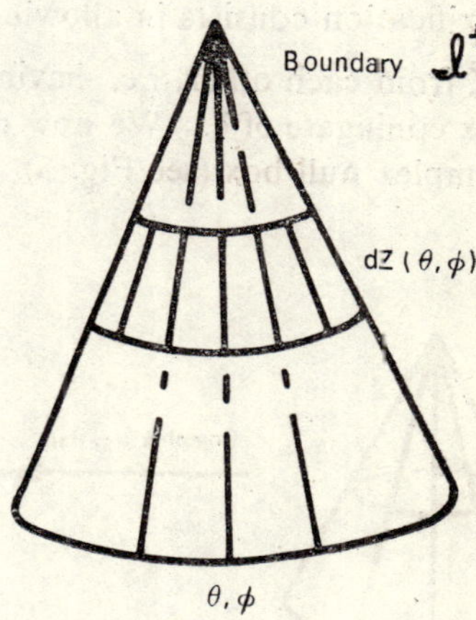

Two Neighboring Null Cone Cuts of Null Infinity

Fig. 4

Null infinity, which is a null surface, can be co-ordinatized by θ and φ (or equivalently by complex stereographic co-ordinates ζ and $\bar{\zeta}$ labelling the generators and the affine distance along the generators. If we have two neighbouring good cuts, the affine distance between them (along each generator) can be written as a function of θ and φ or explicitly as $dZ(\theta, \varphi)$ (see Fig. 4). The Minkowski distance between the two interrior points represented by the two good cuts is now given by

$$ds^2 = \left\{\frac{1}{8\pi}\int \frac{d\Omega}{(dz)^2}\right\}^{-1} = \eta_{ab}\, dx^a\, dx^b$$

where $d\Omega$ is the area element of the unit sphere. By staying near null infinity one can reproduce all the interior local differential geometry.

Can we then make a similar construction in the case of asymptotically flat spaces? Recall that we had asymptotic null cones, (with a shear that behaved as $\sigma = \sigma^0/r^2 + O(r^{-3})$ which intersected null infinity, producing cuts. Two questions arise: In a general asymptotically flat space, are there any asymptotic cones with $\sigma^0 = 0$; their intersections with null infinity producing what would be 'good' cuts? How many good cuts are there and could we define a metric on the space of good cuts? Unfortunately the answer to the first question, do good cuts exist, is no. For the general radiation situation, i.e. with matter (bounded source) on the inside spewing out gravitational radiation, you cannot find any asmptotic null cones which are asymptotically shear-free. It thus appears as if our program ends here. There is, however, a way out of this apparent impasse. Complexification is the answer. What we do is take the real physical space (or really first a neighbourhood of null infinity) and analytically extend it into the complex. In particular real null infinity had the structure of $S^2 \times R$, co-ordinatized by ζ and $\bar{\zeta}$ and the affi-ne length up the generators u—its complexification consists in allowing u to become complex and freeing the ζ and $\tilde{\zeta}$ from each other, i.e. having ζ and $\tilde{\zeta}$ with $\tilde{\zeta}$ no longer being the complex conjugate of ζ. We now draw null infinity not as a null cone but as a complex null box (see Fig. 5).

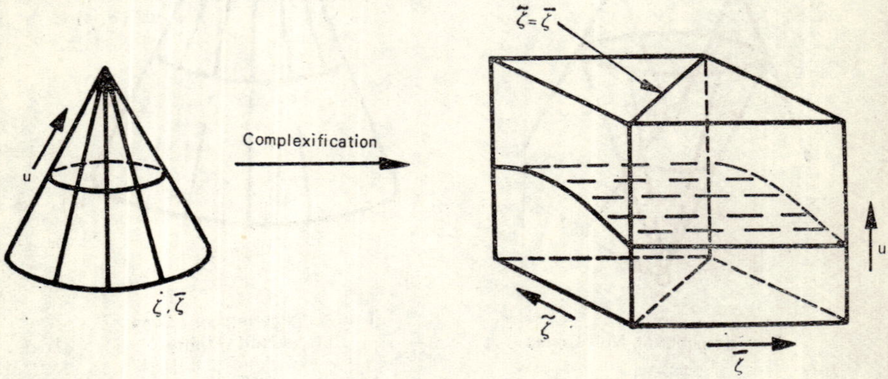

Fig. 5

Remember, everything now is complex. This should really be a six-dimensional picture, but we are drawing the ζ in one direction, the $\tilde{\zeta}$ which is a complex variable in the other direction, and u going up. There is a diagonal, that is where $\tilde{\zeta} = \bar{\zeta}$, and in the complexification one should not go too far off that diagonal. The complexification only really works in some thickened area around that diagonal and in some thickend region around the u being real. We are taking the null infinity of the physical space and assuming all the functions are real analytic, and we analytically extend it into this thickened region of the physical space. It is the same physical space we started with, all the information is there, except we analytically extend all the relevant functions. This is still just the boundary, but if we look back into the inside of the thickened physical space, we can ask our previous questions once more. Are there any good null surfaces? The answer to this yes. There are good null surfaces or good cuts, and in particular there are a four complex parmeters' worth of them. They form a four-(complex) dimensional manifold and a distance between them can be defined. So in the real situation you could not do what we were talking about, but in the complex case you can. This is what I meant by things arising naturally. It was not much of a jump saying that things are real analytic and then simply taking the analytic extension into the complex. In this complex extension of the real physical space, you discover you have a four parameter worth of the good null cones. Although the differential equation that yields the good cuts was a rather nasty equation, it was intuitively clear that you could get four parameter worth of solutions. It was not until recently that Penrose, using the very powerful theorems of sheaf cohomology theory, and the Kodaira deformation theorems, gave a rigorous proof that you do get a four parameter worth of good cuts. The four-(complex) dimensional manifold of asymptotically shear-free complex null surfaces is what is referred to as $\mathscr{H}$-space. (Sometimes it is called heaven for the rather funny reason that all the good cones go to it.) I will just point out as a brief aside that the shear, σ^0, which we have been talking about all along, was really a complex-valued function in the original real space and there was always its complex conjugate function. When we went to the analytic extension, the complex conjugate function no longer was the complex conjngate. It was just another function $\tilde{\sigma}^0$ that was a complex conjugate of the real part but in the complexified physical space no longer the complex conjugate. So you really have two separate functions. You could ask for the good cones that make σ^0 vanish or you could ask for the good cones that made the $\tilde{\sigma}^0$ vanish, and it leads to two separate types of good cones. In other words, it leads to two separate types of $\mathscr{H}$-space. One is an $\mathscr{H}$-space and the other is a conjugate space referred to as $\tilde{\mathscr{H}}$-space. They are actually conjugates of each other. There is no new information in the second one, except what happens is that the self-dual goes into an anti-self-dual. What we have now been able to do is produce from the real physical space by analytically extending it, a new space of these good cones. In the Minkowski space case, it simply reproduded the same space. There is not much likelihood that this is going to produce the original space in the general situation. We can at this point, with a considerable amount of work, prove a series of theorems, which I feel are remarkable.

1. There is a natural metric on the $\mathscr{H}$-space. If two neighbouring good cuts are separated along their generators by the function dZ (as in the Minkowski space case), it can be proved that a holomorphic Rieman-

nian metric is produced by

$$ds^2 = \left\{\frac{1}{8\pi}\int\frac{d\Omega}{(dz)^2}\right\}^{-1} = g_{abc}\,(z^c)\,dz^a\,dz^b$$

the integration taken over the real sphere or a deformation of it. I want to emphasize that this result is not obvious and it takes considerable work to prove it.

2. The second theorem (perhaps more surprising than the first) states that these holomorphic metrics *automatically* satisfy the vacuum Einstein equations. This can be proved directly from the integral defining the metric (though this is a hideous calculation) or by more sophisticated techniques.

3. The third thing that can be proved is that the Weyl tensor is self-dual, i.e.

$$C^*_{abcd} = iC_{abcd}$$

(For the $\tilde{\mathcal{H}}$-space the Weyl tensor is anti-self-dual.)

4. It was recently shown by Ludvigsen that for an $\mathcal{H}$-space an entire series of Hertz-like potentials could be produced. From a tensor with the symmetries of the Weyl tensor ψ_{abcd} one obtains

$$\nabla^a\psi_{abcd} = \psi_{bcd}$$
$$\nabla^c\psi_{abc} = h_{ab}$$
$$\nabla_{[a}h_{b]c} = c_{abc}$$
$$\nabla_{[d}C_{ab|c]} = C_{abcd}$$

It was known in the linear theory of gravity that this set of potentials existed, but to my knowledge it was not realized that they existed in the full non-linear theory, albeit in the self-dual or anti-self-dual cases. The importance of these potentials is that they permit a norm or scalar product to be defined on the space of the holomorphic, self-dual vacuum Einstein metrics.

Further theorems which can be proved concern $\mathcal{H}$-spaces which are themselves asymptotically flat. We can then repeat the argument that led to $\mathcal{H}$-spaces starting from the real solutions, i.e. we can ask for the $\mathcal{H}$-space of an $\mathcal{H}$-space or the $\tilde{\mathcal{H}}$-space of an $\mathcal{H}$-space. The results are that the $\mathcal{H}$-space constructed from a given $\mathcal{H}$-space is identical to the given $\mathcal{H}$-space and the $\tilde{\mathcal{H}}$-space constructed from an $\mathcal{H}$-space is complex Minkowski space, i.e.

$$\mathcal{H}\mathcal{H} = \mathcal{H} \quad \text{and} \quad \tilde{\mathcal{H}}\mathcal{H} = M$$

(I want to point out that there are actually two Minkowski spaces that can be constructed from the $\mathcal{H}$-space—one can use the future null infinity or the past null infinity to do the construction. One obtains two Minkowski spaces that cannot be canonically identified, see Fig. 6.)

This completes the brief account of the major geometric results arising from $\mathcal{H}$-space theory. Let us new turn our attention to what use this is being put to by other workers:

1. Roger Penrose has suggested with a very convincing argument that an $\mathcal{H}$-space should be thought of a non-linear graviton. This interpretation gets strengthened by the existence of the scalar product on the space of $\mathcal{H}$-spaces. One hopes that this will lead to a quantum theory of gravity.

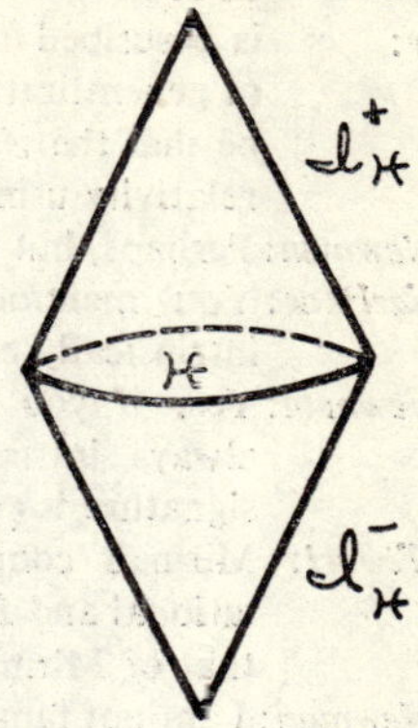

Null Infinity of $\mathcal{H}$– Space

Fig. 6

2. One obtains remarkable equations of motion induced on the $\mathcal{H}$-spnce from the real space, i.e. from physical fields on the real space there are natural mappings to tensor fields on the $\mathcal{H}$-space. These in turn can be used to define centre of mass motion and spin motion. We have no reasonable interpretation of these results.

3. All the results described here can be obtained from twistor theory—in fact some of them were first obtained from it. If one takes twistor theory seriously then the results reported here should be taken seriously.

As a final remark, I want to mention that most of the things said here apply also to gauge or Yang-Mills type theories, and are thus closely related to instanton theory.

References

1. Plebanski, J., *J. Math. Phys.* **16**, 2395 (1975).
2. Plebanski J. and Hacgan, S., *J. Math. Phys.* **16,** 2403 (1975).
3. Finley, J. and Plebanski, J., *J. Math. Phys.* **17,** 585 (1976).
4. Plebanski, J. and Robinson, I., *Phys. Rev. Lett.* **37,** 493 (1976).
5. Boyer, C.P. and Plebanski, J., *J. Math. Phys.* **18,** 1022 (1977).
6. Penrose, R., *GRG*, **7,** 31 (1976).
7. Hansen, R. Newman, E.T., Penrose, R., Tod, P.
8. Penrose, R. in *Quantum Gravity*: *An Oxford Symposium*, C. Isham, R. Penrose, D. Sciama, (Eds.), Oxford University Press (1975).
9. Ko, M., Newman, E.T., Tod, P., in *Asymptotic Structure of Space-time*, P Esprsi to and L. Witten (Eds.), N.Y. Plenum Press (1976).
10. Ko, M., Ludvigsen, M., Newman, E.T., The Theory of $\mathcal{H}$-space, preprint, Submitted to Physics Reports.
11. Ludvigsen, M., GRG **8,** 357 (1977).
12. Newman E.T., in *General Relativity and Gravitation*, G. Shaviv and J. Rosen, (Eds.), Wiley (1975).

Discussions

S G Rajeev: How is this connected with twistors? Is the $\mathcal{H}$-space isomorphic to the twistor space?

E T Newman: There is an intimate connection. Penrose obtains $\mathcal{H}$-space by deforming twistor space, The results are identical to what I described.

S Ramamurty: In Jean's mathematical theory of Electricity and magnetism, a method is described for solving electrostatic problems by using a Riemann space (a generalization of Riemann surface to three dimensions). Can it not be that the $\mathcal{H}$-space theory is a method of solving problems in general relativity using such a method?

E T Newman: Perhaps, but it seems unlikely.

J V Narlikar: You mentioned that the space you get from your approach is intrinsically complex. Is the Minkowski space-time a very special case?

E T Newman: Yes, if you ask for a real subspace of an $\mathcal{H}$-space the metric is always intrinsically complex except for the flat case or when the signature is $(++++)$.

S C Tiwari: Mirman complex physical space gives a theory of unification of gravitational and *E M* interaction. How does your $\mathcal{H}$-space compare with that of Mirman?

E T Newman: I am not familiar with Mirman's work.

C Sivaram: (1) Since you are naturally getting self-dual curvature tensors, won't this be a natural space for describing gravitational instantons? (2) Can you have an analogue of the Atiyah-Singer index theorem here?

E T Newman: (1) Yes, this is what Hawking is working on.
(2) I do not know.

A K Pandey: In electrodynamics there is a method for solving electrostatic problems by complex variables. This is due to an intrinsic connection between the theory of complex variables and Poisson's equation in two-dimensions. In this formalism ($\mathcal{H}$-spaces) such a connection is with Einstein's equation intrinsically. By drawing the analogy further we should take only a real part of the g_{ij} to give physical solutions. Will it be fair in your opinion?

E T Newman: It does not seem to work. Since the equations are non-linear one cannot add or subtract parts.

Cosmology and Astrophysics

Cosmology in the Post-Einstein Era

J.V. NARLIKAR

Introduction

No review of the remarkable developments in cosmology in post-Einstein era can do justice to the subject, if it does not first provide the background of the origin and growth of modern cosmology during Einstein's own lifetime. It was Einstein's pioneering effort that established cosmology firmly as a branch of science. I can do no better than repeat the tribute paid by Einstein's colleague Leopold Infeld[1] on the occasion of Einstein's seventieth birthday:

> Speculations about the Universe in which men live are as old as human thought and as old as art; as old as the view of shining stars on a clear night. Yet it was the general relativity theory which, only thirty years ago, shifted cosmological problems from poetry or speculative philosophy into physics. We can even fix the year in which modern cosmology was born. It was in 1917, when Einstein's paper appeared in the Prussian Academy of Science under the title 'Cosmological Considerations in General Relativity Theory.'

Einstein's paper[2] (op. cit.) attempted for the first time what the modern cosmologist today is still groping for: to provide a mathematical model which simulates the large-scale structure of the universe. To be judged 'successful' the model must describe and explain reasonably accurately the observations of the large-scale structure of the universe. Moreover, it must make predictions which can be tested by future observations.

In 1917, observational astronomy had hardly progressed beyond the confines of our galaxy. There was practically no observational data to really test the 'large-scale' predictions of a cosmological model. Under the circumstances Einstein looked for the simplest possible model of a homogeneous, isotropic and *static* universe. The so-called Einstein-Universe achieves just that goal, but at a cost. Einstein had to modify his field equations to the form

$$R_{ik} - \frac{1}{2} g_{ik} R + \lambda g_{ik} = -\frac{8\pi G}{c^4} T_{ik} \tag{1}$$

where the λ-term on the left hand side provides the force of repulsion which is needed to balance the self-gravity of the universe to produce the static model. The λ-term had to be small enough so as not to affect the local tests of gravity. Einstein showed that it was related to the *mean* density ρ_0 of matter in the universe by a simple formula:

$$\lambda = \frac{4\pi G\rho_0}{c^2} \tag{2}$$

This is the type of relation a cosmologist is always on the look out for. It predicts a definite relation between two quantities which are in principle observable. Unfortunately, even today ρ_0 is only roughly estimated at $\sim 10^{-30}$ g cm^{-3}, if limited to the visible matter in the universe while it has not yet been possible to think of an observation capable of measuring λ.

The λ-term of Einstein, however, lost its *raison d'etre* when Edwin Hubble[3] discovered in the late 1920s the phenomenon of nebular redshift. If Einstein's paper of 1917 is considered to mark the beginning of *theoretical cosmology*, Hubble's paper of 1929 marks the beginning of *observational cosmology*. With the interpretation of the nebular redshift in terms of the models of the expanding universe, the need for obtaining a static model disappeared. The work of Friedmann[4] in 1922 had already shown that it is possible to obtain expanding models of the universe from Einstein's field equations without the λ-term.

Einstein himself readily recognized the apparent superfluousness of the λ-term in the light of this changing situation on the observational front. Again opting for simplicity, he set $\lambda = 0$ and in 1932 he wrote a paper 'On the relation between the expansion and the mean density of the universe', with de Sitter as the co-author.[5] This model, known as the Einstein-de Sitter universe gives the following relation between the mean density ρ_0 and Hubble's constant H:

$$\rho_0 = \frac{3H^2}{8\pi G} \equiv \rho_c \tag{3}$$

This value ρ_c holds a 'limiting status'. If $\rho_0 > \rho_c$ the universe is *closed*, if $\rho \leqslant \rho_c$ it is *open*. Whether the universe is open or closed is an important property of its large-scale structure and a relation like (3) shows how cosmology can make a scientific prediction. The present estimates of ρ_0 and H suggest that $\rho_0 < \rho_c$. However, the possibility of invisible matter, say in the intergalactic space, cannot be ruled out. If we want to insist on a closed model we can always call upon the black-hole bandwagon to supply the 'missing mass' in the form of black holes.

In the last decade of Einstein's life two developments took place which accelerated the development of cosmology both as a physical theory and as an observationally testable theory. The first was the hypothesis of primordial nucleosynthesis of George Gamow[6] and his colleagues Hans Bethe, Ralph Alpher and Robert Hermann and the second was the steady state theory[7] of Hermann Bondi, Thomas Gold and Fred Hoyle. I shall first describe how the steady state theory played a leading role in stimulating observational cosmology.

Observational cosmology

The steady state theory was proposed in 1948 out of dissatisfaction both with the fundamental assumptions of the standard relativistic cosmology and with the unsatisfactory features of some of its observational predictions. The doubts raised with the existing framework were of the following type:

(i) All standard models of relativistic cosmology lead to the deduction that the universe was created in a singular event, the so-called big-bang'. The appearance of a singularity is an undesirable feature of any theory.

(ii) What guarantee have we (except by a self-imposed arbitrary fiat) that the laws of physics operating close to the big bang, when the universe was radically different from its present state, were the same as they are now?

(iii) How was matter created in the primordial event of big bang?

(iv) Why is the age of the universe so small compared to the age of our Galaxy?

The last objection to the big bang model has lost its original power now, since an upward revision of the age of the universe has taken place. The Hubble's constant is now believed to be 5 – 10 times *smaller* than what was estimated by Hubble.[8] However, even with the lowest recognized value it becomes difficult to sustain the closed models of the universe which predict lower ages (see Fig. 1).

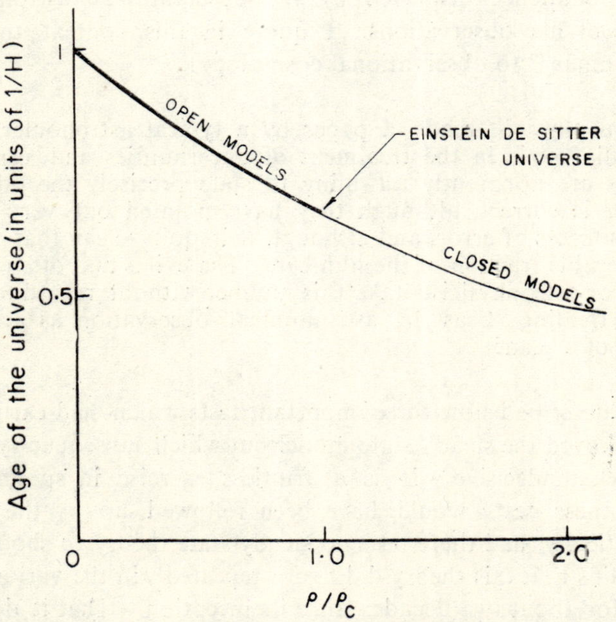

Fig. 1 The age of the Friedmann universe in units of $1/H$, expressed as a function of the ratio of the present mean density ρ and the closure density ρ_c, The open models (shown by the thick curve) give higher ages than the closed models.

The steady state theory got round these objections by postulating an expanding but unchanging universe. This model is described by the de Sitter line element

$$ds^2 = c^2dt^2 - e^{2Ht}\,[dr^2 + r^2\,(d\theta^2 + \sin^2\theta\; d\phi^2)] \tag{4}$$

with $H=$ constant. This universe has no space-time singularity, an infinite age and its unchanging character ensures the validity of the same physical laws at all epochs. The sudden and infinite creation of matter of the big bang models is replaced by a finite and continuous creation of matter which can even be understood by a simple field theory.[9]

Rightly or wrongly the steady state model roused a great deal of controversy. Having taken part in some of these developments I can recall several occasions when the criticism of this theory was either genuinely misinformed or downright prejudiced. While the theoretical foundations of this theory were secure, it was vulnerable to observational tests. By making definite predictions (as opposed to the multi-parameter-dependent predictions of the big bang models) this theory, in the words of one of its originators Hermann Bondi 'stuck its neck out'.

The general hostility to this model and its observational vulnerability prompted many observational astronomers to devise tests to shoot it down. In this process the theory 'died' several times since its birth in 1948, but revived again. The reason for this has, curiously enough, nothing to do with the tenacity of the steady state theory. Rather, it owes its revival to the overconfidence displayed by the observational astronomer in the accuracy of his observations. I quote in this context the reaction of R.P. Feynman[10] to observational cosmology:

> When a physicist reads a paper by a typical astronomer, he finds an unfamiliar style in the treatment of uncertainties and errors.... The authors are apparently unwilling to state precisely the odds that their number is correct, although they have pointed out very carefully the many sources of error, and although it is quite clear that the error is a considerable fraction of the number. The evil is that often other cosmologists or astrophysicists take this number without regard to the possible error, treating it as an astronomical observation as accurate as the period of a planet....

I will describe below three important tests which had earlier claimed to have disproved the steady state model but which subsequently were seen to be somewhat indecisive. It is a fruitless exercise in speculation to ask whether these tests would have been followed up by the observers so enthusiastically, had there been no steady state theory to shoot down. The fact remains that this theory did figure repeatedly in the various cosmological tests for about two decades since its inception. That it does not do so now has nothing to do with the three tests being described below. The death-blow (or a near death-below!) to this theory came from an unexpected direction. But more of that in the following section.

THE REDSHIFT-MAGNITUDE RELATION

The original deduction of the velocity (V)/distance (D) relation

$$V = HD \tag{5}$$

was made by Hubble by plotting the logarithm of the redshift z against the apparent magnitude m for relatively nearby galaxies and by using the relations

$$V = cz \tag{6}$$

$$m = -2.5 \log \frac{L}{4\pi D^2} + \text{const.} \tag{7}$$

where L = luminosity of the galaxy.

In the expanding universe models the redshift receives a direct interpretation as the fractional increase in the expansion factor between the epoch of emission and the epoch of reception of the light wave from the galaxy. Equations (5) to (7) may be considered the flat-space approximations to the exact z-m relation in any cosmology. All big bang cosmologies as well as the steady state theory agree on (5) for nearby galaxies.

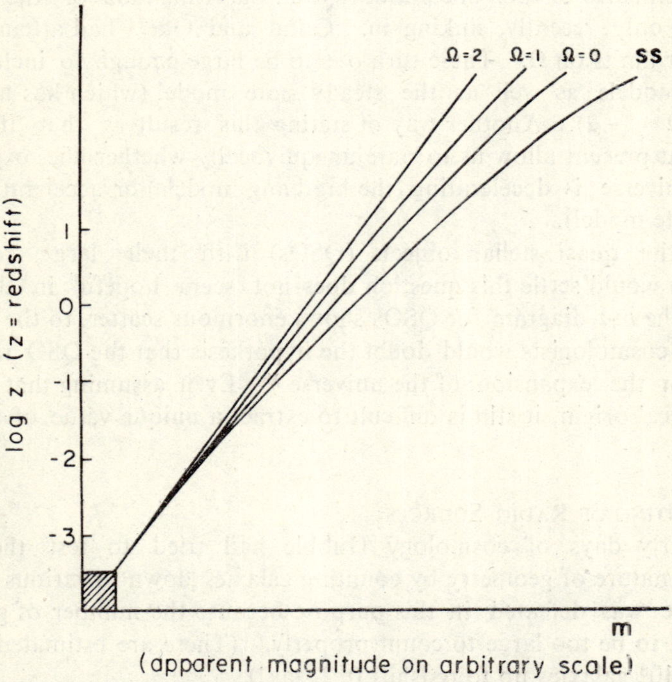

Fig. 2 The schematic m-z relation for three Friedmann models with $\Omega = 2, 1, 0$ and for the steady state (SS) model. Observationally there are not many galaxies with $\log z \gtrsim 0.7$. The shaded rectangle shows the redshift range covered by Hubble.

Figure 2 shows the schematic curves predicted by the various big bang models and by steady state model. The parameter

$$\Omega = \frac{\rho_0}{\rho_c} \tag{8}$$

labels the various big bang curves. Thus $\Omega = 1$ is the Einstein de Sitter model while $\Omega = 2$ represents a closed model. The redshifts of galaxies extend up to $z \sim 0.7$, although there are relatively few points above $z = 0.2$. Notice the shaded rectangle at the bottom which indicates the extent of Hubble's original observations. Modern data cover nearly 500 times more distance in space and time. The question is, whether any particular curve is singled out by the data as giving the most accurate fit?

Allan Sandage[11] and his collaborators at Mt Palomar quote a 'formal' value of Ω close to 2, corresponding to a closed universe which will eventually contract. To what extent does the formal value represent the correct model? Bearing in mind Feynman's remarks one has to be extremely cautious in exercizing judgement since no realistic error estimates are quoted. Do galaxy luminosities evolve with time? To what extent intergalactic absorption is likely to be important for this test? What is the scatter in the standard candle luminosities like? It is difficult to quote realistic estimates of such uncertainties, and the realization of these difficulties is only recently sinking in. Gunn and Oke[12] had attempted to place error limits on Ω. These turn out to be large enough to include all bigbang models as well as the steady state model (which has a formal value of $\Omega = -2$). Another way of stating this result is that the data does not at present allow us to state unequivocally whether the expansion of the universe is decelerating (the big bang models) or accelerating (the steady state model).

That the quasi stellar objects (QSOs) with their large redshifts ($z \leqslant 3.53$) would settle this question does not seem hopeful in the near future. The *m-z* diagram for QSOs shows enormous scatter, to the extent that some cosmologists would doubt the hypothesis that the QSO redshifts arise from the expansion of the universe.[13] Even assuming that z is of cosmological origin, it still is difficult to extract a unique value of Ω from this data.

The Counting of Radio Sources

In the early days of cosmology Hubble had tried to test the non-Euclidean nature of geometry by counting galaxies down to various magnitudes. He was defeated in this purpose because the number of galaxies turned out to be too large to count properly. (There are estimated to be around $\sim 10^8$ galaxies up to redshift 0.2, say!)

The number count test was, however, considered feasible with the advent of radio astronomy. The number of extragalactic radio sources is believed to be much smaller than the number of galaxies. Moreover, the radio astronomers believed (many still do) that the faint radio sources picked up by their instruments are considerably further away than typical galaxies. So whether Euclidean geometry holds out to great distances can be tested by counting radio sources. More specifically, the assumptions to be tested are as follows:

(a) The space-time geometry is Minkowskian.
(b) The radio sources are uniformly distributed in space.
(c) The radio sources are identical in their intrinsic power.

Under these assumptions if S is the flux density (i.e. the amount of radiation crossing unit are normally per unit time per unit frequency range) of a typical source at distance R, and N_0 the number of radio sources brighter than S, then we get $S \propto R^{-2}$ and $N_0 \propto R^3$,

i.e. $$N_0 S^{3/2} = \text{const.} \qquad (9)$$

The radio astronomer plots the ratio N/N_0 against S, where N is the actually observed number of radio sources brighter than S. Pioneering work in the 1950s was done by Sir Martin Ryle and his co-workers at Cambridge and by B. Y. Mills et al. at Sydney. The early surveys have by now been replaced by many more mainly in England (Cambridge). Australia, and the U.S.A. These surveys not only extend to much lower values of S than the early surveys but they also span several frequencies of observation. The N/N_0 plots made from different surveys[14] are shown at different frequencies in Fig. 3.

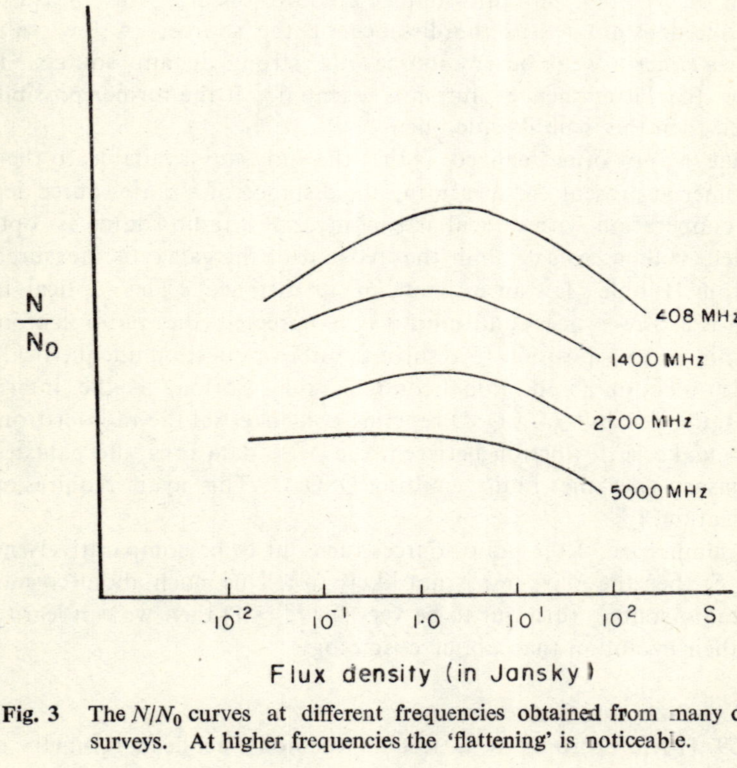

Fig. 3 The N/N_0 curves at different frequencies obtained from many different surveys. At higher frequencies the 'flattening' is noticeable.

Several interesting features emerge from these surveys. At high frequencies there is a fairly good agreement with the Euclidean curve (i.e.

$N = N_0$). At low frequencies there is a departure in that the N/N_0 curve is concave with respect to the S-axis. The difference arises because the lower frequency surveys miss sources with flat spectrum, i.e. with intensity (I)—frequency (ν) relation of the form

$$I(\nu) \propto \nu^{-x}, \quad x \lesssim 0.5 \tag{10}$$

Steep spectrum sources therefore predominate at lower frequencies.

If we believe that radio sources are indeed very far, then redshift effects must predominate at low S. The agreement between N and N_0 at low S then becomes inexplicable unless we postulate that strong evolutionary processes are at work. These conspire with the redshift effect to produce the illusion of Euclidean geometry in space. The evolution refers to the changes in the number density and power of the radio sources with epoch. Evolution is also usually invoked to obtain exact fits with the curves at lower frequencies also. This exercise therefore shifts the emphasis of the test from cosmology to the evolutionary properties of radio sources. Nevertheless, it must be admitted that if epoch dependent evolution is shown to be necessary then this rules out the steady state theory.

Is evolution really necessary to explain the N/N_0 data? The answer depends on whether the radio sources are far or near. The measurement of S alone does not tell us the distance of the source. A low value of S implies *either* a weak nearby source *or* a strong distant source. If we opt for the latter then evolution *is* essential. If the former possibility is accepted then the issue is wide open.

What is not often realized is that the only way available to the radio astronomer at present for measuring the distance of a radio source depends on his cooperation with optical astronomy. If a radio source is optically identified with a galaxy and the redshift of the galaxy is measured, we have from Hubble's law an estimate of its distance. This optical identification is a slow process and until it is completed the radio astronomer would not be in a position to resolve the above question unequivocally.

I also mention an additional complication. So long as the interpretation of the redshift of a QSO remains controversial the radio astronomer should make a distinction between the N/N_0 data for radio galaxies and for quasars (i.e. the radio emitting QSOs). This again requires optical identification.

To summarize, if the radio sources turn out to be comparatively nearby ($z \lesssim 0.5$) then the N/N_0 test is not likely to tell us much about cosmology. If the radio sources turn out to be very far ($z \gtrsim 1$) then we can learn more about their evolution than about cosmology.

Angular Size and Cosmology

In 1959 Hoyle[15] proposed a test of the non-Euclidean geometry of the universe. This involves, in principle, the variation of the angular size α of an object of fixed linear size l as its distance (D) from us is changed. In Euclidean geometry the relation is simple:

$$\alpha = \frac{l}{D} \tag{11}$$

The corresponding relation in the Einstein-de Sitter model shows how α changes with the redshift of the object:

$$\alpha = \frac{lH}{2c} \cdot \frac{(1+z)^{3/2}}{(1+z)^{1/2} - 1} \tag{12}$$

And similar relations can be worked out in the different big bang models and in the steady state model. In the big bang models, as in the special case described in (12), α has a *minimum* at a finite value of z which depends on the parameter Ω. The larger the value of Ω the closer is the minimum, as shown in Fig. 4. This curious effect is a manifestation of the non-Euclidean geometry of space-time. Or, it may be interpreted as the consequence of the matter in the universe acting as a gravitational lens. I am sure this result would have delighted Einstein, had it come in his lifetime.

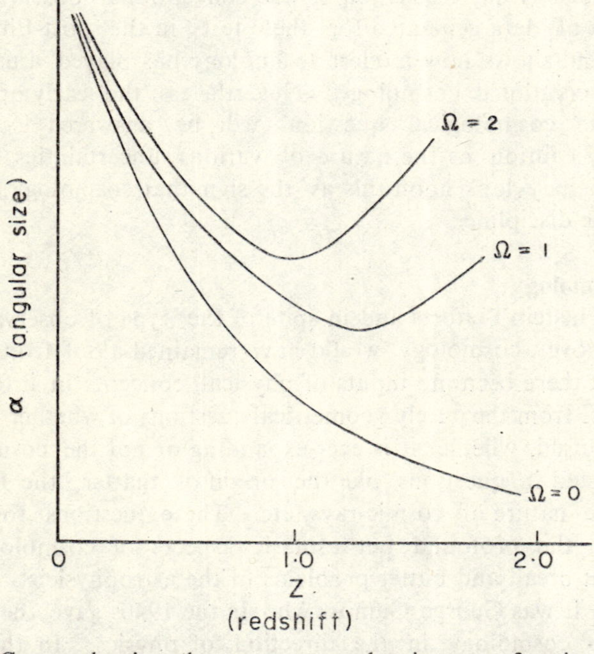

Fig. 4 Curves showing the apparent angular size α as a function of redshift z for Friedmann cosmologies with varying values of Ω. The existence of the minimum of α is in practice masked by various non-cosmological factors.

Hoyle had suggested observations of radio sources for this test. Unfortunately, as a practical proposition for deciding which cosmological model is right this test is not so promising. First, the theoretical result stated above becomes suspect if the universe has inhomogeneity as discussed by Roeder.[16] Secondly, as mentioned earlier, the radio astronomer cannot measure z directly and hence has to rely on S to provide a measure of distance. As the intrinsic powers of radio sources vary over a range of $\gtrsim 10^5$, the $S \leftrightarrow z$ correspondence has a lot of scatter. Finally, the radio sources are not spherical in shape but tend to have linear structure. So α stands for the separation between the two components of a typical double source. Unfortunately, l varies over $\sim$10 kpc — $\sim$1 Mpc. and this variation, including the effects of projection perpendicular to the line of sight causes a scatter in the values of α.

Recently, Swarup[17] and Kapahi[18] have done an extensive analysis of the angular size-flux density relation and have concluded that the observed relation is explicable only if an evolution in l implying $l \propto (1 + z)^n$, $n \sim 1$ is introduced. Whether an evolution is really needed is still open to debate. It can be shown, for example, that a correlation between l and P can also produce results in satisfactory agreement with the data, without requiring evolution[19]. The trouble is that the test is still too vague to yield a clear-cut result.

This concludes my discussion of the conventional observational tests. The amount of data generated for these tests in the post-Einstein era is impressive and shows how modern technology has played a useful role to promote observational cosmology. Nevertheless, the early optimism that the 'ultimate cosmological question' will be answered is now being tempered by caution as the nature of various uncertainties is becoming clearer. We may look upon this as the sign that cosmology is maturing as a scientific discipline.

Physical cosmology

In spite of Einstein's efforts and in spite of the type of observational tests described above, cosmology would have remained aloof from the rest of physics, had there been no inputs of physical concepts in it from time to time. Apart from the purely geometrical questions of whether the universe is open or closed, whether it is ever-expanding or not the cosmologist has been interested in questions like the origin of matter, the formation of galaxies, the nature of cosmic rays, etc. These questions form the vital link between the profound but esoteric aspects of cosmology and the mundane but bread and butter problems of the astrophysicist.

Probably it was George Gamow who in the 1940s gave the first significant turn to cosmology in the direction of physics. In the big bang universe the matter density ρ and the radiation density u in the absence of any mutual interaction depend on the scale factor according to the formulae

$$\rho \propto \frac{1}{S^3}, \quad u \propto \frac{1}{S^4} \tag{13}$$

Thus, however small u may be in comparison with ρ *now*, there must have been an early enough phase in the life of the universe when radiation was more dominant than matter. Thus in the very early stages the universe must have possessed a very high radiation temperature.

Gamow seized upon this important consideration to argue that in the first few moments after the big bang the radiation of the universe had been high enough to set off thermonuclear reactions. The outcome of these reactions would be to synthesize matter from a primordial brew of elementary particles like protons, electrons, neutrinos, photons, etc., into 'higher forms' like the nuclei of elements. The work of Hans Bethe a few years earlier had demonstrated the feasibility of

$$H^1 \rightarrow He^4 \tag{14}$$

within the interior of the sun. Gamow, Alpher, Hermann and Bethe attempted the more ambitious task of producing all the nuclei in the first few moments of the life of the universe.

It is now known that this type of approach achieves only a limited goal. Light nuclei like D and He can indeed be produced this way, but the primordial nucleosynthesis cannot surmount the barrier of unstable nuclei in the atomic number range 5-8. The important work a decade later by E.M. Burbidge, G.R. Burbidge, W.A. Fowler and F. Hoyle[20] showed that stellar interiors provide ideal sites for nucleosynthesis. It should be mentioned, by the way, that this work was also inspired by cosmology. It used to be argued that the Gamow process of nucleosynthesis could not be invoked in the steady state cosmology because of its lack of a high temperature epoch. The stellar nucleosynthesis process was designed to show that a high temperature cosmic epoch is not really necessary. This had two consequences, one positive the other negative. On the positive side, this is an example of how cosmological consideration can generate important astrophysical contributions.

The negative consequence was to divert attention of the cosmologists from the primordial nucleosynthesis theory. And probably for this reason one important guess made by Gamow, Alpher and Hermann[21] remained in the background—the guess that as a relic of the early hot era we should now be seeing a low temperature radiation background.

During 1963–64 several workers took renewed interest in primordial nucleosynthesis: Zeldovich[22] in Moscow, Hoyle and Tayler[23] in Cambridge and Peebles[24] in Princeton. Hoyle, for long regarded as the prime opponent of 'the big bang' was impressed by the fact that the estimates of the ratio of Helium to Hydrogen from various astronomical observations were significantly higher ($\sim 25\%$ by mass) than what could be explained by stellar nucleosynthesis ($\sim 10\%$ by mass). If these estimates were to be regarded as 'universal' they implied that the matter in the universe had passed through a high temperature ($\sim 10^{10}$ °K) in the past. If one does not want the big bang one has to postulate supermassive objects ($M \gtrsim 10^6 M_\odot$) in whose centres high temperatures might be possible. The work of Tayler and Hoyle was, however, directed more towards re-doing Gamow's early calculations with the fresh inputs available from nuclear and particle physics. They found that the observed high abundance of Helium can be comfortably explained in the big bang scenario. Later work of Wagoner, Fowler and Hoyle[25] confirmed this result as well as the earlier expectation that for heavier nuclei (carbon and upwards) stellar nucleosynthesis is definitely needed.

Peebles had reached similar conclusions. His work also brought out another of Gamow's early prediction that a relic of the primordial hot era should be the background radiation at a temperature a few degrees Kelvin. His colleague R.H. Dicke[26] at Princeton had arrived at a similar conclusion from his own discussions of cosmology. Dicke also realized that if this radiation was in the microwave region it could be measured from a ground-based antenna. At his suggestion his experimentalist colleagues P.G. Roll and D.T. Wilkinson began to set up an experiment for the detection of the background radiation.

However, they were overtaken by events. The background radiation at 7.3 cm was detected unexpectedly by two physicists at the Bell Telephone Laboratories, A.A. Penzias and R.W. Wilson.[27] The results of Penzias and Wilson were reported in the spring of 1965, some six months before the results of Roll and Wilkinson.[28] Penzias and Wilson had in fact detected this radiation in 1964 but as they were unaware of the cosmological background, they were puzzled about its origin. It was only when they compared notes with the Princeton group that they realized the enormous significance of their observation. The paper of Penzias and Wilson in the Astrophysical Journal was preceded by that from Dicke and others[29] giving the cosmological interpretation.

In retrospect it is surprising that such as important prediction of

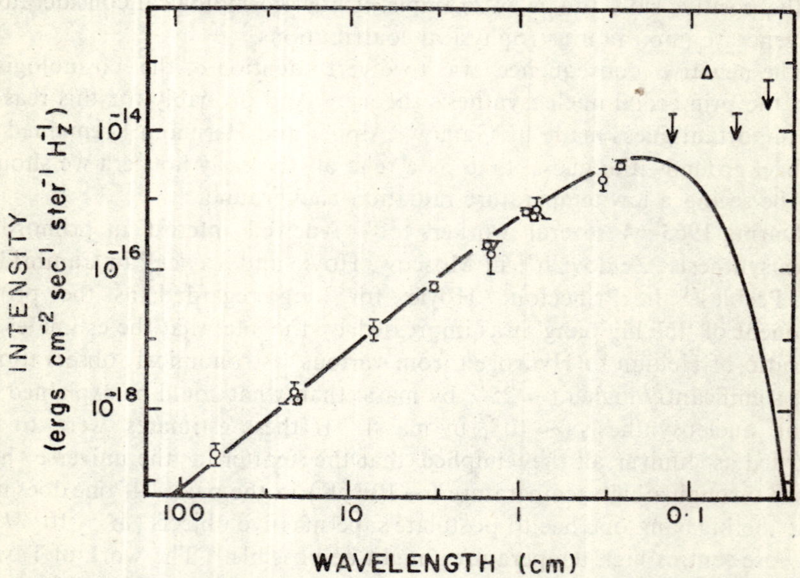

Fig. 5 The curve drawn through the observed points of the microwave background radiation gives a good fit to a black-body curve of temperature 2.7 °K.

Gamow remained untested for so long. It is even more surprising that the Princeton group was also unaware of this prediction! By hindsight it is possible to argue that theoretically and observationally the existence of the background radiation could have been established a decade earlier than it actually was.

Subsequent to 1965, however, the work on the measurement of the background radiation progressed rapidly. Fig. 5 shows the measurements at different wavelengths together with a 'best fit' black body curve of 2.7° K. The measurements at wavelengths shorter than $\sim$1 mm have to be done above the atmosphere and here the accuracy of the observations available to date is far from satisfactory. The short wavelength side of the black body curve still remains to be confirmed.

The black body interpretation has given a tremendous boost to the big bang concept of the universe. It has also inspired cosmological theorists to be bold and apply the local physics to primordial conditions. And thus has emerged in the last dozen years considerable literature on physical cosmology which attempts to answer questions such as: 'When were hadrons formed?', 'In what sequence were elementary particles combined into nuclei?', 'When did atoms and molecules form?', 'How did galaxies come into existence?', 'Did black holes dominate the primordial era?', etc. While many of these theories are highly speculative, the fact that they draw heavily upon the current knowledge of physics indicate how much cosmology has become an integral part of physics.

Non-standard cosmology

Anyone who reads the highly persuasive account of the big bang universe as given by Steven Weinberg[30] will come away with the impression that the problem of the origin of the universe is now more or less solved and that only a few details need to be filled in. If this impression were confined to the layman it would not be so important. That it is now being accepted by the run-of-the-mill physicist is disturbing, for it has the danger of removing the critical scientific ability so essential to the cosmologist if he is to function effectively as a scientist. Against such a complacent attitude Hoyle[31] has remarked in the course of his talk at the Vatican Conference in 1970: "I think it is very unlikely that a creature evolving on this planet, the human being, is likely to possess a brain that is fully capable of understanding physics in its totality, I think this is inherently improbable in the first place, but, even if it should be so, it is surely wildly improbable that this situation should just have been reached in the year 1970."

Einstein himself never asserted that his general theory of relativity was the ultimate theory of gravity. His attempts to unify the basic interactions indicate his aspiration to find a more comprehensive framework for physics. His interest in Mach's principle, his dissatisfaction with the foundations of quantum mechanics or his flirtation with the λ-term are examples of his unconventionality in the quest of solutions to the mysteries of the physical world.

Similar motivation has inspired several research workers to depart from the standard big bang picture of the universe and to look for alternative cosmological theories. I have already discussed the steady state cosmology as an example of this 'non standard' group of cosmologies. I mention below a few more examples of non-standard cosmologies which have come up after the death of Einstein. Even to the most conservative standard big bang theorist a review of non-standard cosmologies should be of interest if only to judge how the standard cosmology compares with them.

The appearance of large number of ratios between typical units in cosmology and elementary particle physics had been noticed even in Einstein's lifetime. For example:

$$\text{(i)}\quad \frac{e^2}{Gm_p m_e} \sim 10^{40}, \quad \text{(ii)}\quad \frac{c/H}{e^2/m_e c^2} \sim 10^{40} \tag{15}$$

Of these, (i) denotes the ratio of the electrostatic force between an electron and a proton to their gravitational force, while (ii) is the ratio of the 'radius' of the universe to the 'radius' of the electron.

Is the appearance of similar large numbers in different contexts above a coincidence? Dirac[32], in 1937 had interpreted this coincidence in terms of an epoch-dependent property of the universe. Thus the figure 10^{40} in (ii) could be considered the 'age' τ of the universe in atomic units, and (i) could be looked upon as giving an epoch-dependent gravitational constant ($G \propto \tau^{-1}$). In 1973 Dirac[33] developed these ideas further and his cosmological theory makes use of two time-scales. The atomic time-scale τ governs the atomic processes where as the macroscopic time-scale t may be used to write the gravitational equations in the standard Einstein form. However, since a non-linear transformation connects the two time-scales the Dirac cosmology differs from relativistic cosmology in a non-trivial way. For example, the gravitational constant decreases at a rate

$$\frac{\dot{G}}{G} = -\alpha H \tag{16}$$

where α is of order unity.

The variability of the gravitational constant with epoch is a consequence of other cosmologies too. For example, the Brans-Dicke theory[34] and the Hoyle-Narlikar cosmology[35] make somewhat similar predictions. Both these theories make Mach's principle as their starting point. In the Brans-Dicke theory the gravitational constant G is replaced by a scalar field ϕ which measures the relative strengths of inertial and gravitational forces so that $G \sim \phi^{-1}$. The gravitational field equations of this theory include ϕ and its derivatives in addition to the Einstein-like terms. The recent measurements of the solar system tests seem to rule out the effect of ϕ except in a very weak way. The cosmological solutions of the Brans-Dicke theory are somewhat similar to the standard big bang models with the difference that the effective gravitational constant decreases with epoch.

In the Hoyle-Narlikar theory the Machian concept of inertia is built into the theory right at the beginning by defining the mass of a particle as a scalar quantity satisfying a wave equation with sources in the rest of the particles in the universe. Besides its Machian origin, the theory is conformally invariant and in a special conformal frame called the Einstein frame it looks like general relativity. Recently it has been shown that the cosmological singularity of general relativity arises because the conformal function needed for going over to the Einstein frame vanishes on the zero mass hypersurfaces.[36] The theory can also be modified easily to introduce a 'fourth' cause of redshift which might be required to describe the anomalous redshifts of quasars[37].

On the physical side a significant departure from the standard cosmology is found in the matter-antimatter symmetric cosmology[38] of Alfven and Klein. This cosmology attempts to explain how from a primordial era of symmetry, matter and antimatter got separated so that we now live in a large region ($\gtrsim$ size of cluster of galaxies) dominated by matter.

Limitations of space and time severely restrict this discussion of non-standard cosmologies (see, however, Narlikar and Kembhavi[39] for a comprehensive review). These cosmologies have to be judged by the elegance of their theoretical framework and by their success or failure on the observational front. As an example of the latter, consider a recent discussion by Gunn and Tinsley[40] of standard cosmologies. Taking into account the various observational inputs available, these authors conclude that to be within the various observational limits it may become necessary to reinstate the λ-term in Einstein's field equations.

Concluding remarks

To the conventional cosmologists this review of cosmology in the post-Einstein era will appear tilted towards a non-conformist point of view. I have deliberately adopted this attitude because I feel that on an important occasion like the Einstein centenary the subject of cosmology should be presented not as a 'closed book' but as an area brimming with activity and with many unsolved problems and controversies. To this end, I have attempted to convey the remarkable strides made by the subject in the comparatively short span of less than a quarter century after the death of the man who was so intimately connected with its early development.

I also take the liberty of exposing my personal bias by making a couple of long term predictions about cosmology. It is my expectation that in due course the microwave background will receive a purely astrophysical interpretation which will divest it of its present cosmological significance. As for the big bang I expect that it will turn out to be a relatively local explosion of high temperature and density rather than the primary event describing the origin of the universe. And, with the tremendous growth rate of information in extragalactic astronomy continuing in the future, I do not think that one will have to wait until Einstein's second centenary for this to come about!

References

1. Infeld, L., in *Albert Einstein, Philosopher-Scientist*, P.A. Schilpp, (Ed), The Library of Living Philosophers, Inc., Illinois, p. 475 (1949).
2. Einstein, A., *Pruss. Akad. der Wiss. Sitzber*, part 1, 142 (1917).
3. Hubble, E.P., *Proc. Nat. Acad. Sci.* (USA), **15**, 169 (1929).
4. Friedmann, A., *Zeits. Phyc.*, **10**, 377 (1922).
5. Einstein, A. and de Sitter, W., *Proc. Nat. Acad. Sci.* (USA), 18, 213 (1932).
6. Gamow, G., *Phys, Rev.*, **70**, 572, (1946); Alpher, R.A., Bethe, H. and Gamow, G., *Phys. Rev.*, **73**, 803, (1948); Alpher, R.A. and Herman, R., *Nature*, **162**, 774 (1948).
7. Bondi, H. and Gold, T., *Mon. Not. R. astron. Soc.*, **108**, 252, (1948); Hoyle, F., Ibid., p. 372.
8. Hubble, E.P. and Humason, M., *Proc. Nat. Acad. Sci.* (USA) **20**, 264 (1934).
9. Hoyle, F. and Narlikar, J.V., *Proc. Roy. Soc.*, **A273**, 1 (1973).
10. Feynman, R.P., *Lectures on Gravitation*, (Caltech) p. 13.3 (1971).
11. Sandage, A.R. and Tamman, G.A., *Astrophys. J.*, **194**, 559 (1974).
12. Gunn, J.E., and Oke, J.B., *Astrophys. J.*, **195**, 255 (1975).
13. Field, G.B., Arp, H.C. and Bahcall, J N., *The Redshift Controversy*, Benjamin (1973).
14. Wall, J.V., *Radio Astronomy and Cosmology*, D.L. Jauncey (Ed.), D. Reidel p. 57, (1977).
15. Hoyle, F., *Paris Symposium on Radio Astronomy*, Braccwell (Ed.), Stanford, p. 529 (1959).
16. Roeder, R.C., *Radio Astronomy and Cosmology*, D.L. Jauncey, (Ed.), D. Reidel, p. 146 (1977)
17. Swarup, G., *Mon. Not. R. astr. Soc.*, **172**, 501 (1975).
18. Kapahi, V.K., Ibid., p. 513.
19. Narlikar, J.V., and Chitre, S.M., *Mon. Not. R. astr. soc*, **180**, 525 (1977).
20. Burbidge, E.M., Burbidge, G.R., Fowler, W.A. and Hoyle, F., *Rev. Mod. Phys.* **29**, 547 (1957).
21. Alpher, R.A. and Hermann, R., *Phys. Rev.*, **75**, 1089 (1949); Gamow, G., *Kong Dansk. Ved. Sels.* **27**, 10 (1953).
22. Zeldovich, Ya.B., *Soviet Phys.—Uspekhi* **6**, 475 (1964).
23. Hoyle, F. and Tayler, R.J., *Nature* **203**, 1108 (1964).
24. Peebles, P.J.E., *Astrophys. J.*, **142**, 1317 (1965).
25. Wagoner, R.V., Fowler, W.A. and Hoyle, F., *Astrophys. J.*, **148**, 3 (1967).
26. Dicke, R.H., *Phys. Rev.*, **125**, 2163 (1962).
27. Penzias, A.A. and Wilson, R.W., *Astrophys. J.* **142**, 419 (1965).
28. Roll, P.G. and Wilkinson, D.T., *Phys. Rev. Litt.* **16**, 405 (1966).
29. Dicke, R.H., Peebles, P.J.E., Roll, P.G. and Wilkinson, D.T. *Astrophys. J.*, **142**, 414 (1965).
30. Weinberg, S., *The First Three Minutes*, Basic Books (1977).
31. Hoyle, F., in *Study Week on Nuclei of Galaxies*, D.J.K. O'Connell, (Ed.), North Holland, p. 757 (1970).
32. Dirac, P.A.M., *Nature* **139**, 323 (1937).
33. Dirac, P.A.M., *Proc. Roy. Soc.* **A.333**, 403 (1973).
34. Brans, C. and Dicke, R.H., *Phys. Rev.*, **124**, 925 (1961).
35. Hoyle, F. and Narlikar, J.V., *Action at a Distance in Physics and Cosmology*, Freeman (1974).
36. Kembhavi, A.K., *Mon. Nat. R. astr. Soc.* **185**, 807 (1979).
37. Narlikar, J.V. *Ann. Phys.*, **107**, 325 (1977).
38. Alfven, H., *Rev. Mod. Phys.*, **37**, 652 (1965).

39. Narlikar, J.V. and Kembhavi, A.K., *Fundamentals of Cosmic Physics* (to be published).
40. Gunn, J.E. and Tinsley, B.M., *Nature*, **257**, 454 (1976).

Discussions

J Pachner: (1) There are three measurable quantities in cosmology, the mean mass density, the Hubble constant, and the deceleration parameter. In the steady state cosmology the deceleration parameter is -1. The recent observations show that this parameter is positive (with a reasonable probability). I consider this as a decisive argument against the steady state cosmology.

(2) There are some modifications of the steady state cosmology (McCrea, Pachner, N. Rosen) but since they introduce some cosmic field, I consider them too arbitrary (one can easily develop several cosmologies by assuming an ad hoc cosmic field).

J V Narlikar: I do not agree with you that the deceleration parameter q has been accurately determined. There are several sources of 'errors' of astrophysical origin which make a realistic determination of q very difficult.

The cosmic field used by Hoyle and me was the 'simplest' possible modification: it was a zero rest mass scalar field. Thus by choosing simplicity as a criterion one can counter the criticism of its being ad hoc.

C Sivaram: How would you account for the initial helium abundance in steady state theories?

J V Narlikar: Ordinary stellar nucleosynthesis is inadequate to account for the He/H ratio as high as is observed. One has to postulate the existence of supermassive objects ($M = 10^6 M_\odot$) for this.

N D Hari das: Do you think the recently discussed Hawking phenomena applied either to mini black holes or to cosmological metrics (if you do not believe in black holes) can be utilized as a source of black body radiation for *SS* theories?

J V Narlikar: To make this work you have to produce a theory of continuous creation of mini black holes and to demonstrate that their integrated radiation is thermal and of 2.7°K in a steady state. Off hand, I don't think this can be done.

N Panchapakesan: The data points on the 3°K curve do not seem to imply a bending of the curve. Is the black body nature of radiation still in doubt?

J V Narlikar: The data points on the short wavelength side of the 'peak' at 1 mm are few and not very accurate (in my opinion). They are, however, consistent with 2.7°K black body.

Spontaneous Symmetry Breaking and the Origin of the Universe

R. BROUT, F. ENGLERT and E. GUNZIG

Introduction

This report is a discussion of some of the aspects of a recent suggestion we have made concerning the origin of the universe, which arose in our attempt to avoid the big-bang singularity[1] encountered in Friedman type universes. We shall first summarize our arguments implying that avoidance of the big-bang singularity requires the creation of matter out of vacuum[1], and then discuss in detail the latter subject. This discussion leads to a new interpretation of our previous results which may have far-reaching consequences.

In our previous work we suggested that the mechanism of material creation was cooperative in character. An infinitesimal bit of matter seeds a space-time varying gravitational field which through quantum mechanics results in more matter production which..., the whole leading to the primitive fireball. Usual cosmological models take over from this point on: we have in mind models which begin with a very hot primitive universe (hot refers to the scale of elementary particle masses), such as sketched in S. Weinberg: *The First Three Minutes*.

The attempt to realise this programme in Ref. 1, has led to an interesting conceptual problem. In the formalism of section 3 and Appendix 2, of that paper it was found surprisingly that production of particles with mass took place without appeal to space-time varying gravitational field. Thus, in effect we came upon a production mechanism which seems to have nothing to do with our initial suggestion of matter-gravity cooperation. In the present note we interpret this mechanism in terms of a phase transition in which the "edge of the universe" is the boundary wall between two phases. The matter field outside the universe is characterized, by zero mass whereas inside, the phase transition has caused mass to arise. At first sight, this phase transition could be described by a conventional type of symmetry breakdown[2]: however, the formalism suggests a more general symmetry breaking mechanism in which the gravitational field itself arises at the same time that the universe comes into existence. In either case the phase transition would imply that the fundamental agent of symmetry breaking in physics is the creation of the universe itself.

Lest we be misunderstood, one last remark is in order: We do not wish to imply from the above that the cooperative mechanism which we initially proposed has been proven false. It is rather that our first attempts to realize it seem to point towards the existence of a phase transition, that

is a modification of the vacuum state. This modification of the vacuum state sets up the cooperative process.

Cosmological kinematics

We recall some essential elements of cosmological kinematics. We assume Einstein's equation without a cosmological constant. Consistent with present data, we assume that the universe is open. *Straightforward* extrapolation to the distant past then leads to the big-bang singularity. (By big-bang we literally mean the mathematical singularity and *not* the fireball).

The underlining of *straightforward* is deliberate. More precisely, the big-bang arises in any perfect fluid model of the gravitational source with the pressure having its usual mechanical interpretation [$p \geqslant 0$]. We argue below that a way to avoid the singularity is that in the early universe, the pressure is negative. The laws of energy and momentum conservation then imply that mass must be produced in this region of space-time. We emphasize this important conclusion: *the singularity of an open universe can be avoided through the spontaneous creation of matter.* For this reason we have had recourse to search for the origin of the universe as a quantum phenomenon. For it is only in the context of quantum field theory that the creation of matter can at present be conceived.

Our approach offers the solution to another vexing problem, that of causality. In the straightforward extrapolation, the existence of the singularity at a finite time in the past inevitably gives rise to horizons. Our proposition to avoid the big-bang automatically eliminates these.

We first review the argument which leads to the big-bang for $p \geqslant 0$. We begin with the conventional perfect fluid model of cosmology expressed in terms of the Robertson-Walker metric. For an open homogeneous isotropic universe one has

$$T_{\mu\nu} = (\sigma + p)\, u_{\mu\nu} u - p g_{\mu\nu} \tag{1}$$

$$ds^2 = d\tau^{*2} - R^2(\tau^*) \left[\frac{dr^2}{1+r^2} + r^2 d\Omega^2 \right] \tag{2}$$

$T_{\mu\nu}$ = energy-momentum tensor, σ = proper mass density, p = pressure, τ^* = proper synchronous time; $R(\tau^*)$ = cosmological scale factor, r, $\Omega(\theta, \psi)$ comoving co-ordinates, u_μ = unit velocity vector.

It is convenient and conventional to parametrize R essentially by its logarithm

$$R = \tau e^{\lambda/2} \tag{3}$$

where τ is defined through

$$d\tau^* = e^{\lambda/2}\, d\tau \tag{4}$$

In these terms the relevant Einstein's equations are condensed into a single differential equation for λ

$$e^{-\lambda-2a}\,\frac{d\lambda}{du}\left[\frac{d\lambda}{du}+4\right]=(4/3)\,\tau_0^2\,K\sigma \tag{5}$$

and the important conservation law

$$d/du\,(\sigma R^3) = -\,p\,d/du\,(R^3) \tag{6}$$

where u is the logarithmic time, $\tau = \tau_0 e^u$, and τ_0 an arbitrary scale.

The criterion for the absence of a particle horizon[3] is

$$\int_0^{\tau^*}\frac{d\tau^x}{R} = \int_{\tau_{\min}}^{\tau}\frac{d\tau}{\tau} = \infty \tag{7}$$

which is clearly met when $\tau_{\min} = 0$. The finiteness of $\sigma(0)$ then implies that $\lambda\,(\tau = 0)$ and $d\lambda/d\tau$ $(\tau = 0)$ are finite. So that $R(\tau = 0) = 0$; hence σR^3 $(\tau = 0) = 0$ and it then follows from (6) that $p\,(\tau) < 0$ at least in the neighbourhood of $\tau = 0$. Thus the absence of horizons and finiteness of $\sigma(0)$ imply the negativity of pressure in the early universe. From (5) and (6), it follows immediately that $d^2\lambda/d\tau^2 < 0$ if the pressure $p \geqslant 0$, in the case of an expanding universe. Hence the shape of the curve $\lambda\,(\tau)$ in familiar realizations $(p = 0,\ p = \sigma/3)$: $\lambda\,(\tau)$ is a monotonically increasing function of τ with negative curvature everywhere; its vertical asymptote which appears at some finite value of τ corresponds to the big-bang singularity. This is clearly avoided by letting $p < 0$ in a finite region for $\tau \geqslant 0$, thus converting curve (I) to curve (II) (Fig. 1).

From eqn. (6) we see that $p < 0$ implies the production of mass for an expanding universe. It is also clear that $p < 0$ in the early universe gives rise to an inflection in function $\lambda(\tau)$ to avoid the singularity in λ.

Matter Creation

Turning to the theory of quantum production, we consider a scalar field in interaction with gravitation. The action is

$$S = \frac{1}{2}\int d^4x\,\sqrt{-g}\left\{g^{\mu\nu}\,\partial_\mu\Psi\partial_\nu\Psi - m^2\Psi^2 + \frac{R}{6}\,\Psi^2 - K^{-1}\,R\right\} \tag{8}$$

R is the Einstein scalar curvature. The coupling to gravity is non-minimal due to the term $(R/6)\,\Psi^2$. This has been included so that $T^\mu_\mu = 0$ for $m = 0$. Our point of view is that conformal symmetry must be respected in preparation for a more fundamental theory. For example, if one takes as fundamental constituents, fermions and Yang-Mills fields then minimal coupling to these fields would respect $T^\mu_\mu = 0$ in the massless limit. In exemplifying a fundamental approach with scalar fields we assume that this condition must be maintained. It is useful to make the Weyl transformation

$$g_{\mu\nu} = \check{g}_{\mu\nu}\left(\frac{K}{6}\phi^2\right),\ \Psi = \psi/(K/6)^{1/2}\,\phi \tag{9}$$

$$S = \frac{1}{2}\int d^4x \sqrt{-\check{g}}\left\{\left[\check{g}^{\mu\nu}\,\partial_\mu\psi\partial_\nu\psi - (K/6)\,m^2\phi^2\psi^2 + \frac{\check{R}}{6}\,\psi^2\right] - \left[\check{g}^{\mu\nu}\,\partial_\mu\phi\partial_\nu\phi + \frac{\check{R}}{6}\,\phi^2\right]\right\} \tag{10}$$

As mentioned the comological part of $g_{\mu\nu}$ is its scale, thus in cosmological equations, the $\check{g}_{\mu\nu}$ of eqn. (10) is the Minkowski metric. A remarkable theorem then emerges. From the manifest Lorentz invariance of (10) the statement of energy momentum conservation reads

$$\partial_\mu\,[T_{\mu\nu}(\phi) + T_{\mu\nu}(\psi)] = 0 \tag{11}$$

where $T_{\mu\nu}$ are conventional Lorentz energy-momentum tensors. By definition: $T_{\mu\nu}(\phi) = -\{\partial_\mu\phi\,\partial_\nu\phi - \frac{1}{2}\check{g}_{\mu\nu}\,\partial_\sigma\phi\,\partial^\sigma\phi - \frac{1}{6}\,(\partial_\mu\partial_\nu - \check{g}_{\mu\nu}\,\check{\Box})\,\phi^2\}$;
$$T_{\mu\nu}(\psi) = +\left\{\partial_\mu\psi\,\partial_\nu\psi - \tfrac{1}{2}\check{g}_{\mu\nu}\partial_\sigma\psi\partial^\sigma\psi - \tfrac{1}{6}(\partial_\mu\partial_\nu - \check{g}_{\mu\nu}\check{\Box})\,\psi^2 + \tfrac{1}{2}\check{g}_{\mu\nu}\,\frac{K}{6}\,m^2\phi^2\psi^2\right\}.$$
The all-important point is the minus sign in front of $T_{\mu\nu}(\phi)$. This permits the creation of matter ex-nihilo. One can begin with a situation $T_{\mu\nu}(\phi) = 0$, $T_{\mu\nu}(\psi) = 0$ for $t < 0$ and yet have each separately non-zero in the future, maintaining the sum equal to zero. The expanding universe, manifested in space-time variations of ϕ [in our previous notation $\phi = (6/K)^{1/2}\,e^{\lambda/2}$] supplies in energy source for the creation of matter.

In our first efforts to exploit this circumstance we have set ϕ equal to a classical field, whilst treating ψ quantum-mechanically. We ask for the production of ψ-particles due to space-time variations of ϕ. It is supposed that the universe nucleates about some (arbitrary) origin ($r = 0$, $t = 0$). The Lorentz invariance of (10) then tells us that the response will be a function of τ only where $\tau^2 = t^2 - r^2$.

These considerations prompted us[1] to look for non-trivial solutions (i.e. those for which massy matter is engendered) in the forward light cone of the origin. It turns out that for $Km^2 \ll 1$ the space-time variations of λ are so slight for $0 \leqslant \tau \ll (1/(Km^2)^{1/2})\;1/m$ that λ can be essentially treated as constant in this region. Heisenberg's equations of motion are then exactly integrable in this region and one finds in the interval $[1 < (m\tau) \ll (Km^2)^{-1/2}]$ that σ is a constant $\sim m^4$. In order to perform this calculation a boundary condition on the surface $\tau^2 = 0$ has to be specified. In the above calculation, the boundary condition was the 'natural' one in the formalism. Namely, the analysis is naturally cast in terms of the eigenfunctions of the Laplacian corresponding to displacements on the unit hyperboloid $\tau^2 = 1$. These are given by Gegenbauer functions; they are indexes by the (continuous) quantum number k in what follows. Our boundary condition was that $\langle N(k;\,\tau = m^{-1})\rangle = 0$ where expectation values refer to a Heisenberg state, called 'vacuum'.

This calculation has now been made more precise than in Ref. 1 and an exact solution to the problem has been found. In what follows, the calculation has been extended to the light-cone itself and the boundary condition is $\langle N(k;0)\rangle = 0$. (The divergence as $\tau_0 \to 0$ in Ref. 1 was an artifact of the approximation.)

We have calculated the causal Green's function $G(x_1 - x_2)$ of the ψ-field for x_1, x_2 both inside the light cone of the origin. Defining $\tau_i = (X_i^{02} - \mathbf{X}_i^2)^{1/2}$ and working in the co-ordinate system $X_1^0 = \tau_1$, $X_2^0 = \tau_2 \cosh \xi$, one finds

$$G(x_1, x_2) = \frac{1}{\tau_1\tau_2 \sinh \xi} \int_0^\infty dk \sin k\xi \; G_k(\tau_1, \tau_2) \tag{12}$$

$$G_k = \{\theta(u_2 - u_1)\, J_{ik}(e^{u_1})\, J_{-ik}(e^{u_2}) + \theta(u_1 - u_2)\, J_{-ik}(e^{u_1})\, J_{ik}(c^{u_2})\} \frac{\pi}{\sinh \pi k} \tag{13}$$

where $\tau_i = m^{-1} e^{u_i}$; $J_{ik}(e^u)$ is the Bessel function of imaginary index ik.

One finds the number of particles in mode k at time τ by forming

$$\lim_{\tau_1, \tau_2 \to \tau} \overrightarrow{\Omega}_k(\tau_1)\, G_k(\tau_1, \tau_2)\, \overleftarrow{\Omega}_k^+(\tau_2) = \langle N_k \rangle \tag{14}$$

where
$$\Omega_k = \frac{\partial_u + iK}{\sqrt{2K}}; \; K = (k^2 + m^2\tau^2)^{1/2} \tag{15}$$

From (12), (13), (14) the calculation of the particle density at time τ is easily performed and one finds $n \sim m^3$ is expected.

Interpreting these results we find that the causal Green's function (13) is precisely what would have been calculated by the functional integration technique where the Lagrangian is formally made to vanish outside the light cone ($\tau^2 = 0$, $t > 0$). Clearly, in the context of the type of theory which we wish to construct, where the past describes empty space (hence filled with quantum vacuum fluctuations) this construction is untenable. The boudary condition $\langle N_k(\tau = 0)\rangle = 0$ is therefore not physical. Nevertheless, the calculation in question may be relevant to the solution of the following physical problem; the action (10) is characterized by the vanishing of the interaction term $Km^2\phi^2\psi^2$ outside the light cone, and $m^2\phi^2$ takes on a constant value inside. The jump in $m^2\phi^2$ would then give rise to the production of massy particles inside the light cone.

The reasoning behind this conjecture is based on the following argument. First consider that m vanishes everywhere. In the usual formalism where action exists over all Minkowski space, one has

$$G_{m=0}(x_1, x_2) = \frac{i}{4\pi^2} [(x_1 - x_2)^2 + i\varepsilon]^{-1} \tag{16}$$

which in our co-ordinate system reads

$$G_{\substack{m=0 \\ \text{usual}}}(x_1, x_2) = \frac{i}{4\pi^2} \frac{1}{2\tau_1\tau_2[chu - ch\,\xi + i\varepsilon]} \tag{17}$$

This is to be contrasted with (13) evaluated with $m = 0$ where $J_{ik} \sim \tau^{ik}$

$$G_{m=0}(x_1, x_2) = \frac{i}{4\pi^2} \frac{\xi}{\tau_1 \tau_2 \, sh\, \xi} \frac{1}{u^2 - \xi^2 + i\varepsilon} \tag{18}$$

One checks that (17) gives rise to a boundary condition $\langle N_k(\tau = 0)\rangle = [e^{2\pi k} - 1]^{-1}$. As $\tau \to \infty$ this gives rise to a particle density dropping off as τ^{-3}. Thus for massless particles the difference between (17) and (18) leads to negligible effects deep inside the light cone ($\tau \to \infty$).

For massy particles within the light cone, the essential point to note is that for $\tau > m^{-1}$, the $J_{ik}(k \neq 0)$ approach τ^{ik}. Hence G_m connects smoothly $G_{m=0}$ within a Compton wavelength of the light cone. Thus if we suppose, that the matter field is essentially massless for $\tau < m^{-1}$ and develops massy character for $\tau > m^{-1}$, then the Green's function of eqn. (13) would be expected to furnish a reasonable approximation to the solution to the problem in the region $\tau > m^{-1}$.

We are at present examining the detailed mathematics of such a model. A non-trivial problem which must be settled is concerned with re-normalization of the infinities which necessarily arise in the calculation of the mass density T^{μ}_{μ}. We think that the solution to this problem resides in setting to zero the re-normalized cosmological constant in a manner which preserves conformal invariance[4]. Indeed, we have checked that this procedure eliminates spurious contributions to T^{μ}_{μ} in usual Minkowski field theory.

We are thus led to the suggestion that early particle creation is set up by a phase transition in which $m^2\phi^2$ increases from zero to a finite value in an interval m^{-1} across the light cone.

There are two solutions to $m^2\phi^2 = 0$ corresponding to two avenues of approach for subsequent research. The fiirst thing that comes to mind is $m = 0$, $m \neq 0$ is then the expectation value of a scalar field. This would suggest that a field theoretical conventional symmetry breaking mechanism is at work. The scenario for the unfolding of the very early universe in such an approach would resemble the droplet model.[5] Particle production would be the quantum effect induced by the droplet. This would then have to be further modified by taking into account interaction with the gravitational field.

The more exciting alternative is that $m^2\phi^2 = 0$ implies $\phi = 0$ (thereby giving up general relatively outside the universel). We may analyse the meaning of this solution by reference to the action (10) in the context of spontaneously broken conformal symmetry (s.b.c.s.). Namely, in this interpretation ϕ is the Goldstone meson: the dilaton; in s.b.c.s. we refer to the hypothesis set out in Ref. 4 wherein conformal symmetry is a true symmetry of nature in spite of its violation in renormalization procedures. Its spontaneous violation would then imply the existence of a Goldstone boson. There is a fundamental sign ambiguity in the effective Lagrangian of this phenomenological field ϕ. If the minus sign is chosen, then (10) is its effective action. Unitarity considerations then give rise to the remaining components of gravitational metric tensor.[4] Thus in this approach the gravitational field is a phenomenological field which arises in response to s.b.c.s., that is which comes into existence at the same time as the universe.

It is clear in this approach that the process of generating mass and the gravitational field are part and parcel of the same problem. The parameter which describes symmetry breakdown is necessarily dimensionless (owing to conformal symmetry). This parameter is Km^2 and we would expect to determine it from some sort of self-consistent condition to be of O(1). Thus we expect the characteristic mass of the particles to be created in the very early universe to be the Planck mass of $O(K^{-1/2})$. Such a consideration might have been sufficient to reject this proposition a few years ago. But the grand unified scheme[5] is becoming increasingly attractive. In this approach the fundamental splitting which divides the weak and strong interaction sectors turns out to be of $O(K^{-1/2})$. Several arguments concord to give this estimate. In particular the scheme requires vector fields—the so-called lepto-quarks—whose mass is $O(K^{-1/2})$. The attractive possibility then arises that the creation of the universe, gravitation and elementary particle physics as we know it are different facets of one and the same physical phenomenon. An addtional attractive possibility is that a very early fireball composed of lepto-quarks could conceivably decay into quarks and leptons in a manner which is asymmetric with respect to charge conjugation symmetry owing to the interference between the proton decay matrix element and CP violating amplitudes.[6]

Appendix

We have made further calculations concerning the production of particles. We are now in a position to state a number of firm conclusions, some of which are in contradiction with our previous published work on the subject, (due to mathematical subtleties which had originally escaped our attention).

1. In the absence of interaction of the matter field with the gravitational field, there is no asymptotic production, One finds

$$\lim_{\tau \to \infty} \sigma(\tau) \sim \tau^{-3}$$

2. It is therefore necessary to have recourse to the interaction with the gravitational field to produce particles asymptotically. We have examined this problem in the case where the cosmological gravitational field is taken to be de Sitter in character from the outset.

In the notation of Ref. 1, this means we have set $e^{\lambda} = [1 - (\tau/\tau_{\infty})^2]^2$; $\tau_{\infty} = \sqrt{12/k\sigma_A}$ where σ_A is some mass density which must be calculated self consistently.

We have proven that in this case there is asymptotic production, i.e. $\lim_{\tau \to \tau_\infty} \sigma(\tau)$ exists. Furthermore, for $m\tau_\infty \gtrsim 0(1)$ the asymptotic value of σ so obtained is consistent with the input value σ_A, *if and only if a condition on the mass is met*, viz., $km^2 = 0(1)$. The exact number cannot be determined by the model, but a (not very wide) range of values of km^2 is admitted.

In principle one could also study the case of $0 < m\tau_\infty \ll 1$ but here our model of treating gravitation as a classical field becomes grossly incorrect, since the condition $m\tau_\infty \ll 1$ implies $k\sigma/m^2 \gg 1$.

It remains to perform the complete self-consistent calculation to show how the gravitational field evolves from zero to the de Sitter limit. On the basis of these results we are encouraged in our conjecture which links the breakdown of grand unification symmetry with the creation of the universe.

The above calculations have been carried out with the collaboration of J.M. Frère, P. Nardone, P. Spindel and C. Truffin, and will be reported in detailed publications elsewhere.

References

1. Brout, R., Englert, F., Gunzig, E., *Ann. Phys.* (N.Y.) **115**, 78 (1978); Brout, R., Englert, F., Gunzig, E., "The causal universe" *G.R.G.* Preprint.
2. Englert, F., Brout, R., *Phys. Rev. Lett.*, **13**, 321 (1964); Higgs, P.W., *Phys. Rev. Lett.* **12**, 132 (1964); *Phys. Rev. Lett.* **13**, 508 (1964).
3. Rindler, W., *Mon. Not. Roy. astron. Soc.* **116**, 662 (1956).
4. Englert, F., Truffin, C., Gastmans, R., *Nucl. Phys.* **B117** 407 (1976).
5. Coleman, S., "The use of Instantons" Lectures delivered at the 1977 International School of Subnuclear Physics, Ettore Majorana.
6. Yoshimura, M., *Phys. Rev. Lett.* 41 281 (1978).

Discussions

J V Narlikar: Around 1965–66 Hoyle and I had considered a model of the universe in which a part of the steady state universe has creation cut off. This part then expands as a Friedmann universe. This model was dictated by astrophysical considerations. I am happy to see a connection with particle physics in your work.

J Pachner: Since the McCrea's paper (1954) it is well known that the negative constant pressure can replace the *C*-field of Hoyle's steady state cosmology. The subsequent papers by Pachner and N. Rosen show that a negative pressure depending on the radius of universe can yield a singularity-free oscillating universe. However, the weak point of these cosmologies is an ad hoc introduction of the negative pressure, I consider the cosmology developed by the speaker as an important step towards understanding the origin of the actual universe because the entrance of the negative pressure is an integral, physically justified part of the theory. Have you been able to compute the ratio of helium to hydrogen atoms in the universe?

E Gunzig: One of the open questions related to this work is, of course, connected with the detailed physics of the 'primaeval soup' here described by the matter field. Another open question concerns the detailed physical phenomena occurring around the joining point between the creation period and the free expansion one. The answer to the question will possibly be found somewhere in the framework of these two problems, but this is for the moment unknown.

QSOs as Probes of the Universe*

JUDITH J. PERRY

Professor Narlikar has given a thorough and delightful review of the theoretical bases of the principal methods of observational cosmology. These methods hold the promise that we shall some day be able to determine the structure of the universe. This promise has been the motivation behind years of work by dedicated and talented cosmologists, yet we seem as far away from our goal today as we were many years ago. The purpose of this talk is to give a brief insight into some of the practical reasons underlying this frustration and to discuss the particular, and peculiar, role played by QSOs in this game.

The goal of observational cosmology is the determination, through the use of astronomical observations, of the geometrical structure and evolution of the universe. Ever since the early work of Hubble[1] established a redshift-distance relationship for nearby galaxies, the most commonly used method of observational cosmology has been the construction of redshift-magnitude (m, z) diagrams. In principle a direct, simple and elegant method (as outlined by Prof. Narlikar), the construction of (m, z) diagrams is in practice so riddled with difficulties as to render it only marginally useful.[2] If we assume that the universe is both isotropic and homogeneous, then it can be shown that the geometrical structure of that universe is described by a Friedmann-Robertson-Walker metric.[3] Neither the assumption of isotropy nor of homogeneity has been established beyond reasonable doubt[4] and further observational tests of both assumptions are needed. However, in order to proceed, we shall here follow the usual practice and assume both isotropy and homogeneity. The metric (FRW) is then

$$r_0(z) = \frac{1}{H_0 q_0^2 (1+z)^2}\{q_0 z + (q_0 - 1)(\sqrt{2q_0 z + 1} - 1)\} \qquad (1)$$

In this case the goal of observation reduces to that of determining H_0 and q_0.

If the universe were filled with a set of identical astronomical objects—'standard candles'—this would be a trivial task. The measurement of the diminution of the light from such objects as functions of z would directly measure $1/r_0^2(z)$, and from the curves of r_0 against z one could determine H_0 and q_0 directly. However, the population of the universe as we observe it is diverse, and all of these diverse objects are believed to evolve in

*This lecture draws extensively on material in Ellis and Perry[2], Perry, Burbidge and Burbidge[7] and in Dyson, Falle and Perry.[11]

time, i.e. to change their properties as a function of z.

Let us first consider the optical observations of galaxies with a view toward the construction of real (m, z) diagrams. Opiical images contain the information $I_\nu(\alpha)$, the specific intensity at an angular position α at the frequency ν. The relationship between the angles at the detector and linear distances r at the source is illustrated in Fig. 1 and satisfies the equation

$$r = \alpha r_0(z) \tag{2}$$

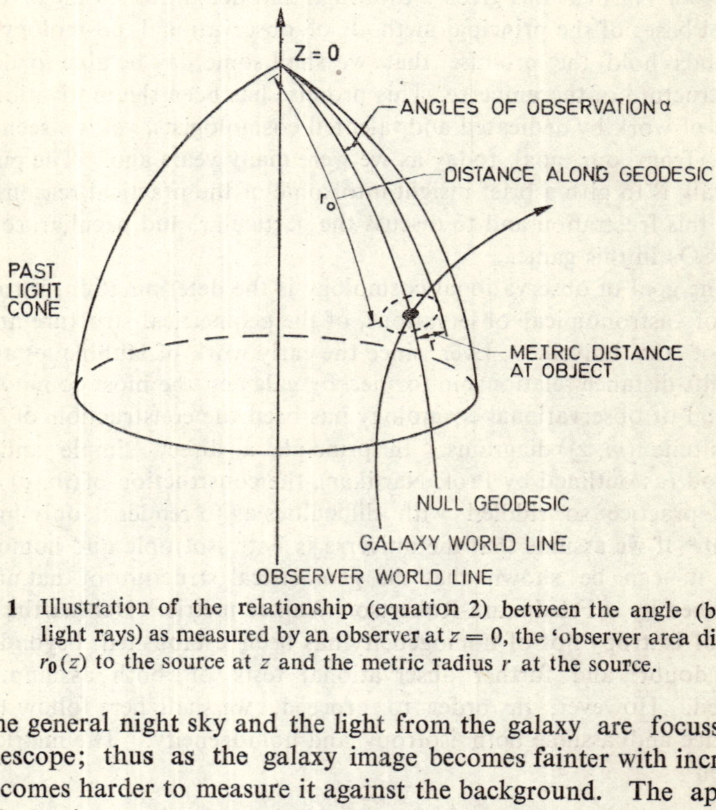

Fig. 1 Illustration of the relationship (equation 2) between the angle (between light rays) as measured by an observer at $z = 0$, the 'observer area distance' $r_0(z)$ to the source at z and the metric radius r at the source.

Both the general night sky and the light from the galaxy are focussed by the telescope; thus as the galaxy image becomes fainter with increasing r_0 it becomes harder to measure it against the background. The apparent size of the image is that angle A_{ap}, at which the observed intensity of radiation from the source falls to the limiting apparent brightness which can be detected in the presence of the general background (Fig. 2). Disney[5] has pointed out that a primary selection criterion for the galaxies which are studied may just be A_{ap}. As is illustrated in Fig. 2, for galaxies of the same total luminosity if A_{ap} is large the surface brightness may be so low that the galaxy is not chosen for studies, whereas if A_{ap} is small the galaxy may not be distinguishable from a star.

Once a galaxy has been selected for observations, usually on the basis of its appearance on a photographic survey plate, what can then be measured, and how does that measurement relate to the actual properties of the galaxy in its own resi frame?

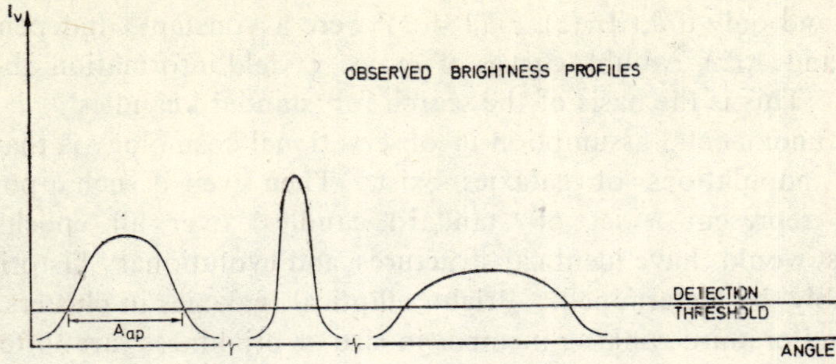

Fig. 2 Schematic illustration of galactic brightness profiles. The apparent angular size of the image, A_{ap}, as shown, is that angle at which the image brightness falls to that of the background—the detection threshold. The central profile represents a centrally condensed galaxy of the sort whose image may be mistaken for that of a star on low resolution plates; the right most that of a large diffuse galaxy which may be missed in optical searches of plates (see Disney[5]).

For distant galaxies, whose images are faint, the usual measurement made at present is of the integrated specific intensity, i.e. the specific flux within a given aperture centred on the image. The specific flux at frequency ν within an aperture of half angle A ($|A| \ll 1$) is*

$$F_\nu(A) = \int_0^A I_\nu(\alpha)\, 2\pi\alpha\, d\alpha \tag{3}$$

If the surface brightness of the source is $B_\nu(r, z)$ then the specific intensity of radiation at the observer is measured to be

$$I_\nu(\alpha, z) = B_{\nu(1+z)}(r, z) \tag{4}$$

Here the redshift dependence of B reflects the possibility of source evolution and r is the 'metric radius' or distance measured at the source from its centre, corresponding to the observed angle α. α and r are related through eqn. (2). Accordingly, the specific flux received from the source within an aperture of half angle A is

$$F_\nu(A, z) = \frac{h_\nu(Ar_0(z), z)}{r_0^2(z)(1+z)^3} \tag{5}$$

where

$$h_\nu(\beta, z) \equiv \int_0^\beta B_{\nu(1+z)}(r, z)\, 2\pi r\, dr \tag{6}$$

Now the magnitude $m_\nu(A)$ of the galactic image is

$$m_\nu(A) = -2.5 \log_{10} F_\nu(A) + \text{const} \tag{7}$$

or $$m_\nu(A) = 5 \log_{10} r_0(z) - 2.5 \log_{10} \{h_\nu(Ar_0(z), z)/(1+z)^3\} \tag{8}$$

*We will limit our discussion to spherical sources, in which case the image is circular. In case where ellipticity is significant a similar analysis can be carried out where $r^1 (\equiv \{x^2/a^2 + y^2/b^2\}^{1/2})$ replaces $r (\equiv \{x^2 + y^2\}^{1/2}.)$ In that case eqn. (6) becomes a two-dimensional integral.

Thus, if and only if $h_\nu(Ar_0(z), z)/(1+z)^3$ were a constant, independent of both z and $r_0(z)$, would curves of m_ν vs. z yield information about $r_0(z)$ directly. This is the basis of the search for 'standard candles'.

The fundamental assumption in observational cosmology is that homogeneous populations of galaxies exist. Then even if such a population did not represent a set of 'standard candles' over all epochs, z, its members would have identical structures and evolutionary histories could presumably be determined. Bright elliptical galaxies in clusters, usually the second or third ranking member in size or brightness are often taken to be members of such a set. However, the extent to which these galaxies do exhibit standard structures and luminosities at any given z, particularly large z, is not yet known. The assumption that they do is built into all methods at use at present for determination of H_0 and q_0. Developing methods to test these assumptions is therefore imperative.

The assumptions about this 'standard' galactic brightness structure, made for reasons of simplicity and for lack of detailed data, are outlined below. The galaxy brightness function is usually written

$$B_\nu(r, z) = B_0(z)\, \mathcal{J}(\nu, z)\, f\left(\frac{r(z)}{a(z)}\right) \tag{9}$$

where the spectral function, $\mathcal{J}$, which gives the proportion of the intensity in each frequency band at any epoch z, is assumed constant across the face of the galaxy. Since colour gradients are, in fact, observed in nearby galaxies, this is a quite weak assumption. $\mathcal{J}$ is normalized so that

$$\int_0^\infty \mathcal{J}(\nu, z)\, d\nu = 1$$

The radial variation of the brightness is assumed to be characterized by an amplitude which may evolve in time, $B_0(z)$, and a normalized radial function whose functional form is assumed not to evolve, $f(\beta)$. The radial function is normalized so that $f(0) = 1$. Here $\beta(z) = r(z)/a(z)$, where a is a scaling radius. It has been generally assumed that the scaling radius, a, does not evolve with z. Since this assumption has never been justified we will write $a = a(z)$. With these assumptions, h_ν becomes:

$$h_\nu(Ar_0(z), z) = B_0(z)\, \mathcal{J}(\nu(1+z), z)\, a^2(z) \int_0^{Ar_0(z)/a(z)} f(\beta)\, 2\pi\beta\, d\beta \tag{10}$$

and the magnitude is then

$$m_\nu(A) = 5 \log_{10} \Theta(z) - 7.5 \log_{10}(1+z) - 2.5 \log_{10} \{B_0(z)\, \mathcal{J}(\nu(1+z), z)\, g(A\,\Theta(z))\} \tag{11}$$

where

$$\Theta(z) \equiv r_0(z)/a(z) \tag{12}$$

and

$$g(\gamma) \equiv \int_0^\gamma f(\beta)\, 2\pi\beta\, d\beta \tag{13}$$

Note that the magnitude depends on the cosmologically interesting quantity $r_0(z)$ only through the ratio $\Theta(z)$; there is a direct inverse square dependence on Θ plus the dependence through g. If and only if

$$[B_0(z)\, \mathcal{I}(\nu(1+z), z)\, g(A\, \Theta(z))] = \text{const.} = [B_0(0)\, \mathcal{I}(\nu, 0)\, g(A\, \Theta(0))]$$

will curves of m_ν vs. z yield information about $\Theta(z)$ directly. Even if the assumption of the uniformity of the set of galaxies were true, it would also be true that (a) because of source evolution $B_0(z_1) \neq B_0(z_2)$; (b) because measurements are made at standard frequencies in the *observer's* frame $\mathcal{I}(\nu(1+z_1), z_1) \neq \mathcal{I}(\nu(1+z_2), z_2)$; and (c) $g(A_1\, \theta(z_1)) = g(A_2\, \theta(z_2))$ only if $A_1\, \theta(z_1) = A_2\, \theta(z_2)$, i.e. the apertures, A_1 and A_2 must be chosen based on an assumed knowledge of the very function $\theta(z)$ which the test is supposed to determine. This latter difficulty could only be overcome by direct observation if galaxies had sharp edges or a brightness maximum at some definite metric radius. Since such clear metric indicators do not exist, it is immediately obvious that no (m_ν, z) relationship can yield information on $\theta(z)$ without independent knowledge of $f(\beta)$ at all z. Every attempt to develop methods for determination of $\theta(z)$ is thus forced to incorporate assumptions about $f(\beta)$, assumptions which are then not checked at large z in the standard methods used.

Because of the aforementioned difficulties a set of 'corrections' have been introduced in order to attempt to relate the observable magnitudes $m_\nu(A)$ to the theoretical concept m. These are:

(1) The 'aperture correction', which attempts to correct the error introduced by measuring different physical extents of galaxies at different z. If we define $g(A\,\theta(z)) = G(A, z)\, g(r_m)$, where r_m is a metric radius in a nearby "standard" galaxy, $G(A, z)$ is the aperture correction.

(2) The K-correction attempts to account for the fact that measurements made by an observer at ν correspond to flux emitted at $\nu(1+z)$ at the source. If we define

$$\mathcal{I}(\nu(1+z), z) = K(\nu, z)\, \mathcal{I}(\nu, z=0),$$

then $K(\nu, z)$ is the K-correction. Evaluation of the K-correction requires knowledge of both the ultraviolet spectrum at $z = 0$, and the evolution of the spectral distribution function with z.

(3) The evolutionary correction attempts to account for the evolution of the surface brightness with z. If we define $B_0(z) = E(z)\, B_0(0)$, then $E(z)$ is the evolutionary correction factor.

In terms of these corrections, the flux may be written as

$$F_\nu(A,z) = \frac{K(\nu, z)\, E(z)\, G(A,z)}{(1+z)^3}\, \frac{B_0(0)\, \mathcal{I}(\nu,0)\, g(r_m)}{\theta^2(z)} \qquad (14)$$

and the magnitude as

$$m_\nu(A,z) = 5 \log_{10} \theta(z) + 7.5 \log_{10} (1+z) - 2.5 \log_{10} K(\nu, z)\, E(z)\, G(A, z) + \text{const.} \qquad (15)$$

It is now clear that without *independent* astrophysical information as to the z dependence of $\mathcal{J}$, B_0 and $f(\beta)$ the (m_v, z) diagram *cannot* be used to determine $\theta(z)$, nor can it be inverted to determine *KEG* without an independent determination of $\theta(z)$.

The scatter in the galactic magnitude data, even after application of all the corrections, has been too large to allow a reliable determination of H_0 and q_0. (It is argued[2] that the magnitude-redshift diagram is not the best method to use to attempt to determine H_0 and q_0. We have proposed that (a) direct comparison of the theory should be made, e.g., eqn. (5) with actual measurements and (b) differential measurements using several apertures per galaxy are to be preferred to current measurement techniques.) Despite the difficulties with the (m, z) diagrams, however, the hope remained that if large redshift objects could be found, these would allow separation of the curves for different values of q_0 and would in addition provide direct evidence as to the evolution of astronomical bodies and the mass density of the universe.

The discovery of QSOs in the early sixties was therefore greeted with a tremendous enthusiasm by observational cosmologists. Although QSO redshifts ranging up to 3.5 have been measured for over 800 objects (measured redshifts of galaxies hardly exist for $z > 0.5$) the hope that (m, z) diagrams for QSOs would determine the metric has not been realized. The scatter in the apparent magnitude at any given redshift z (approximately three to four magnitudes) has led more to a debate over the origin of the redshifts than to the direct use of the QSOs for cosmological purposes. The attempt to understand the objects themselves, and to build models capable of explaining the apparently enormous energy output from relatively tiny volumes has dominated QSO research in recent years. In fact, in the proceedings of the Copenhagen Symposium[6] held in summer of 1977, only about 20 out of 400 pages deal in any way with the cosmological use of QSOs. The only paper dealing with a (m, z) relationship[6a] focused not on the determination of the metric but on the question of whether such a relationship could be established at all for any subset of QSOs.

Must we therefore abandon all efforts to map the universe in light of the difficulties present above? Fortunately, the answer to this question appears to be no, although the time scale for success appears to be rather long. Detailed examination of the spectra of many QSOs has revealed rich and complex absorption patterns. This rich absorption appears most frequently in the spectra of QSOs with large emission-line redshifts. The lines are generally very sharp—in fact, they are often at the limit of the spectral resolution, particularly at intermediate resolutions, and when looked at higher resolution strong lines often tend to split up into several components. This extreme narrowness of the lines implies very small internal velocity dispersions in the absorbing gas, consistent with a few times the thermal velocity in gas at a few 10^4 °K. No physically sensible identification of all the lines present in any given spectrum is possible in most cases; generally subsets of lines can be grouped together into redshift systems with redshifts, z_a, quite different from one another, all of which are generally quite different from the emission line redshift, z_e. These absorption line systems are generally blue shifted with respect to the emission line redshift of the object. Were these blue shifts Doppler shifts then they would correspond to velocities ranging from approximately 0.001 to 0.5 of c (the velocity of light).

The interpretation of these differential redshifts has been the subject of considerable controversy ever since they were first discovered.[7] Opinion has polarized about two extreme hypotheses, first that all the absorption is due to intervening matter: galactic haloes, protogalactic clouds, etc. (the intervening hypothesis) or second, that it all arises in clouds or shells in the vicinity of the QSO in material probably ejected from the QSO (the intrinsic hypothesis).

Were most of the absorption due to intervening matter, then the study, of that absorption would be a powerful tool for cosmology. In fact, at the moment, it appears that such studies may be the only way of mapping the universe at large redshifts. So much absorption has already been observed that if it is all intervening it forces us to modify the picture we have, based on local observations, of the structure of galaxies. (It is important to realize that we can as yet set only lower bounds to the amount of absorption present.) In particular, it implies that in the early universe, galaxies had extremely large haloes (perhaps as large as 0.5 *Mpc*).[8] There are also indications that the universe must have a uniformly distributed population of hydrogen clouds which do not show any indication of the clustering typical of galaxies.[9] Because of the importance of these and other conclusions for our understanding of galactic structure and evolution, it is obviously essential to determine what proportion, in fact, of the absorption systems are intrinsic to the QSOs. Until recently no plausible models for intrinsic absorption existed. This fact led many to suppose that all the narrow-line absorption must be intervening, even those systems with absorption line redshifts approximately equal to that of the emission lines.

In considering the problems posed by the assumption that the absorption is intrinsic, we must first ask what the physical properties of the absorbing gas need be in order to give rise to the observed absorption-line systems. The observations yield directly (for all objects) the absorption-line redshifts, number of systems, minimum column densities per ion, and a minimum set of co-existing ions. Particularly for high redshift QSOs, in which absorption appears in essentially every object, the absorbing gas must cover 4π steradians about the central object if the absorption is intrinsic. The relative velocity, if the difference in redshifts is due to the Doppler effect is

$$v_{\text{rel.}} = \beta c = \left[\frac{(1+z_e)^2 - (1+z_a)^2}{(1+z_e)^2 + (1+z_a)^2}\right] c \qquad (16)$$

The mass and kinetic energy of the absorbing shell is then:

$$M \approx 4\pi r^2 m_P N_H \qquad (17)$$

$$T \approx 2\pi r^2 m_P N_H v_{rel}^2. \tag{18}$$

where r is the distance of the absorbing gas from the source, N_H is the column density of hydrogen nuclei, and m_p is the proton mass. r cannot be determined directly from observation, but must be inferred indirectly. In about 40% of the absorption line systems lines from ions with fine structure in their ground state are observed. Where the ground state of an ion has such fine structure, absorption lines out of the upper excited level can be observed—if the level is populated. Either collisions, excitation of high-lying states by UV photons followed by cascades, or direct infrared absorption can populate such levels. Thus the relative strengths of the 0-volt and excited lines can be used to determine either the density or the local flux of exciting radiation. In general, based on such arguments, upper limits to the density in the absorbing gas of about 10^2 - 10^3 cm^{-3} have been set. (For a fuller discussion of the problems associated with these arguments See Ref. 7.) Once limits on the density have been set, r can be determined if it is assumed that the ionization state is determined by photo-ionization equilibrium with the excitation due to the non-thermal continuum of the QSO. Based on such arguments distances of the order of 10-100 kpc or more are deduced. The observed column densities are of the order of 10^{17} cm^{-2} (with a few systems having column densities as high as 10^{22} cm^{-2}). The physical thicknesses of the systems are then of the order of only fractions of a parsec. Thus the picture of intrinsic absorption which emerges is one of very thin shells at great distances from the central source, expanding at high velocity with total kinetic energies of perhaps 10^{59} ergs and containing of the order of $10^8 M_\odot$ or more. Such a picture is difficult, if not impossible, to explain on the basis of clouds ejected from the central object, or on the basis of instabilities arising in a continuous wind. In neither of these two models is the observation of material only at such large distances explicable; in both models one expects to observe the absorbing material at all radii. In fact, in the case of continuous winds the instability mechanisms which could give rise to such shells are most effective in regions of high density, i.e. near to the central source. The picture of thin shells of material at distances from the central source orders of magnitude greater than the shell thickness is, however, reminisant of the situation found in supernovae, for instance.

It is now generally believed that QSOs are active nuclei of galaxies; it is in any case unresonable to assume that they are surrounded by vacuum. It is therefore reasonable to consider the effect the winds emanating from the QSO would have on the interstellar gas of any galaxy surrounding the QSO or on whatever ambient gas were there. Models for the central energy producing regions of QSOs[10] fall into three categories; two of which involve dense stellar clusters. In one, radiation is produced by accretion of stellar debris onto a central black hole, in the other by plasma processes triggered by multiple supernovae events. The other alternative is that nuclear processes in a supermassive object or objects produce the radiation field. In all three cases it is plausible [11, 12] to assume that winds carrying kinetic energies of the order of about one-thousandth of the bolometric luminosity of the QSO are emitted. QSOs with absorption-line systems have observed visual luminosities of $10^{46.2}$-$10^{48.2}$ erg/sec.[7] The bolometric luminosities must be at least 5-10 times the visual luminosities. Thus, mechanical energy outflows of the order of 10^{43}-10^{45} erg/sec are reasonable. We now ask what the effect of such an outflow on surrounding gas would be.

For simplicity we shall here consider the surrounding gas to have a uniform density and to be initially stationary.

If the velocity of the wind is V, and the mass-loss rate is $\dot{m}$, then the flow pattern will be as shown in Fig. 3. This flow pattern is set up because V is much greater than the sound speed in the ambient gas. This situation

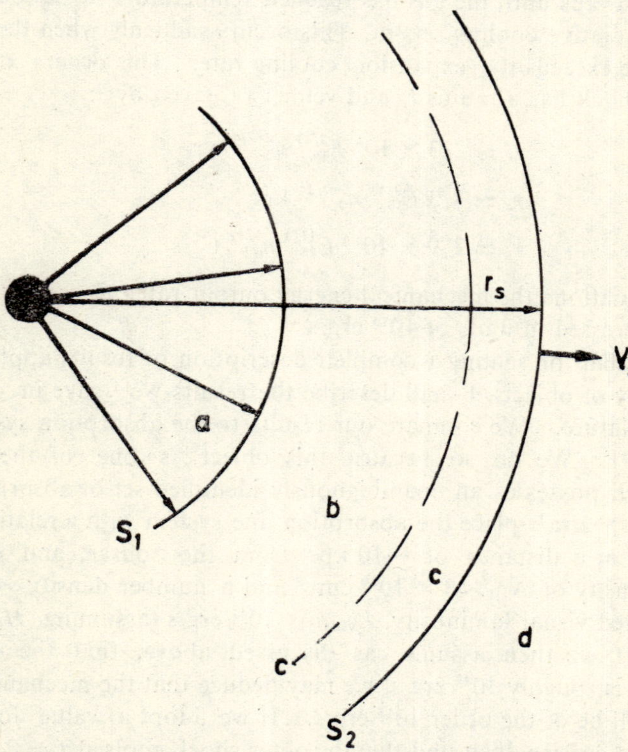

Fig. 3 Flow pattern of a QSO: region (a) the freely outflowing wind; (b) the shocked wind; (c) swept up ambient gas; (d) undisturbed ambient gas. S_1 is an inward facing shock in the wind, c is a contact discontinuity and S_2 is the outwards moving shock in the ambient gas. r_s and v_s are the radius and velocity of the outer shock (after Ref. 11).

is analogous to the interaction of a high speed stellar wind with interstellar gas.[13,14]. In the initial stages of the flow the shocked ambient and wind material are far too hot either to cool or to give rise to absorption lines. The shocked QSO wind gas behaves adiabatically over all time-scales of interest. Radiation pressure plays no role during the adiabatic expansion phase.[12] The characteristic length scale of the flow, $l = [\dot{m}/(v\rho_0)]^{1/2}$, is small compared with the typical scale size of the flow for all reasonable values of the parameters of the problem. Therefore, during the adiabatic phase, the position of the outer shock, r_s, can be described by the similarity law

$$r_s = \alpha \left[\frac{1}{2}\frac{\dot{m}V^2}{\rho_0}\right]^{1/5} t^{3/5} \tag{19}$$

where α is a dimensionless constant which is found from computations[14] to be 0.88. As the flow expands, the material in region c cools adiabatically. Further, as the shock decelerates, the material coming through the outer shock is shock-heated to increasingly lower temperatures. This situation prevails until the gas has reached temperature to about 10^7 °K, at which radiative cooling sets in. This occurs suddenly when the radiative cooling rate exceeds the expansion cooling rate. This occurs at a time t_c when the shock has a radius r_c and velocity v_c given by[12]:

$$t_c \simeq 2.3 \times 10^6\, \dot{E}_{45}^{3/11}\, n_0^{-8/11}\ \mathrm{yr} \tag{20}$$

$$r_c \simeq 3.3\, \dot{E}_{45}^{4/11}\, n_0^{-7/11}\ \mathrm{kpc} \tag{21}$$

$$v_c \simeq 2.9 \times 10^{-3}\, \dot{E}_{45}^{1/11}\, n_0^{1/11}\, C \tag{22}$$

In these equations the mechanical energy output rate, $\dot{E} = \dot{m}V^2/2$, of the QSO is expressed in units of 10^{15} erg/s.

Rather than presenting a complete description of results applicable to a wide variety of objects, I shall describe the results we[11] gave in our recent article in Nature. We compare our results to the absorption system in the QSO 3C 191. We do so because this object is one of the very few QSOs which possesses an unambiguously identified set of absorption line.[7] Studies of spectra[15] place the absorption line system with a relative velocity of 0.0027c at a distance of $\sim$10 kpc from the source, and estimate a column density of $N_{\mathrm{HI}} \sim 3 \times 10^{19}$ cm^{-2} and a number density $\sim 10^3$ cm^{-3}. The observed visual luminosity, L_{vis}, is 10^{47} erg/s (assuming $H_0 = 50$ and $q_0 = 0$). If we then assume, as discussed above, that the bolometric luminosity is roughly 10^{48} erg/s, we may deduce that the mechanical energy outflow will be of the order 10^{45} erg/s. If we adopt a value for n_0 of 1 particle per cm^3 we then find that the outer shock cools at $t_c = 2.3 \times 10^6$ yr, $r_c = 3.3$ kpc, $v_c = 0.0029c$. At this shock speed, the post-shock temperature is 10^7 °K. The gas will now cool rapidly until it reaches a state of ionization and thermal equilibrium with the radiation field of the QSO. In such a situation the equation of state of the gas is[16,17]

$$P = \mathcal{G}\rho y f(y) \tag{23}$$

where
$$y = \frac{L_{\mathrm{vis}}}{(4\pi r^2 \rho c^3)} \tag{24}$$

is the ionization parameter[18,17], $\mathcal{G}$ is the gas constant, and $f(y) \equiv T(y)/(\mu y)$.

The state is critically dependent on the spectrum of the QSO. For flat spectra, $F(\nu) \sim \nu^{-\alpha}$ ($\alpha \sim 1$), the equilibrium temperature is a few times 10^4 °K.

As the gas cools V will initially decrease, with an upper limit to the postshock pressure given by

$$P_s \leqslant \frac{2}{\gamma+1} \rho_0 V_c^2 \tag{25}$$

The number density in the cool material is therefore

$$n \leqslant n_0 \frac{V_c^2}{\Omega y_c f(y_c)} \tag{26}$$

where y_c is the equilibrium value of the ionization parameter. Substituting eqns. (24) and (20) into (26), and using $\dot{E}_{45} \simeq 1$, $n_0 = 1$, $L_{vis} = 10^{47}$ erg/s, we find that $\log_{10} f(y_c) \simeq 14.1$. From Roser's[17] calculations we find that $\log_{10} y_c \simeq -9.5$ for a flat ($\alpha = 1$) spectrum with, e.g., a drop or cut off at about the He (II) ionization limit. For this spectrum, at y_c, C (IV) and Si (IV) are the dominant ions of C and Si. During the subsequent evolution[14] y decreases so that the gas becomes more neutral. Substituting these values for $f(y)$ and y_c into eqn. (26) we find that $n \leqslant 2.3 \times 10^3$ cm^{-3}. The total column density of hydrogen, $N_H \simeq n_0 r_c$ is $\simeq 10^{22}$ cm^{-2}. The column density of neutral hydrogen is $(1 - x) N_H$, where $x = x(y)$ is the degree of ionization of hydrogen. $x(y)$ depends sensitively on the spectrum; for the flat spectrum discussed above, at y_c, $(1 - x) \sim 10^{-4}$, so that $N_{HI} \sim 10^{18}$ cm^{-2}. Since y decreases as the shell moves out, N_{HI} will increase.

All these values are in surprisingly good agreement with observations in view of the crudity of this first model. This agreement can be shown to remain even when we consider non-uniform and non-spherically symmetric ambient density distributions.[12] Because the sound speed in the shocked QSO wind gas is very much higher than the expansion velocity itself, during most of evolution of the flow, the pressure in region 'b' will be spatially uniform. This is the case even if the wind itself is non-isotropic (e.g., if it is caused by multiple explosive events). The inner shock would then be nonspherical, but pressure gradients in region 'b' would still relax in a time scale much less than the dynamical time scales of the flow. Departures from sphericity can then be produced in the outer shock only by non-uniformities in the galactic material.[12]

Once the cool dense shell has formed, it will start to absorb momentum from the radiation field. In the majority of cases the shell will accelerate due to the radiative driving. If this happens, then it is unavoidable that the shell becomes unstable to radiative driven density wave formation (as discussed elsewhere.)[19] The more effective the radiation driving, the higher the velocity and the more unstable is the system. Thus this mechanism seems capable of proving a natural explanation for the multiplicity of high velocity sytems. The flow in this post-cooling phase is very complex, involving multiple secondary shocks which are very diffcult to treat either analytically or numerically. In addition, in contrast to the formation of the shell as described above, the post-cooling flow depends strongly on the parameters and the details of the QSO spectrum. We (J.E. Dyson, S.A.E.G. Falle and I) are now studying this post-cooling phase in detail.

We do not expect that this mechanism can explain all the absorption line systems of QSOs. In fact, if QSOs are at their redshift (z_e) distances, then a significant fraction of the absorption *must* arise in intervening objects. The detailed explanation of the consequences of this model of intrinsic absorption should provide a basis for distinguishing between those systems which are intrinsic and those which are intervening.

References

1. Hubble, E., *Proc. Nat. Acad. Sci.* **15**, 168 (1979); Humason, M.L., Mayall, N.V., and Sandage, A.R., (1956), *Astrophys. J.* **61**, 97.
2. Ellis, G.F.R. and Perry, J.J., *Mont. Not. R. astron. Soc.*, preprint (1979).
3. Robertson, H.P., *Astrophys. J.* **82**. 284 (1935); Walker, A.G., *Proc. Lond. Math. Soc.*, **42**, 90 (1936); Friedmann, A., *Phys. Zetts.*, **10**, 377 (1922).
4. Ellis, G.F.R. 'Proc. 8th Texas Conference on Relative Astrophysics', *Acad. Sci.*, in press (1979).
5. Disney, M.J., *Nature* **263**, 573 (1976).
6. *Proc. Copenhagen Sym. on 'Active Nuclei', Physica Scripta*, **17** (1978).

6a. Bahcall, J.N. and Turner, E.L., Ibid.

7. Perry, J.J., Burbidge, E.M., and Burbidge, G.R., *Proc. Astron. Soc. Pacific*, **90**, 337 (1978).
8. Roberts, D.H., Burbidge, E.M., Burbidge, G.R., Crowne, A.H., Junkharinen, V.T., and Smith, H.E., *Astrophys. J.* **224**, 344 (1978).
9. Boksenberg, A., (1978) private communication.
10. Rees, M J., *Physica Scripta* **17**, 193 (1978).
11. Dyson, J.E., Falle, S.A.E.G. and Perry, J.J., *Nature* **227**, 118 (1979).
12. Dyson, J.E., Falle, S.A.E.G., Perry, J.J., (1979), in preparation.
13. Dyson, J.E. and de Vries, J., *Astron. & Astrophys* **20**, 223 (1972).
14. Falle, S.A.E.G., *Astron. & Astrophys.* **43**, 323 (1975).
15. Williams, R.E., Strittmatter, P.A., Carswell, R.F., and Craine, E.R. *Astrophys. J.* **202**, 296 (1975).
16. Kippenhahn, R., Mestel, L., and Perry, J.J., *Astron. & Astrophys.* **44**, 123 (1975),
17. Röser, H.J., (1979) *Astron. & Astrophys.*, preprint.
18. Kippenhahn, R., Perry, J.J., and Röser, H.J., *Astron. & Astrophys.* **34**, 211 (1974),
19. Mestel, L., Moore, D.W., and Perry, J.J., *Astron. & Astrophys.* **52**, 203 (1976),

Discussions

B Buti: The relativistic winds you are considering has only relativistic speeds or is it relativistically hot also? If it is hot then I am surprised that you do not have beam-plasma instabilities.

J Perry: We are not discussing fully relativistic winds—the velocities of the nuclear outflow are only of the order of several tenths of *c* at most. The wind temperatures are of the order of 10^6 to 10^7 °K. The flow is spherically symmetric to a good approximation, and there is no two stream flow pattern as all the ambient gas is swept up by the forward shock. In essence no 'beam' interaction occurs.

M Sohnius: Could particular spectral features, e.g., the possible accompanying high-ionization hot shells, distinguish intrinsic from intervening absorption?

J Perry: Unfortunately, since the same ionic species occur in the interstellar gas in galaxies as are predicted to occur in both the cool and hot shells of our model, such a simple selection criterion would not work. All workers in the field have hoped that such a selection criteria would be found, but it appears not to be the case. Detailed radio maps in the Hydrogen 21 cm line appear to offer more hope, in the few cases where such a line is observed, of finding a purely oberservation selection between intrinsic and intervening absorption.

D Brill: You have a new classification of QSOs, depending on whether they can be obtained from your model or not. Does one or the other class give you a better Hubble line?

J Perry: Were enough high resolution spectral data available, it would be interesting to attempt such a separation into two classes of QSOs. Unfortunately the data are as yet far too incomplete and inhomogeneous to attempt such classification.

M C Durgapal: In order to explain absorption redshifts z_{ab} and z_{em} it is assumed that the absorption clouds are ejected with very large relative velocities (as high as 0.6c). Why is it that no blue shift has been observed in the absorption spectrum of any quasar?

J Perry: Absorption is predominantly observed in large (z_{em}) redshift QSOs—essentially no absorption has been seen in objects of redshift less than 1. Were most of the absorption to prove to be intrinsic this would be interpreted as an evolutionary effect, the absorption arising only in the early stages of the life of a QSO. This would be consistent with sweeping away the ambient gas necessary to build absorbing shells. However, if our knowledge of the structure and distribution of galaxies is even approximately correct, much of the absorption *must* be intervening. It is then likely that the systems with apparent ejection velocities as high as 0.6c are intervening. As long as no absorption occurs in low redshift objects no blue shifts are to be expected.

Analytical and Numerical Investigation of Gravitational Collapse

J. PACHNER

1. Introduction

From the logical standpoint, the general theory of relativity is an excellent theory of gravitation. While the Newtonian dynamics is based upon three independent laws of nature—the equations of motion, Newton's law of general gravitation, and the equality of gravitational and inertial mass—the equivalence principle is the corner-stone of general relativity, the Newton's law is replaced by Einstein field equations, and the equations of motion are no more an independent law of nature, but the integrability conditions of the field equations. In addition, the gravitational field is propagating with a finite velocity equal to the propagation of electromagnetic phenomena. On the other hand, the three Einstein's tests indicate that the deviations from Newtonian dynamics are extremely small so that the cosmology seemed for many years to be the only domain where the general relativity played an important role.

However, the situation changed when Chandrasekhar (1934) showed that a star with an overcritical mass could not reach a stable configuration after it had exhausted all its nuclear fuel and must collapse into a black hole. The gravitational collapse may be thus another domain where the application of general relativity is indispensable not only because a collapsing star may disappear within a black hole, but also because a collapsing rotating star is a source of gravitational radiation.

The aim of the present investigation is to get a better understanding of the gravitational collapse with the help of analytical methods, and where they fail, to develop numerical methods for the integration of Einstein equations.

2. Analytical approach to a spherically symmetric collapse

The problem of the gravitational collapse of a spherically symmetric body consisting of incoherent matter was investigated by Oppenheimer and Snyder (1939). Unfortunately, they restricted their investigation to a very special case of a uniform initial distribution of the mass density in spite that an integration method for Einstein equations with an arbitrary spherically symmetric initial distribution of incoherent matter had been published by Datt (1938) one year earlier.

Oppenheimer and Snyder found that the body disappeared first within the Schwarzschild radius and then continued its collapse to a point singularity. On the basis of Datt's solution it will be shown that the gravitational collapse of incoherent matter with any spherically symmetric distribution of mass density decreasing monotonically from the centre towards the surface of the body is proceeding so fast at the centre that the density here becomes singular before the whole body, or its interior part, might disappear within the Schwarzschild radius. It is just one term, absent if the mass is uniformly distributed, which prevents the formation of a black hole before the mass density becomes singular at the centre of the body.

In the spherical comoving co-ordinates (r, θ, ϕ, t) a spherically symmetric metric within a body consisting of incoherent matter may be written in the form $(c = G = 1)$

$$ds^2 = e^{2\alpha}dr^2 + r^2e^{2\beta}(d\theta^2 + \sin^2\theta d\phi^2) - dt^2 \tag{1}$$

The unknown functions $\alpha(r, t)$ and $\beta(r, t)$ are determined by Einstein field equations. The field is regular at the origin of co-ordinates if

$$\alpha = \beta, \quad \partial\alpha/\partial r = \partial\beta/\partial r = 0 \quad \text{at } r = 0 \tag{2}$$

It is assumed that the initial distribution of mass density is given,

$$\rho(r, 0) = \rho_i(r) \quad \text{at } t = 0 \tag{3a}$$

that the mass is at rest,

$$\partial\alpha/\partial t = \partial\beta/\partial t = 0 \quad \text{at } t = 0 \tag{3b}$$

and that the co-ordinate r and the radial distance from the origin are identical at this moment,

$$\alpha = 0, \quad \text{at } t = 0 \tag{3c}$$

Only if a spherically symmetric body consists of incoherent matter, Datt's (1938) analytical solution may be applied. His equations were simplified by Pachner (1965) by a suitable change of notation and used for the classification of all solutions of that class. In the case under consideration these equations get a still simpler form by the substitutions

$$k\sigma = w(2 - w) \text{ and } r^3\psi = vw(2 - w)$$

The functions α, β in eqn. (1) are then given by

$$e^{2\alpha} = (1 - w)^{-2}\,[(\partial/\partial r)(vF)]^2 \tag{4}$$

and

$$r^2 e^{2\beta} = (vF)^2 \tag{5}$$

Because of the initial condition (3b)

$$0 \leqq w(r) \leqq 1$$

and the function $F(r, t)$ is defined by the eqn. 6

$$tq(r) + \eta(r) = \arccos F^{1/2} + [F(1 - F)]^{1/2} \tag{6}$$

where

$$q(r) = [w(2 - w)]^{1/2}/v \tag{7}$$

The mass density is determined by the equation

$$8\pi\rho = \{(d/dr)\,[vw(2 - w)]\}/\{v^2F^2(\partial/\partial r)(vF)\} \tag{8}$$

The preceding equations contain three arbitrary functions $v(r)$, $\eta(r)$, $w(r)$, corresponding to the free choice of the initial distribution of mass density, of its initial velocity, and of the scale of the co-ordinate r. The imposed initial conditions (3a, b, c) require

$$dw/dr = 4\pi\rho_i v - (1 - \tfrac{1}{2}w)(w/v) \tag{9a}$$

$$\eta(r) = 0 \tag{9b}$$

$$dv/dr = 1 - w \tag{9c}$$

For $\eta(r) = 0$ and $t = 0$, eqn. (6) yields $F(r, 0) = 1$. If the function $w(r)$ is eliminated from eqn. (9a, c) and $v(r)$ is replaced with β by eqn. (5), the resultant equation takes the familiar form (the subscripts 1 indicate the partial derivatives with respect to r)

$$\beta_{11} + 3(\beta_1/r)(1 + \tfrac{1}{2}\beta_1 r) + (1 - e^{-2\beta})/2r^2 = 4\pi\rho_i$$

of the constraint equation

$$R_4^4 - \tfrac{1}{2}R\,\delta_4^4 = 8\pi T_4^4$$

for the metric (1) with α given by the initial condition (3c).

The metric is Minkowskian in the infinitesimal neighbourhood of $r = 0$ if

$$v(0) = 0, \quad w(0) = 0 \tag{10}$$

Eqns. (4) and (8) are thus reduced to

$$e^{2\alpha} = \{F - t\,\Phi(r)\,(1/F - 1)^{1/2}\}^2 \tag{11}$$

and

$$\rho(r, t) = \rho_i(r)\, e^{-\alpha}/F^2 \tag{12}$$

where

$$\Phi(r) = (4\pi\rho_i/q) - (3q/2) \tag{13}$$

If the mass is distributed uniformly, $\rho_1(r) = \rho_0 = \text{const.}$, then the particular integral of (9a, c) satisfying (10) is

$$v(r) = (1/q_0)\sin(rq_0) \tag{14a}$$

and

$$w(r) = 1 - \cos(rq_0) \tag{14b}$$

with

$$q_0 = (8\pi\rho_0/3)^{1/2} \tag{14c}$$

Hence if

$$\Phi(r) = 0$$

$$q(r) = q_0 = \text{const.} \tag{15}$$

and the mass density remains uniformly distributed

$$\rho(t) = \rho_i/F^3 \tag{16}$$

At $t = 0$ the particles of incoherent matter start to fall freely from a finite distance, while in Oppenheimer-Snyder solution they already have a certain velocity (they start to fall freely from infinity at $t = -\infty$). This inessential difference manifests itself in different times necessary for the formation of singular mass density at $r = 0$, but the difference in $\rho(t)$ disappears when t approaches this moment, for, if $tq_0 \to \pi/2$, the right-hand side of (6) with $\eta = 0$, expanded into a power series, yields

$$F(t) = (3/2)^{2/3}(\pi/2 - tq_0)^{2/3} \tag{17}$$

in agreement with Oppenheimer-Snyder for τ given by

$$\tfrac{3}{2} r_0^{1/2} R_b^{-3/2} \tau \to 1 \tag{18}$$

As is well known, the whole body disappears within its Schwarzschild radius before the density becomes singular at $r = 0$.

To understand the essential difference in the process of collapse in the case of initial mass density monotonically decreasing from the centre of the body, the equations valid in a close neighbourhood of $r = 0$ have been deduced under the assumption

$$4\pi\rho_i(r) = \bar{\rho}_0 + \tfrac{1}{2}\bar{\rho}_0'' r^2 + \ldots \tag{19a}$$

In this case

$$v(r) = r - (\bar{\rho}_0/9)\, r^3 + (\bar{\rho}_0^2/270 - \bar{\rho}_0''/50)\, r^5 + \ldots \tag{19b}$$

$$w(r) = (\bar{\rho}_0/3)\, r^2 - (\bar{\rho}_0^2/54 - \bar{\rho}_0''/10)\, r^4 + \ldots \tag{19c}$$

$$q(r) = q_0\, [1 + (3/20)\, (\bar{\rho}_0''/\bar{\rho}_0)\, r^2 + \ldots] \tag{19d}$$

$$\Phi\, (r) = (3\bar{\rho}_0/50)^{1/2} (\bar{\rho}_0''/\bar{\rho}_0) r^2 + \ldots \tag{19e}$$

Since $\bar{\rho}_0'' < 0$, the function $q(r)$ is decreasing from the centre. As a consequence, more distant the particles are at $t = 0$ the more delayed is their arrival at $r = 0$. The steep decrease of the cycloid (6) when $tq(r) \to \pi/2$ acts as a powerful amplifier of the steady decrease of the function $q(r)$ (with the amplification factor $|\, dF/d(tq)\,| = (1/F - 1)^{1/2}$), resulting in effectively slowing down the more distant particles and a highly non-uniform collapse. The mass density is increasing much faster at $r = 0$ than the radius of the body is decreasing. It is just the term involving $\bar{\rho}_0''$ which changes the constant q_0 to the function $q(r)$ and thus the whole picture of gravitational collapse. That term is missing in Oppenheimer-Snyder solution.

In order to get the whole picture of collapse, eqns. (9a, c) are to be integrated numerically, using with advantage the standard fourth order Runge-Kutta method. The function $q(r)$ ane $\Phi(r)$ are computed simultaneously by eqns. (7) and (13). It is important to realize that the function $\Phi(r)$ equals the difference of two relatively large numbers. Eqn. (19e) indicates that it starts from 0 at $r = 0$ and then takes the negative values. If the error bound in Runge-Kutta method is not chosen sufficiently small. the numerical integration may result in erroneous (positive) values of $\Phi(r)$. For any chosen $t > 0$ the non-linear eqn. (6) with $\eta = 0$ is to be solved by Newton's method (for t small eqn. (6) is reduced first to its parametrical form and then solved by Newton's method). Finally the functions α, β and ρ are computed by eqn. (5), (11) and (12). The whole computation is programmed for a computer, the program computes also the first and second derivatives of α and β with respect to t.

The computation has been carried out for a body of initial radius

$R_0 = 1.75\ R_\odot$ with the mass $M = 3.47 M_\odot$ distributed by the function

$$\rho_i = 50.0 \exp\,(-15.0(r/R_0)^4)\ (\text{g/cm}^3) \tag{20}$$

The density does not vanish but takes negligible values in the region outside the model. The geometrical radius R of a concentric layer with the co-moving r has been determined by the formula

$$R(r, t) = \int_0^r e^\alpha dr \tag{21}$$

and compared with the radius computed from the circumference

$$R_{cf}(r, t) = re^{\beta(r, t)} \tag{22}$$

Both the radii, R and R_{cf}, are equal where the curvature of space is negligible. The body is collapsing smoothly, its density increasing faster at the layers with a smaller co-ordinate r, until at $T = \pi/2q(0)$ it becomes singular at $r = 0$. At any moment and at any layer, the Schwarzschild radius is smaller than the radii R and R_{cf}, but of the difference between the Schwarzschild radius and the radii R and R_{cf} the smaller is the value of the comoving co-ordinate r (while for a uniform distribution, $\rho = \text{const.}$, such as in Oppenheimer-Snyder model, the difference is the smaller the closer is the layer to the surface; at the moment when the body disappears within a black hole the Schwarzschild radius equals to R_{cf} at the surface of the body).

Table 1 exhibits the results of computation at moment T. Note that all the quantities in it are scalars; the results are thus independent of the choice of the co-ordinate system. The absolute error bound in Runge-Kutta method has been chosen 10^{-15} between the origin and the first grid point, and 10^{-8} in the remaining part of the integration domain. The Schwarzschild radius $R_s(R_T)$ corresponds to the mass within the radius R_T. Table 1 also contains the mass density ρ_{CT} computed by Newtonian dynamics.

Table 1 shows that the ultimate distribution of mass density falls extremely steeply from infinity to values insignificantly higher than the initial ones at radii close to the surface of the body. The ratios R_T/R_S and R_{cft}/RS indicate that both radii $R(r, T)$ and $R_{cf}(r, T)$ are at all layers a few orders of magnitude greater than the corresponding Schwarzschild radii.

Table 1 is just one of several examples that demonstrate the inadequacy of Oppenheimer-Snyder model as a standard picture of gravitational collapse (Penrose, 1972). The Penrose-Hawking theorem (Penrose 1965, Hawking and Penrose 1970, and Hawking and Ellis 1973), which states that "the gravitational collapse of a massive star to a black hole is inevitable if the deviations from spherical symmetry are not too great at the moment of formation of a closed trapped surface", cannot thus be applied as an argument for the existence of black holes, because in an actual star

Table 1 Distribution of the mass density ρ along the geometrical radius R at the initial moment (subscript 'i' and at the moment $T = 297.1307188$ s when the central density has become singular (subscript 'T')

R_i (km)	R_T (km)	R_{cfT} (km)	ρ_i (g/cm^3)	ρ_T (g/cm^3)	ρ_T / ρ_i	ρ_{CT}/ρ_i	R_s (km)	R_T / R_S	R_{cfT}/R_s
0.0000E0	0.0000E00	0.0000E00	5.0000E01	∞	∞	∞	—	—	—
3.0220E03	8.1535E–4	9.6602E–4	5.0000E01	5.3242E20	1.0648E19	1.0354E19	6.8293E–7	1.1939E03	1.4145E03
6.0440E03	9.2699E–3	9.8708E–3	5.0000E01	3.3760E18	6.7520E16	6.9452E16	5.4634E–6	1.6967E03	1.8067E03
9.0660E03	4.1068E–2	4.1932E–2	5.0000E01	1.3973E17	2.7946E15	2.8064E15	1.8439E–5	2.2272E03	2.2741E03
1.5110E04	2.6976E–1	2.7066E–1	5.0000E01	2.3739E15	4.7478E13	4.7496E13	8.5366E–5	3.1600E03	3.1706E03
3.0220E04	3.4338E00	3.4348E00	5.0000E01	9.2868E12	1.8574E11	1.8574E11	6.8293E–4	5.0280E03	5.0295E03
7.5550E04	9.8831E01	9.8835E01	4.9995E01	6.0919E09	1.2185E08	1.2185E08	1.0670E–2	9.2625E03	9.2625E03
1.5110E05	1.2531E03	1.2531E03	4.9925E01	2.3907E07	4.7886E05	4.7886E05	8.5311E–2	1.4689E04	1.4689E04
7.5550E05	3.5907E05	3.5906E05	1.9580E01	6.2983E01	3.2167E00	3.2166E00	7.4158E00	4.8420E04	4.8419E04
1.5110E06	1.3957E06	1.3956E06	1.5295E–5	1.5470E–5	1.0114E00	1.0114E00	1.0293E01	1.3560E05	1.3559E05

R_S is the Schwarzschild radius corresponding to the mass inside the radius R_T. The values of the ratios R_T / R_S and R_{cfT}/R_s indicate that the corresponding Schwarzchild radius lies deep inside the layer of radius R_T and R_{cfT}. The density ρ_{CT} at the moment T have been computed classically. The usual computer notation, such as 3.0220E03 = 3.0220×10^3 and 8.1535E–4 = 8.1535×10^{-4}, is used throughout.

the extremely high pressure gradient in the nearest neighbourhood of its centre may prevent the formation of a closed trapped surface.

3. Analytical approach to a non-symmetric collapse

A body consisting of incoherent matter cannot be considered as an adequate model of a collapsing massive star. To obtain a more realistic model the incoherent matter is to be replaced by an ideal fluid whose physical behaviour is defined by the energy momentum tensor ($c = G = 1$, the signature of the metric is $+ 2$)

$$T_\mu^\nu = \rho e^\chi g_{\mu\alpha} u^\alpha u^\nu + p\delta_\mu^\nu \tag{23}$$

and by the equation of state

$$p = p(\rho, \varepsilon) \tag{24}$$

which interrelates the proper rest mass density ρ, the proper internal energy per unit mass ϵ, and the pressure p. The e^χ is the proper enthalpy per unit mass

$$e^\chi = 1 + \varepsilon + p/\rho \tag{25}$$

and u^μ the components of the four-velocity which have to satisfy the normalization equation

$$g_{\mu\nu} u^\mu u^\nu = -1 \tag{26}$$

The ideal fluid has to obey the equation of continuity

$$(\rho u^\nu);_\nu = 0 \tag{27}$$

the conservation law of energy momentum tensor

$$T^\nu_{\mu;\nu} = 0 \tag{28}$$

and six independent Einstein equations

$$R_{ik} = 8\pi(T_{ik} - \tfrac{1}{2}Tg_{ik}) \tag{29}$$

At any moment the free choice of Cauchy data is restricted by four constraint equations

$$\tfrac{1}{2}I^4_\mu \equiv R^4_\mu - \tfrac{1}{2}R\delta^4_\mu - 8\pi T^4_\mu = 0 \tag{30}$$

The semicolon denotes the Riemannian derivatives, the comma indicates partial derivatives. Greek indices run from 1 to 4, Latin indices from 1 to 3.

If eqn. (23) is substituted into (28) and differentiated, the conservation law can be written with the help of (27) in the form

$$T^\nu_{\mu;\nu} \equiv (\rho u^\nu)(e^\chi g_{\mu\alpha} u^\alpha)_{;\nu} + \delta^\nu_\mu p_{,\nu} = 0 \tag{31}$$

In general, the conservation laws (27) and (31) can be integrated only in connection with the field equations after four co-ordinate conditions have been chosen. However, there exists a class of co-ordinate systems, that of comoving co-ordinates, in which they can be integrated separately from the field equations. In a particular subclass of comoving co-ordinates the integration is trivial. The choice of these co-ordinates has a further important advantage that the quantities ρ, ε, p appearing in equations are measured in a local co-ordinate system at rest.

The comoving co-ordinates are defined by three co-ordinate conditions

$$u^i = 0 \tag{32a}$$

The fourth contravariant component u^4 is given by eqn. (26):

$$u^4 = (-g_{44})^{-1/2} \tag{32b}$$

If eqns. (32a, b) together with (25) are substituted into (31) the $\mu = 4$ component takes the form

$$d\varepsilon/d\rho = p/\rho^2 \tag{33}$$

This is an equation familiar from elementary thermodynamics as defining adiabatic thermodynamic processes. This result is, of course, a consequence of the accepted definition (23) of the energy momentum tensor (Tolman, 1934).

Now it is easy to verify equation

$$\rho,_{\mu} = \rho e^{\chi} \chi,_{\mu} \tag{34}$$

With its help eqn. (31) reduces for $\mu = i$ to

$$T^{\nu}_{i;\nu} \equiv (-g_{44})^{-1/2} \rho \, [g_{i4} e^{\chi} (-g_{44})^{-1/2}],_4 + \rho e^{\chi} (\partial/\partial x^i) \ln [e^{\chi} (-g_{44})^{1/2}] = 0 \tag{35}$$

The integration of this equation becomes trivial if the fourth co-ordinate condition is chosen in the form (Pachner, 1971)

$$(\partial/\partial x^i) \ln [e^{\chi} (-g_{44})^{1/2}] = 0 \tag{36}$$

The g_{44} component is determined from this equation, while a formula for g_{i4} components follows from the integration of (35) These four components may be expressed by a single equation

$$g_{\mu 4} = - A_{\mu} A_4 e^{-2\chi} \tag{37}$$

where $$A_i = A_i (x^1, x^2, x^3), \; A_4 = A_4 (x^4) \tag{38}$$

The co-ordinate conditions (32a) and (36) restrict the allowed co-ordinate transformations to

$$\bar{x}^i = \bar{x}^i \; (x^1, x^2, x^3) \tag{39}$$

and $$\bar{x}^4 = \bar{x}^4 \, (x^4) \tag{40a}$$

The arbitrary function $A_4 (x^4)$ may be reduced to ± 1 by a mere scaling of the co-ordinate x^4 without any loss of generality, but then the transformation of x^4 is restricted to a linear transformation

$$\bar{x}^4 = \pm x^4 + \text{const.} \tag{40b}$$

The arbitrary functions $A_i \; (x^1, x^2, x^3)$ are determined by the inital distribution of rotation. One of these three functions can be reduced to zero without loss of generality.

In the comoving co-ordinates the equation of continuity can easily be integrated:

$$\rho = |g_{44}/g|^{1/2} \psi(x^1, x^2, x^3) \tag{41}$$

where

$$g = \det g_{\alpha\beta} \tag{42a}$$

The function $\psi(x^1, x^2, x^3)$ is determined by the initial distribution of the mass density.

By algebraic operations that do not change the value of the determinant g its g_{4i} elements can be reduced to zero and the determinant itself expressed by the formula

Table 2 Comparison of the mass density computed analytically, ρ_a, with the density ρ resulting from the numerical integration at different moments t and at different values of the comoving radial co-ordinate r within a body consisting of incoherent matter

Steps	$t(s)$	r	ρ_a (g/cm³)	ρ_a/ρ_i	ρ/ρ_a
1000	296.675 3253	0.00	1.1794E06	2.3588E04	1.0002
		0.01	1.1794E06	2.3588E04	1.0001
		0.50	6.2232E01	3.1783E00	1.0000
		1.00	1.5468E–5	1.0113E00	1.0000
1400	297.055 4023	0.00	1.4077E08	2.8154E06	1.0013
		0.01	1.4069E08	2.8138E06	1.0013
		0.50	6.2914E01	3.2132E00	1.0000
		1.00	1.5470E–5	1.0114E00	1.0000
1700	297.118 2187	0.00	5.0953E08	1.0191E09	1.0079
		0.01	5.0772E09	1.0154E08	1.0079
		0.50	6.2972E01	3.2161E00	1.0000
		1.00	1.5470E–5	1.0114E00	1.0000
2000	297.128 6110	0.00	1.7904E11	3.5808E09	1.0485
		0.01	1.7532E11	3.5064E09	1.0480
		0.50	6.2981E01	3.2166E00	1.0000
		1.00	1.5473E–5	1.0114E00	1.0000

The initial density is denoted by ρ_i. Note that the ratio ρ/ρ_a indicating the error of the numerical integration increases faster at the centre where $|\partial\rho/\partial t|$ always has a maximum at the given moment t that at the outer layers. Since the pressure gradient slows down the local increase of the density, one may expect lower errors within a body consisting of ideal fluid. The usual computer notation, such as 2.4071E08 $= 2.4071 \times 10^8$, is used throughout.

$$g = -|g_{44}|\gamma \tag{42b}$$

where
$$\gamma = \det \gamma_{ik} \tag{43}$$

and
$$\gamma_{ik} = g_{ik} - g_{i4}g_{k4}/g_{44} \tag{44}$$

The γ_{ik} have been deduced here by purely algebraic operations. Landau and Lifshitz (1962) deduced them on the basis of physical considerations as the covariant components of a three-dimensional positive finite metric determining the geometrical properties of space (with contravariant components $\gamma^{ik} = g^{ik}$). The equation of continuity (41) takes now a simpler form

$$\rho = \psi/\gamma^{1/2} \tag{45}$$

Once the equation of state is given and eqn. (33) integrated, the internal energy ε and the pressure p are known functions of the mass density determined by the equations of continuity (41). With the help of eqns. (33) and (34) the derivatives of χ occurring in the field equations and constraint equations may be expressed by the formulas

$$\chi,_{\mu} = (\rho,_{\mu}/\rho)^{E} \tag{46a}$$

$$\chi,_{\mu\nu} = (\rho,_{\mu\nu}/\rho)^{E} + (\rho,_{\mu}/\rho)(\rho,_{\nu}/\rho)^{F} \tag{46b}$$

where the functions E, F of the mass density are defined by

$$E = e^{-\chi}(dp/d\rho), \tag{47a}$$

$$F = e^{-\chi}\rho(d^{2}p/d\rho^{2}) - E(1+E) \tag{47b}$$

The gravitational collapse is then described by six independent Einstein's eqns. (29) for six unknown g_{ik}. The free choice of the initial data is restricted by four constraints (30). Unfortunately, if the collapsing body consists of ideal fluid, the Einstein equations can be integrated analytically only in the simplest, but astrophysically meaningless, case of a spherically symmetric collapse with a uniform distribution of the mass density.

However, in the applied system of comoving co-ordinates it is possible to deduce analytically a conservation law of the vorticity vector defined by the formula (Gödel, 1949)

$$\Omega^{\mu} = \tfrac{1}{2}(-g)^{-1/2}\,\varepsilon^{\mu\lambda\alpha\beta}\,u_{\lambda}u_{\alpha,\beta} \tag{48}$$

wher $\varepsilon^{\mu\lambda\alpha\beta}$ is the totally skew-symmetric tensor density

$$\varepsilon^{\mu\lambda\alpha\beta} = \pm 1, 0, \quad \varepsilon^{1234} = +1 \tag{49}$$

In the comoving co-ordinates the covariant components of the four-velocity take the values

$$u_{\mu} = g_{\mu 4}u^{4} = -A_{\mu}e^{-\chi} \tag{50}$$

The functions A_{μ} are thus components of a covariant vector. By comparison with (37) and (44), the $g_{\mu 4}$ and γ_{ik} may now be defined by:

$$g_{\mu 4} = -u_{\mu}u_{4} \tag{51}$$

$$\gamma_{ik} = g_{ik} + u_{i}u_{k} \tag{52}$$

Without loss of generality the component A_4 is henceforth assumed to be equal to 1. With the help of (42b) the vorticity vector may be expressed as

$$\Omega^{\mu} = e^{-\chi}\,\gamma^{-1/2}\,C^{\mu} \tag{48a}$$

with the contravariant time-independent vector density

$$C^{\mu} = \tfrac{1}{2}\,\varepsilon^{\mu\lambda\alpha\beta}\,A_{\lambda}A_{\alpha,\,\beta} \tag{53}$$

The components Ω^{μ} depend, of course, on the applied system of co-ordinates. A conservation law valid in any co-ordinate system must have the form of a scalar equation involving the absolute magnitude of the vorticity vector

$$|\,\Omega\,| = (g_{\mu\nu}\,\Omega^{\mu}\Omega^{\nu})^{1/2} \tag{54}$$

The eqn. (48) shows that the vorticity vector is proportional to the curl of the four-velocity; the vectors Ω^{μ} and u_{μ} are thus orthogonal:

$$u_{\mu}\Omega^{\mu} = 0 \tag{55}$$

The component Ω^4 may therefore be expressed in terms of other three components, and, with regard to (51) and (52), as

$$|\,\Omega\,| = (\gamma_{ik}\,\Omega^{i}\,\Omega^{k})^{1/2} \tag{56}$$

The substitution from eqn. (48a) yields the conservation law

$$e^{\chi}\,|\,\Omega\,|/(\gamma_{ik}\,\breve{C}^{i}\breve{C}^{k}/\gamma)^{1/2} = 1 \tag{57}$$

in which the three-dimensional time-independent vector $\breve{C}^{i}$, covariant under general three-dimensional co-ordinate transformation (39) on a hypersurface $x^4 = \text{const.}$, is given by the formula immediately following from eqn. (53):

$$\breve{C}^{i} = \tfrac{1}{2}\,\varepsilon^{iab}\,A_{a,b} \tag{58}$$

The $\varepsilon^{iab} = \pm 1, 0$, $\varepsilon^{123} = +1$, is the totally skew-symmetric three-dimensional tensor density.

In an axially symmetric field also exhibiting a reflection symmetry with respect to the equatorial plane, the vorticity vector, expressed in the cylindrical co-ordinates $(z, \bar{r}, \phi, t)$, has only non-vanishing component Ω^1 along the axis of symmetry and at the equatorial plane. In this special case eqn. (57) takes a simple form

$$e^{\chi}\,|\,\Omega\,|/(\gamma_{11}/\gamma)^{1/2} = |\,\breve{C}^{1}\,| \tag{59a}$$

Since the components γ_{12} and γ_{13} here vanish, the ratio (γ_{11}/γ) is proportional (cf. eqn. (45)) to the mass density ρ_T per area perpendicular to the axis of rotation. The change of ρ_T results thus from the contraction, or expansion, of the mass element dm in the direction transverse to the axis of rotation, but not along the axis of rotation. The conservation law (59a) reduces to

$$e^{\chi}\,|\,\Omega\,|/\rho_T = \text{const.} \tag{59b}$$

Since the time co-ordinate t corresponding to $A_4 = 1$ is related to the time τ by the formula (Landau and Lifshitz, 1962)

$$d\tau = e^{-\chi}\, dt, \tag{60}$$

the $|\Omega|$ is to be interpreted as the vorticity vector with respect to the co-ordinate time t, and $e^{\chi}|\Omega|$ as the vorticity vector with respect to the proper time τ of a local observer at rest. There is only one time in Newtonian dynamics, the absolute time. Therefore, in Newtonian dynamics $e^{\chi} = 1$; the conservation law (59b) has then the form of the conservation law of angular momentum of Newtonian dynamics.

Note that the conservation law (57) and its special form (59b) have been deduced without applying the field equations. They are thus valid in any gravitational theory based upon the principle of general covariance, constant speed of light, and Riemannian geometry.

Hawking and Ellis (1973) consider as "reasonable to expect some loss of angular momentum during the collapse because of braking by magnetic forces and because of gravitational radiation". From the special case (59b) it seems that the angular momentum is not affected by gravitational radiation.

4. Numerical approach to a spherically symmetric collapse

In the spherically symmetric collapse the number of independent Einstein's equations (29) is reduced from six to two and the number of constraints from four to two. With regard to eqn. (37) the metric may be written in the form

$$ds^2 = e^{2\alpha}\, dr^2 + r^2 e^{2\beta}(d\theta^2 + \sin^2\theta\, d\phi^2) - e^{-2\chi}\, dt^2 \tag{61}$$

where e^{χ} is the proper enthalpy per unit mass defined by eqn. (25).

The unknown functions α, β are determined by the Einstein field equations

$$\begin{aligned}\alpha_{44} = {} & 4\pi\rho e^{-2\chi}(e^{\chi} - 2p/\rho) - \alpha_4(\alpha_4 + 2\beta_4 + \chi_4) + \\ & + 2e^{-2\alpha-2\chi}\{\beta_{11} + (\beta_1/r)(2 + \beta_1 r) - (\alpha_1/r)(1 + \beta_1 r) + \\ & + \tfrac{1}{2}\chi_1(\alpha_1 + \chi_1) - \tfrac{1}{2}\chi_{11}\}\end{aligned} \tag{62}$$

$$\begin{aligned}\beta_{44} = {} & 4\pi\rho e^{-2\chi}(e^{\chi} - p/\rho) - \beta_4(\alpha_4 + 2\beta_4 + \chi_4) + \\ & + e^{-2\alpha-2\chi}\{\beta_{11} + 2(\beta_1/r)(2 + \beta_1 r) - (1 + \beta_1 r)(\alpha_1 + \\ & + \chi_1)/r + (1 - e^{2\alpha-2\beta})/r^2\}\end{aligned} \tag{63}$$

in which the subscripts 1, 4 indicate the derivatives with respect to r and t, respectively.

The equation of continuity takes the simple form

$$\rho = \exp\{\psi(r) - \alpha - 2\beta\} \tag{64}$$

with $\psi(r)$ given by the initial distribution of the mass density.

The constraint equations are

$$I_1^4 \equiv -4e^{2\chi}\{\beta_{41} + (1 + \beta_1 r)(\alpha_4 - \beta_4)/r + \beta_4\chi_1\} = 0 \tag{65}$$

$$I_4^4 \equiv 2e^{2\chi}\beta_4(2\alpha_4+\beta_4) - 2e^{-2\alpha}\{2\beta_{11} + 3(\beta_1/r)(2 + \\ + \beta_1 r) - 2(\alpha_1/r)(1+\beta_1 r) + (1 - e^{2\alpha-2\beta})/r^2\} - \\ - 16\pi\rho(e^{\chi} - p/\rho) = 0 \qquad (66)$$

At $r = 0$ these equations contain indeterminate terms 0/0 which must be evaluated analytically. The field is regular if $\alpha = \beta$ at $r = 0$ at any moment, and both functions, as well as the mass density, are even functions in r. The time evolution at $r = 0$ is the described by the field equation (satisfying both constraint (65) and (66))

$$\alpha_{44}(O, t) = \beta_{44}(O, t) = -4\pi\rho e^{-2\chi}(e^{\chi} + 2p/\rho)/3 - \\ - \alpha_4^2(1 - 3E) - e^{-2\alpha-2\chi}(\rho_{11}/\rho)E \qquad (67)$$

with E given by eqn. (52).

The Einstein equations for a spherically symmetric collapse were numerically integrated for the first time by May and White (1967) with the help of a finite difference method. In the numerical example (p.245ff) they assumed a uniform initial distribution of the mass density and a non-vanishing pressure. With this choice it is not only impossible to satisfy the boundary condition $p = 0$ at the surface of the body (Lichnerowicz, 1955), but what is far more serious, a uniform redistribution of the mass density implies a zero pressure gradient and a smooth collapse to a point singularity no matter which equation of state has been chosen (the pressure without pressure gradient just accelerates the collapse, but it cannot change its uniformity). The bounce, shock waves, etc., observed by May and White, are thus numerical instabilities of the applied integration method.

In the process of developing a stable numerical method for the integration of Einstein equations the author and D.A. Swayne, his research associate at that time, soon realized that the main source of numerical instability was the Lagrange formula for numerical differentiation (Pachner and Teshima, 1973). Therefore the finite difference method used by May and White (1967) and later by Smarr (1977) was rejected, because the derivatives with respect to the spatial co-ordinate are computed in it by three-point Lagrange differentiation formula. The expansion of the function that has to be differentiated into a series of orthogonal polynomials, combined with the least-square technique (Ralston, 1965), improved the stability when the initial data were very smooth functions (Pachner, 1975), but further research showed that this approach certainly was not the solution of the problem, for when the initial distribution of the mass density was chosen similar to that of actual stars, growing oscillations in the numerical results were observed soon after the start of integration, not dissimilar to the gravitational waves in Smarr's (1977) numerical integration. The comparison with the analytical data for incoherent matter confirmed the suspicion that those oscillations were generated by numerical instabilities in the applied numerical methods.

The problem was solved after the author had realized that the constraint eqns. (65) and (66) might be replaced by the following three first order differential equations

$$v_1 = e^{\alpha}(1 - w) \tag{68}$$

$$w_1 = e^{\alpha}\{v[4\pi\rho(e^{\chi} - p/\rho) - (\alpha_4 + \tfrac{1}{2}\beta_4)\beta_4 e^{2\chi}] - w(1 - \tfrac{1}{2}w)/v\} \tag{69}$$

$$\beta_{41} = e^{\alpha}(1 - w)(\alpha_4 - \beta_4)/v - \beta_4 y_1 E \tag{70}$$

with the initial values

$$v(0, t) = 0,\ w(0, t) = 0 \tag{71}$$

and

$$\beta = \ln (v/r) \tag{72}$$

$$y = \ln \rho \tag{73}$$

and the field eqns. (62) and (63) by

$$\beta_{44} = -4\pi\rho e^{-\chi} + \beta_4(\alpha_4 - \beta_4) + e^{-\alpha-2\chi} w_1/v + \\ + E\{\beta_4(\alpha_1 + 2\beta_4) - e^{-2\alpha-2\chi} y_1(v_1/v)\} \tag{74}$$

$$\alpha_{44} = -2\beta_{44} - 4\pi\rho e^{-2\chi}(e^{\chi} + 2p/\rho) - \\ - \alpha_4^2 - 2\beta_4^2 + E\{(\alpha_4 + 2\beta_4)^2 - e^{-2\alpha-2\chi}[y_{11} + \\ + y_1(2v_1/v - \alpha_1 + y_1(1 - E + F/E))]\} \tag{75}$$

Note that the constraint equations now contain on the right hand side only one derivative with respect to r, viz, y_1, and that

$$\lim_{r\to 0} w/v = 0 \qquad \lim_{r\to 0} (\alpha_4 - \beta_4)/v = 0$$

The constraint equations can thus be integrated numerically with a very high accuracy by the standrd fourth order Runge-Kutta method. They are used not only for computing the admissible initial data, but also for correcting the function β and its derivative β_4 at each step of the integration of field eqns. (74) and (75). The cubic spline is used for interpolation and for computation of the derivatives y_1, y_{11}, α_1. The derivatives v_1, w_1 occuring in the field equations are the most valuable by-products of the integration of constraint equations. First a Euler method was used for the integration of the field equations; its stability was good, but the CPU-time was excessive, because the method required a very small interval to obtain a good accuracy. Therefore, it was replaced by a new predictor-corrector method for second order differential equations (fourth order strongly stable predictor, fourth order strongly stable corrector, the interval of integration adjusted at each step, with an optional second control of the interval, and with a built-in starter). This predictor-corrector method is much faster, stable, accurate, and since the step is much larger, the lesser number of steps diminishes the accumulation of numerical errors. The subroutine of the equation of state can easily be adjusted to any equation of state of the form (24). Further two, optional, subroutines compute the functions and their first and second time derivatives, and the mass density by analytical formulas of sec. 2 at (a) any moment given by the time evolution program, and (b) the moment when the central mass density becomes singular, to test the accuracy of numerical integration in case $p = 0$. A small auxiliary programe generates (and saves) a file containing the cubic spline matrix. It has to be applied only when the spatial distribution of grid points is changed. In all these programs a dimensionless system of units is used, but the results are converted to c.g.s. system of units.

The program is written in FORTRAN in double precision and has been run on Xerox Sigma 9 of the University of Regina. For 51 grid points the required memory extent is $21.5K$ words. The auxiliary program for the cubic spline requires $22.8K$ words.

In the examples below a star is supposed to be of the initial radius $R_0 = 1.75R_\odot$ with the mass $M = 3.47M_\odot$ distributed by the function (20). The density becomes thus negligibly small outside the model. The geometrical radius R of a concentric layer with the co-moving co-ordinate r is computed by the formula (21) and is compared with the radius R_{cf} computed from the circumference by the formula (22).

In order to test the accuracy of the applied numerical methods, the pressure p was assumed to be zero. The results of numerical integration are compared with the analytical ones of See. 2 Table 2. Total CPU time for 2000 steps was 122.96 minutes. The integration domain was covered by 51 points.

Since these results were very encouraging, the computer program was applied to the investigation of the gravitational collapse of a star consisting of ideal fluid. The equation of state was assumed in the form

$$p = (\gamma - 1)\,\varepsilon\rho \tag{76a}$$

It describes the behaviour of a non-degenerate Fermi gas. After substitution into eqn. (33) and integration,

$$\varepsilon = C_0\rho^{\gamma-1}, \quad p = C_0(\gamma - 1)\,\rho^{\gamma} \tag{76b}$$

C_0 being an integration constant and γ the exponent of the adiabatic process. Hence

$$p/\rho = C_0(\gamma - 1)\,\rho^{\gamma-1} \tag{77a}$$

According to eqn. (25)

$$e^{\chi} = 1 + [\gamma/(\gamma - 1)]\,(p/\rho) \tag{77b}$$

and according to eqns. (47a, b)

$$E = e^{-\chi}\,(p/\rho), \tag{77c}$$

$$F = -E\,(E + 2 - \gamma) \tag{77d}$$

In both models $\gamma = 5/3$. In model I the integration constant C was adjusted to the Harrison-wheeler equation of state (Harrison, Thorne, Wakano, Wheeler, 1965) in the region of low densities ($\simeq 50\text{g/cm}^3$)

$$C_0 = C_1 = 5.46 \times 10^{11}\ \text{cm}^4\ \text{g}^{-2/3}\ \text{sec}^{-2} \tag{78a}$$

In model II it was adjusted in the region of very high densities ($\simeq 10^{12}$ g/cm^3)

$$C_0 = C_2 = 1.44 \times 10^{10} \text{ cm}^4 \text{ g}^{-2/3} \text{ sec}^{-2} \tag{78b}$$

Since the collapse in both models is highly non-uniform, with the most violent increase of the mass density occurring in a very small region around the centre, it is important to put a sufficient number of grid points into the central part of the star, so that the functions can be interpolated with necessary accuracy by the cubic spline. On the other hand, the interval between the grid points may be quite large in the remaining region. The results of the numerical integration are exhibited in Figs. 1 and 2. The maximum density in both models is finite, but as one may expect, it is far greater and concentrated to a smaller region in model II where the pressure does not increase as fast as in model I. However, once the pressure gradient has been built up in model II, the repulsion is stronger and faster with fewer osciliations in the central region than in model I. For the equation of state (76a) with $\gamma = 5/3$ and C given by (78a, b), the pressure gradient is thus able to reverse the gravitational collapse to a steady expansion of the star. One may assume that the numerical errors are responsible for the continuity of expansion when $t \to \infty$.

The equation of state used in the two numerical examples is very simple, but the computer program is written in such a way that any equation of state of the form (24) may be applied. The computer program with a properly chosen equation of state, taking into account a possible re-ignition of nuclear fuel in the region of the maximum compression of the fluid, can enable a detailed study of the gravitational collapse of massive spherically symmetric stars which may not only ultimately decide the question whether the black hole may be considered as the last stage of the evolution of massive non-rotating stars, but also help us to understand the whole process of a supernova explosion; the explosive nucleosynthesis, the origin of most chemical elements, the origin of cosmic rays, the generation of neutron stars (Weaver 1978, Wilson 1978, Arnett 1978).

In the conclusion of this section it has to be emphasized that the excellent accuracy of the numerical integration is a consequence of the existence of the constraint eqns. (68), (69) and (70) with the help of which the field equations have been transformed to the simple form eqns. (74), (75) and of the application of the fourth order predictor-corrector method for the integration of the field equations.

5. Numerical approach to an axially symmetric collapse

In order to determine the intensity of gravitational radiation emitted during a gravitational collapse, it is necessary to extend the investigation to axially symmetric problems.

In his first attempts (Pachner and Teshima, 1973 and Pachner, 1975) the author used a system of cylindrical co-ordinates, but when it has turned out that a spherically symmetric collapse is highly non-uniform, the whole

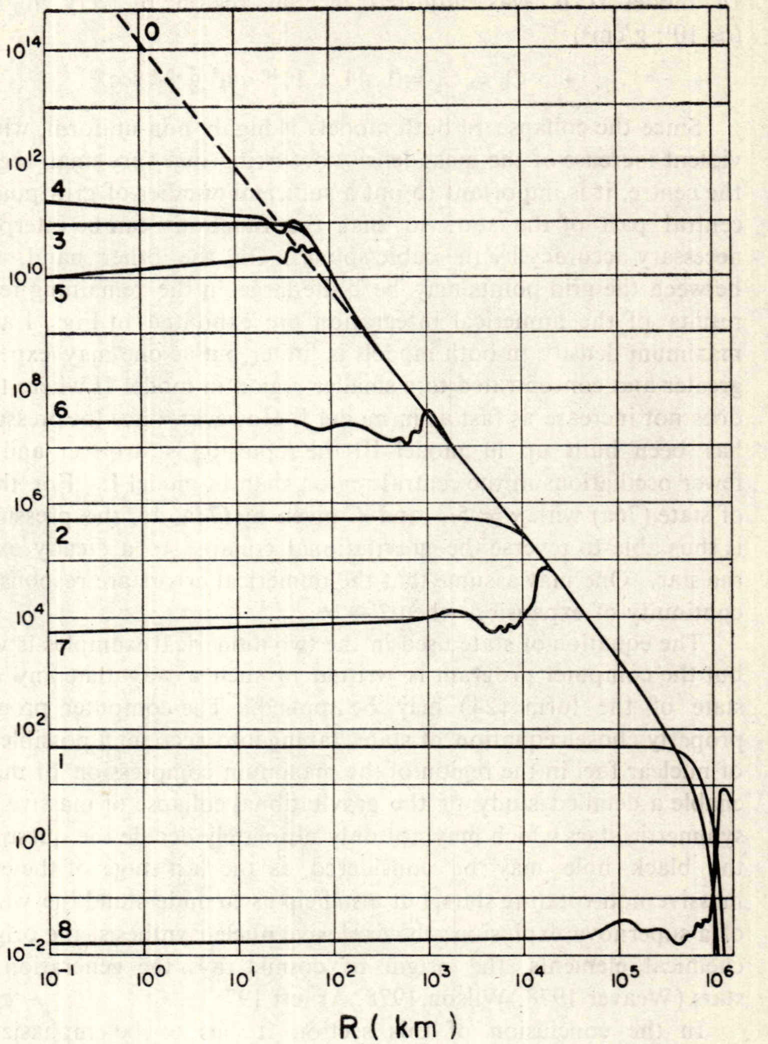

Fig. 1 Distribution of the mass density ρ along the geometrical radius R with time as parameter in Model I, Curve 1: $t = 0.0$ second. Curve 2: $t = 295.868\,5519$ second. Curve 3: $t = 297.106\,5419$ second. Curve 4: $t = 297.109\,6357$ second. Curve 5: $t = 297.111\,9729$ second. Curve 6: $t = 297.143\,7704$ second, Curve 7: $t = 297.537\,0907$ second. Curve 8: $t = 326.074\,7613$ second. The dashed curve 0 corresponds to the moment $T = 297.099\,2827$ second when the central density in the body consisting of incoherent matter has become singular. Total CPU $\simeq$ 582 minutes, 1690 steps, $t = 362.484\,3265$ second. Note: (i) The collapse follows the curve 0 for the collapse of incoherent matter until the pressure gradient has been built up (at $R \simeq 30$ km) to reverse it to an expansion. (ii) The oscillations in density are generated where the inner, expanding, part collides with the still contracting outer part of the model. (iii) The smoothness of other parts of the curve indicates the absence of numerical instability.

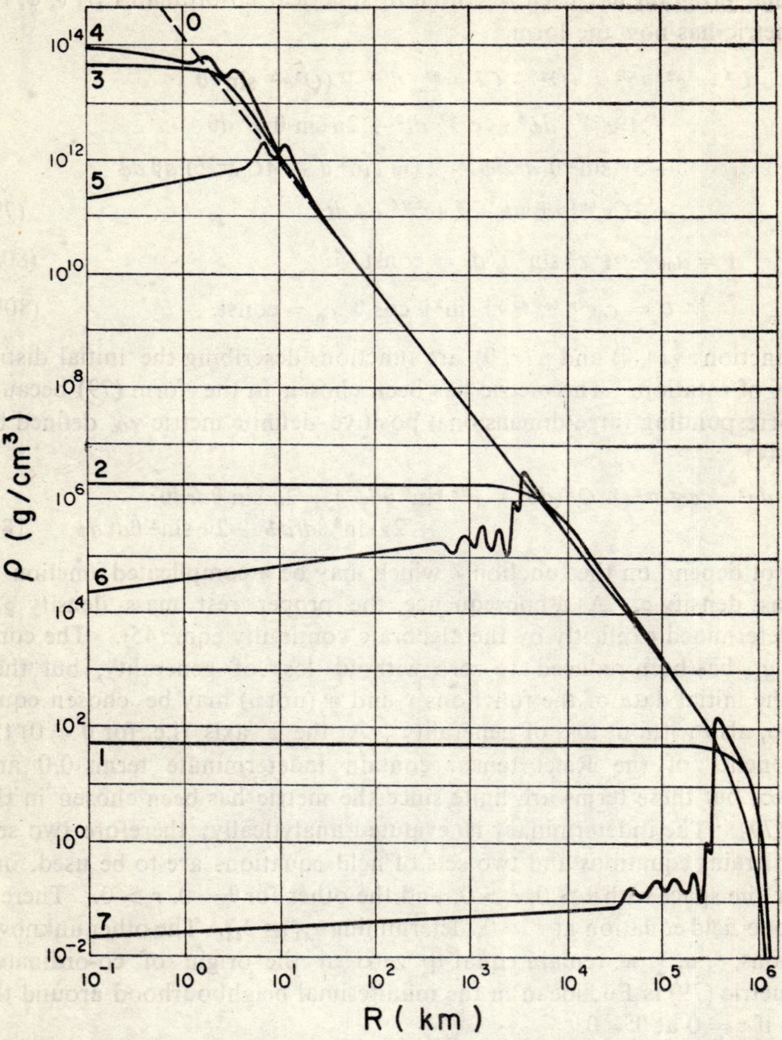

Fig. 2 Distribution of the mass density ρ along the geometrical radius R with time as parameter in Model II. Curve 1: $t = 0.0$ second. Curve 2: $t = 296.468\ 1542$ second. Curve 3: $t = 297.096\ 2612$ second. Curve 4: $t = 297.096\ 4043$ second. Curve 5: $t = 296.096\ 6941$ second. Curve 6: $t = 297.142\ 1926$ second. Curve 7: $t = 301.738 - 6654$ second. The dashed curve 0 corresponds to the moment $T = 297.099\ 2827$ second when the central density in the body consisting of incoherent matter has become singular. Total CPU = 562.6369 minutes, 1634 steps, $t = 755.505\ 0000$ second. Note: (i) The collapse follows the curve 0 for the collapse of incoherent matter until the pressure gradient has been built up (at $R \simeq 2.5$ km because of the lower pressure) to reverse it to an expansion. (ii) The oscillations in density, greater than that in Model I because of the lower pressure, are generated where the inner, expanding, part collides with the still contracting outer part of the model. (iii) The smoothness of other parts of the curves indicates the absence of numerical instability.

computer program has been rewritten in spherical co-ordinates (r, θ, ϕ, t). The metric has now the form

$$\begin{aligned} ds^2 = e^{2\alpha} dr^2 + (Q^2 - C^2 e^{-2\chi}) d\theta^2 + (Q^2 e^{2\varepsilon} \sin^2\theta \\ - A^2 e^{-2\chi}) d\phi^2 - e^{-2\chi} dt^2 + 2u \sin\theta \, dr \, d\theta \\ + 2v \sin^2\theta \, dr \, d\phi + 2 (w \sin^3\theta - AC e^{-2\chi}) d\theta \, d\phi \\ - 2Ce^{-2\chi} d\theta \, dt - 2Ae^{-2\chi} d\phi \, dt \end{aligned} \tag{79}$$

where $A = a_0 e^{2\nu(r,\theta)} r^2 \sin^2\theta, \; a_0 = \text{const.}$ (80a)

$$C = c_0 e^{2\mu(r,\theta)} r^4 \sin^3\theta \cos\theta, \; c_0 = \text{const.} \tag{80b}$$

The functions $\nu(r, \theta)$ and $\mu(r, \theta)$ are functions describing the initial distribution of rotation. The metric has been chosen in the form (79) because the corresponding three-dimensional positive definite metric γ_{ik} defined by eqn. (44)

$$\begin{aligned} d\sigma^2 = e^{2\alpha} dr^2 + Q^2 (d\theta^2 + e^{2\varepsilon} \sin^2\theta d\phi^2) + 2u \sin\theta dr d\theta \\ + 2v \sin^2\theta dr d\phi + 2w \sin^3\theta d\theta d\phi \end{aligned} \tag{81}$$

does not depend on the function χ which may be a complicated function of the mass density ρ. As a consequence, the proper rest mass density ρ is now determined explicitly by the algebraic continuity eqn. (45). The componet g_{14} has been reduced to zero without loss of generality, but then only the initial data of the functions v and w (not u) may be chosen equal to zero, also without loss of generality. At the z axis (i.e. for $\theta = 0$) the components of the Ricci tensor contain indeterminate terms 0/0 and $\infty - \infty$, but these terms are finite since the metric has been chosen in the form (79). The indeterminacy is evaluted analytically; therefore two sets of constraint equations and two sets of field equations are to be used, one set for the space with $\theta > 0$, $r > 0$, and the other for $\theta = 0$, $r > 0$. There is only one field equation at $r = 0$, determining $\alpha_{44} \equiv \beta_{44}$. The other unknown functions, ε, u, v, w, remain equal to zero at the origin of co-ordinates. The metric (79) is Euclidean in the infinitesimal neighbourhood around the z-axis if $\varepsilon = 0$ at $\theta = 0$.

The Ricci tensor is evaluated by the following formulas involving only algebraic operations:

$$R_{\mu\nu} = H_{\mu\nu} + L_{\mu\nu} \tag{82a}$$

where $H_{\mu\nu} = \frac{1}{2} g^{\alpha\beta} (g_{\mu\alpha,\nu\beta} + g_{\nu\beta,\mu\alpha} - g_{\mu\nu,\alpha\beta} - g_{\alpha\beta,\mu\nu})$ (82b)

$$L_{\mu\nu} = g^{\alpha\beta} g^{\gamma\delta} [(\gamma, \mu\alpha)(\delta, \nu\beta) - (\gamma, \mu\nu)(\delta, \alpha\beta)] \tag{82c}$$

The Christoffel symbols of the first kind are defined by the equation

$$(\varepsilon, \mu\nu) = \tfrac{1}{2} (g_{\varepsilon\mu,\nu} + g_{\varepsilon\nu,\mu} - g_{\mu\nu,\varepsilon}) \tag{82d}$$

Similarly as in the spherically symmetric problem of Sec. 4, the constraint equations correct at each step of integration of the field equations' four functions, Q, β_4, u_4, v_4. The derivative χ_{41} occurring in the Ricci tensor must be reduced with the help of eqns. (45) and (46b) to an expression involving β_{41}, u_{41}, v_{41}. Since the Einstein equations are quasi linear, the system of four constraint equations can now be reduced by algebraic operations (by Gaussian elimination) to three first order differential equations for β_{41}, u_{41}, v_{41}, and one second order differential equation for Q_{11}. The second derivative Q_{11} is replaced with the first derivative S_1 by the equation

$$Q_{11} = \alpha_1 Q_1 - e^{\alpha} S_1 \tag{83}$$

and to this system of four explicit first order differential equations for the functions β_4, u_4, v_4, S the fifth first order differential equation

$$Q_1 = e^{\alpha}(1 - S) \tag{84}$$

is added for the unknown function Q which is related to β occuring in the field equations by the equation

$$\beta = \ln(Q/r) \tag{85}$$

Because of the symmetry with respect to the z-axis and the reflection symmetry with respect to the equatorial plane, the integration domain may be restricted to one quadrant $0 \leqq \theta \leqq \pi/2$. The domain is covered by $(1 + n \times m)$ grid points, one point the origin of the co-ordinate system, n points at each straight line $\theta =$ const., and m points at one quadrant of each circle $r =$ const. The n intervals Δr between the grid points may be chosen arbitrarily, the $(m - 1)$ intervals $\Delta\theta$ are equal $(\Delta\theta = \pi/2\,(m - 1))$. The cubic spline is used for the computation of derivatives with respect to r. In order to compute the derivatives with respect to θ, each function at the quadrant $r =$ const. is numerically expanded into a Fourier series and then differentiated. The matrices for the cubic spline and for Fourier expansion and differentiation are computed in a small auxiliary program (core 33.3K words) and saved in corresponding files. This auxiliary program has to be applied only when the spatial distribution of grid points is changed.

The system of five first order partial differential constraint equations may be considered as a system of $5 \times m$ ordinary simultaneous differential equations and integrated in the r-direction by Runge-Kutta method. The system of six second order partial differential field equations may be considered as a system of $(1 + 6 \times n \times m)$ ordinary differential equations and integrated by the fourth order predictor-corrector method of Sec. 4.

In order to minimize the requirement on the necessary memory extent, the variables for the integration of constraint equations are stored as vectors, and some of them (for instance, the Christoffel symbols) as scalars. The variables for the integration of the field equations (i.e. six unknown functions and their first and second derivatives with respect to t) must be stored, of course, as matrices. In this way the memory extent of the computer program written in FORTRAN for the most general axially symmetric problem with 121 grid points is 46.9K words.

In order to save computer time, two more computer programs have been developed, one for axially symmetric problems with rotation in which the choice of the initial distribution of rotation is slightly restricted by putting in the metric (79) the function $C = 0$ (all the vorticity filaments lie at $t = 0$ in the planes $\phi =$ const., at $t > 0$ they remain at the hypersurfaces $\phi =$ const. (no more planes)), and the other for axially symmetric problems without rotation (i.e. $A = C = 0$ in the metric (79). It can be proved that the axially symmetric problems without rotation reduce to four unknown functions α, Q, ε, u, and the corresponding constraint equations to four-first-order differential equations for the functions Q, S, β_4, u_4. The required memory extent in these two programs for 121 grid points is 42.4K words, and 30.2K words, respectively.

Since the Ricci tensor is far more complicated in any axially symmetric program than in a spherically symmetric one, it is more difficult to avoid the occurrence of numerical instability. Nobody should trust any numerical integration of spherically or axially symmetric Einstein equations, unless it has been proved that the integration method is stable by running the program with spherically symmetric initial data and zero pressure and comparing the results with the analytical ones of Sec. 2. Without such a proof there is a real danger that the oscillations in the numerical results which are nothing else than numerical instability are declared for shock waves (spherical spmmetry) or for gravitational radiation (axial symmetry).

At present all three axially symmetric programs developed by the author are still numerically unstable. In the initial steps of the time evolution there is an excellent agreement between the numerical and analytical results, but then the numerical errors in the integration of constraint equations generate fast increasing oscillations in the direction of r-co-ordinate. The author believes that he will succeed to stabilize the numerical computation.

Acknowledgement
In the former stages of the development of numerical integration method for time-dependent Einstein equations (with metric written in cylidrical co-ordinates) Dr. D.A. Swayne of the University of Regina and Mr. R. Hoffman of the University of Texas at Dallas participated in the development of the numerical methods (D.A.S.) and in programming (D.A.S. and R.H.). The participation of D.A.S. in discussions of some problems of numerical analysis and of programming as well as running the programs described in this paper is acknowledged and highly appreciated.

The research project is supported in part by National Research Council of Canada. However, without the understanding of Prof. Ivor Robinson and generous support by the University of Texas at Dallas during my sabbatical year 1975/76 spent at this university, it never would have been possible to accomplish the research described in the present paper.

References

1. Arnett, W.D., Physical Aspects of Collapse (an invited talk at Texas Symposium on Relativistic Astrophysics, to be published in Ann. New York Acad. Sci.).
2. Chandrasekhar, S., *The Observatory* **57** pp. 373, 377 (1934).
3. Datt, B., Phys. Zeits· **108**, 341 (1938).
4. Gödel, K., *Rev. Mod. Phys.* **21**, 447 (1949).
5. Harrison, B.K., Thorne, K.S., Wakano, M., and Wheeler, J.A., *Gravitation Theory and Gravitational Collapse* (Chicago University Press), pp. 108ff (1965).
6. Hawking, S.W., and Ellis, G.F.R. *The Large-scale Structure of Space-Time* (Cambridge University Press), pp. 301–302 (1978).
7. Hawking S.W., and Penrose, R., *Proc. Roy. Soc.* (London) **A314**, 529 (1970).
8. Landau, L.D., and Lifshitz, E.M., *The Classical Theory of Fields*, 2nd ed., Pergamon Press, pp. 341ff (1962).
9. Lichnerowicz, A., *Theories Relativistes de la Gravitation et de l'Electromagnetisme*, Masson et Cie., Paris, p. 63 (1955).
10. May M.M., and White, R.H., *Meth. Comp. Phys.* **7**, 219 (1967).
11. Oppenheimer, J.R., and Snyder, H. *Phys. Rev.* **56**, 455.
11. Pachner, J., *Bull. Astron. Inst. Czech.* **17**, 105 (1965).
12. ———, G.R.G. **1**, 281 (1971).
13. ———, *Ann. N. Y. Acad. Sci.* **262** 141 (1975a).
14. ———, in *General Relativity and Gravitation*, G. Shaviv and J. Rosen, (Eds.), Wiley, p. 143 (1975b).
15. Pachner, J., and Teshima, R., *Can. J. Phys.* **51**, 743 (1973).
16. Penrose, R., *Phys. Rev. Lett.* **14**, 57 (1965).
17. Penrose, R., *Nature* **236** 377 (1972).
18. Ralston, A., *A First Course in Numerical Analysis*, McGraw-Hill, pp. 228ff (1965).
19. Smarr, L., *Ann. N.Y. Acad. Sci.* **302**, 569 (1977).
20. Tolman, R.C., *Relativity, Thermodynamics and Cosmology*, Oxford: Clarendon Press, p. 222 (1934).
21. Weaver, T., 'Evolution and Explosion of Massive Stars' (an invited talk at Texas Symposium on Relativistic Astrophysics, to be published in Ann. N.Y. Acad. Sci.) (1978).
22. Wilson, J., 'Collapse and Supernovae', an invited talk at Texas Symposium on Relativistic Astrophysics, to be published in Ann. N.Y. Acad. Sci.

Discussions

R Cowsik: How long does the program run to solve supernova explosion problems?

J Pachner: In the original version of the program between 9 and 10 hours when a relatively slow Xerox Sigma 9 computer is used. In the improved version the time is less than 5 hours (on the same computer).

S M Chitre: We should perhaps bear in mind that the realistic equations of state are far more complicated. In fact, the fate of the collapse is now believed to be crucially dependent on the form of the equation of state. Is it not, therefore, essential to use the sophisticated equations of state in the numerical problems of gravitational collapse?

J Pachner: For a quantitatively correct description of a gravitational collapse (and a supernova explosion) the use of a sophisticated equation of state is absolutely necessary. The exhibited examples with simple equations of state (a non-degenerate Fermi gas with $\gamma = 5/3$) just show that the applied method of numerical integration of spherically symmetric Einstein equations is stable for a long period to describe the gravitational collapse and the subsequent expansion of the star.

Black Holes for Bedtime

C.V. VISHVESHWARA

"Thinukidanu Phaniraya Ramayanada bharadali"
"The King of Snakes groaned under the weight of the Ramayanas"
—From the Mahabharatha by Kumara Vyasa.*

In the year 1756 Siraj-ud-daula, the Nawab of Bengal tried to create a black hole by squeezing a large amount of matter, in the readily available human form, into a rather small volume. The experiment was a failure and the political consequences were disastrous. But the name black hole came into existence. Only in the last few years have we come to realize that the Nawab's method of manufacturing black holes is feasible after all, but the compressions required are of cosmological proportions. These are the primordial black holes which may have nothing to do with politics.

About two hundred years later, the term black hole was introduced into astrophysics. Since then it has come to represent many things in popular imagination: symbol of Dantesque doom, panacea for astrophysical ills, inexhaustible source of energy, free ticket to time travel, gateway to other universes which may not exist, explanation for the missing mass in the universe, the missing solar neutrinos, the missing link—in general anything that is missing. Fortunately, behind all the folklore and fantasy, fallacy and fiction, stands the black hole of the sane scientist, which is a product of Einstein's general theory of relativity and a manifestation of high space-time curvature.

The first general relativistic black hole—the Schwarzschild black hole—was born immediately after the advent of the theory itself. But it lay dormant for nearly half a century collecting dust in old books and obscure articles. This was ironical in view of the fact that as early as in the thirties mass limits for stable stellar configurations had been derived and the dynamics of spherical collapse leading to the formation of a black hole had been amply demonstrated. Nevertheless, in the past decade, intense research—perhaps unprecedented in the history of relativity—has clarified

*The reason for his choosing the Mahabharatha rather than the Ramayana for rendering in the Kannada language, says the poet Kumara Vyasa, is that the various versions of the latter have already burdened heavily the seven-hooded Adisesha, the King of Snakes, who supports the earth. Perhaps so have the articles on black holes. The present review is then a minor perturbation, an admissible one it is hoped, on the already existing mass of literature.

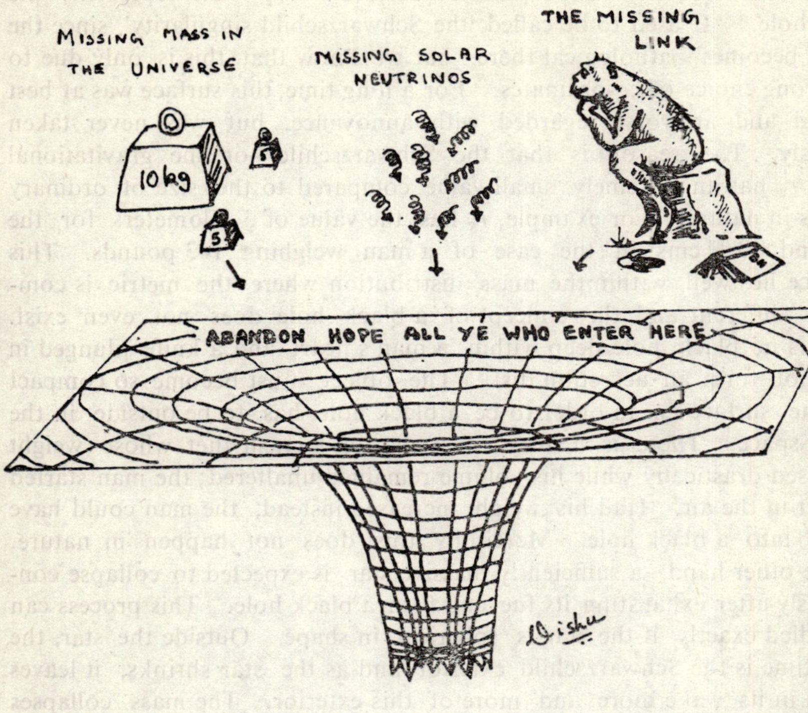

the properties of black holes and their relevance to astrophysics to a great extent. The subject is vast and the space available here limited. We shall not attempt therefore to paint the portrait of a black hole with "pimples, warts, and all"—especially since the black hole, as we shall see, cannot have them. Only a thumbnail sketch will be aimed at. Let us then begin at the beginning.

In the beginning, Einstein created the space-time and the line element. And the space-time was flat and empty. Einstein introduced matter and gravitation. The space-time became curved. And the line element was given by the metric tensor which had ten components in general. These ten functions of space and time also incorporated relevant parameters like the mass and angular momentum of the source of gravitation. And they told the entire gravitational tale of the particular situation on hand.

The Schwarzschild empty space line element, which incorporates the black hole, has the form:

$$ds^2 = \left(1 - \frac{2m}{r}\right) dt^2 - \left(1 - \frac{2m}{r}\right)^{-1} dr^2 - r^2 (d\theta^2 + \sin^2 \theta \, d\phi^2).$$

This is valid outside a spherically symmetric source of mass m ($G = C = 1$ as usual, so that mass is expressed in units of length). Just with the introduction of the Newtonian potential m/r into the metric, a change that seems trivial, we leave the safe moorings of a flat space-time and enter the field of a black hole. The spherical surface $r = r_g = 2m$ represents the black hole.* It used to be called 'the Schwarzschild singularity', since the metric becomes pathological there, but we know that this is only due to the wrong choice of co-ordinates. For a long time, this surface was at best ignored and at worst regarded with annoyance, but was never taken seriously. The reason is that the Schwarzschild—or the gravitational radius r_g has an extremely small value compared to the size of ordinary objects in nature. For example, r_g has the value of 3 kilometers for the sun and 10^{-23} cms in the case of a man weighing 143 pounds. This distance lies well within the mass distribution where the metric is completely different and the concept of a black hole does not even exist. There is no black hole deep within a man's heart and a knife plunged in does not form an accretion disk. The object must become so compact that the surface r_g, in order to be a black hole, has to be outside in the empty space. There is a story of a man on a crash diet whose weight decreased drastically while his volume remained unaltered; the man started to float in the air. Had his weight increased instead, the man could have turned into a black hole. Mercifully this does not happen in nature. On the other hand, a sufficiently massive star is expected to collapse continuously after exhausting its fuel and form a black hole. This process can be studied exactly if the star is spherical in shape. Outside the star, the space-time is the Schwarzschild exterior and as the star shrinks, it leaves behind in its wake more and more of this exterior. The mass collapses into the Schwarzschild radius thereby liberating the black hole and finally hits the singularity at the origin.

What exactly is a black hole? A systematic study of this question in regard to both the Schwarzschild and the Kerr black holes was undertaken in the late sixties. We start the discussion by noting some very basic properties of the Schwarzschild metric. The metric components are independent of time. Furthermore, there are no terms like $dt\, d\phi$ in the metric. The latter observation corresponds to the fact that the source is not rotating and consequently no rotation is built into the space-time either. The time-independence of the metric shows the existence of a very special vector field which spells out the inherent symmetry of the space-time. This is known as the Killing vector which in the usual Schwarzschild coordinates has the form $\xi^a = (1, 0, 0, 0)$, namely the direction of time. Motion along the vector is tantamount to the flow of time which brings no change in the space-time picture. We can define sources and observers with four-velocities along the trajectories of the Killing vector at each spatial point. These will then be the static observers who do not

*There seems to be some confusion of terminology in the literature. Sometimes the entire spacetime region containing the surface $r = 2m$ as well as the singularity at $r = 0$ is referred to as the black hole. On the other hand terms like the "area of a black hole" imply reference only to a surface. In this article the black hole is consistently identified with the Schwarzschild surface $r = 2m$ or its equivalent in any other spacetime.

move; only time passes on as they sit around and watch. However, the Killing vector is timelike only outside the black hole. It is lightlike inside. Consequently, the static observers are defined only up to the black hole. With no reference to any particular coordinate system, we can characterise the Schwarzschild black whole as the surface on which the timelike Killing vector becomes null making it thereby the so-called 'static limit'.

An immediate consequence of the above feature pertains to the gravitational redshift with respect to static sources and observers. As in any other branch of physics, the existence of time symmetry leads to a conserved quantity. Along a geodesic, which represents the natural particle motion inherent to a given space-time, the scalar $\xi_a p^a$ is constant. Here p^a is the four-momentum tangential to the geodesic. At the same time, an observer with a four-velocity u^a measures the energy E of the geodesic particle as $u_a p^a$. Applying these two considerations to a photon travelling along a null geodesic from a static source to a static observer, one readily obtains the ratio of the observed and source frequencies as

$$\frac{\nu_0}{\nu_s} = \frac{(\xi_a \xi^a)_s^{1/2}}{(\xi_a \xi^a)_0^{1/2}} = \frac{(1 - 2m/r_s)^{1/2}}{(1 - 2m/r_0)^{1/2}} \, .$$

As the source approaches the black hole, the observed frequency tends to zero or the gravitational redshift becomes infinite. The Schwarzschild black hole is an infinite redshift surface for static sources.

We now come to the definitive property of the black hole. It is a null surface and hence acts as a one-way membrane. That is, if the length of the normal to the surface is computed using the metric tensor, the length turns out to be zero. Null surfaces are quite familiar to us in everyday life even in flat space-time. Every wavefront is a null surface. Such a surface is tangential to the local light cone. Timelike trajectories which

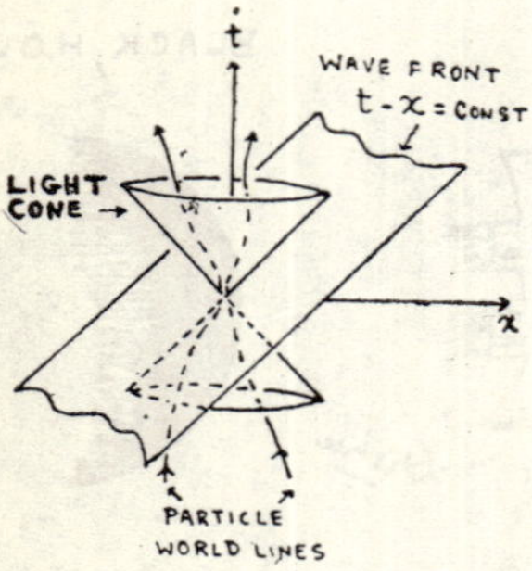

have to be confined to within the coue necessarily cross the wavefront in one direction and cannot cross it again in the reverse direction. In other words, once the wavefront has crossed a material particle, the latter cannot overtake the front and cross it the other way round. Here the wavefront travels in space with the elapse of time. And it is at spatial infinity at some time or the other. The black hole, being a null surface, is similar to the wavefront although it is devoid of any physical content such as the electromagnetism associated with the wavefront. The light cone is again tangential to it everywhere. A particle falling in, having crossed it in one direction, cannot turn back and recross it in the opposite sense. It acts thus as a one-way membrane. However, unlike the wavefront, it does not 'travel' in time and does not extend to spatial infinity. This is a reflection of the gravitational pull exerted by the mass that has fallen in, thereby creating the black hole. To summarise, the Schwarzchild black hole happens to be the surface on which the timelike Killing vector has become null and is hence the static limit; furthermore, it is a null surface and consequently behaves like a static one-way membrane. Since not even photons, let alone material particles, can escape from within this surface, no event occurring in this region can communicate itself with the outside world. The surface is therefore referred to as an 'event horizon'.

The true picture of the black hole, free of co-ordinate pathologies, emerged with the discovery of the Kruskal co-ordinates. The radial co-ordinate r and the time t are combined in a somewhat complicated manner to produce the new co-ordinates u and v in terms of which the line element reads

$$ds^2 = f^2(dv^2 - du^2) - r^2(d\theta^2 + \sin^2\theta d\phi^2),$$

where $f^2 = (32m^3/r) \exp(-r/2m)$. The metric is now regular all the way down to the singularity at the origin. Everywhere the light cones are similar to those of the flat space-time. It can be readily seen that the light cone is tangential to the black hole. Inside, both incoming and 'outgoing' rays hit the singularity, so much so a particle that has entered the black hole is fated to do the same and get crushed. The picture of stellar

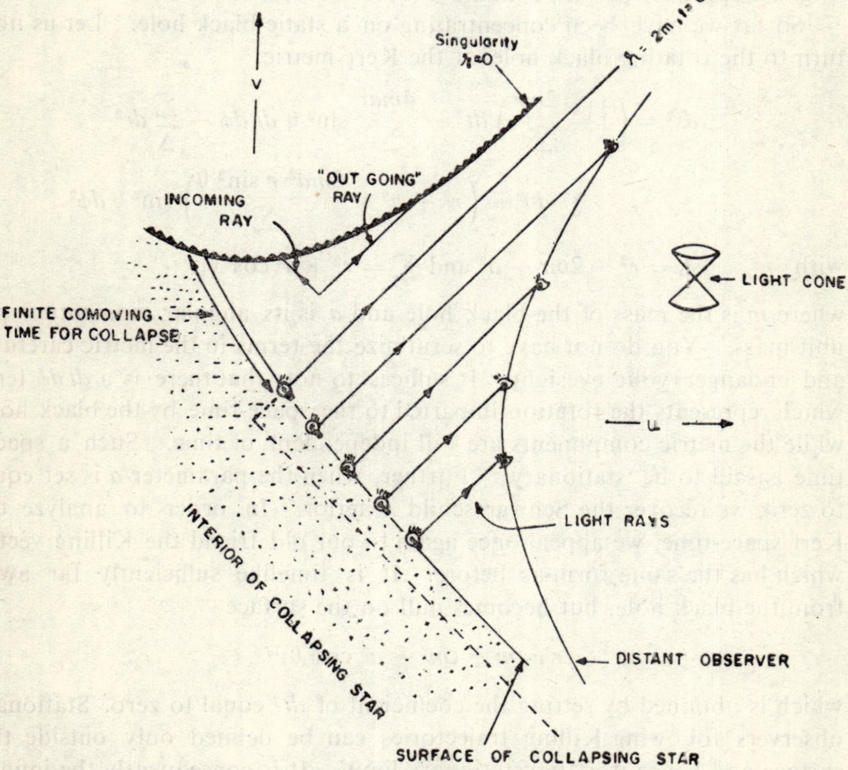

KRUSKAL DIAGRAM

collapse is also clear. Intervals of light pulses emitted from the surface are progressively dilated as seen by an outside observer whose trajectory is a hyperbola in the u-v plane. The surface of the star seems to take infinite amount of the observer's time to reach the Schwarzschild radius. A similar effect is that the colour of the light reaching the observer shifts more and more to red and the intensity dies down exponentially. Old stars never die; they just fade away. The proper infall time of the surface, however, is finite for the entire collapse. In addition to displaying all this in a straightforward manner, the Kruskal diagram becomes the proper framework in which many phenomena are studied, as the metric no longer suffers from the spurious divergences at the Schwarzschild radius.

Orbits around the black hole and their stability have been worked out in detail. These computations are of significance to astrophysics. Another effect of some importance is the tidal distortion of extended objects in the vicinity of a black hole. Gruesome descriptions have been given of this effect on a man falling into a black hole. Since at the Schwarzschild surface the tidal force is proportional to $1/m^2$, one is advised to fall into a massive black hole, if fall one must, in order to enjoy the trip before gravity takes its final toll. This would happen as the singularity at the origin is approached where the tidal forces become infinite.

So far we have been concentrating on a static black hole. Let us now turn to the rotating black hole of the Kerr metric:

$$ds^2 = \left(1 - \frac{2mr}{\Sigma}\right) dt^2 - \frac{4mar}{\Sigma} \sin^2\theta \, dt \, d\phi - \frac{\Sigma}{\Delta} dr^2$$

$$- \Sigma \, d\theta^2 - \left(r^2 + a^2 + \frac{2ma^2 \, r \sin^2\theta}{\Sigma}\right) \sin^2\theta \, d\phi^2$$

with $\Delta = r^2 - 2mr + a^2$ and $\Sigma = r^2 + a^2 \cos^2\theta$,

where m is the mass of the black hole and a is its angular momentum per unit mass. You do not have to scrutinize the terms in the metric carefully and endanger your eyesight. It suffices to note that there is a $dt\,d\phi$ term which represents the rotation imparted to the space-time by the black hole, while the metric components are still independent of time. Such a space-time is said to be 'stationary'. Further, when the parameter a is set equal to zero, we recover the Schwarzschild exterior. In order to analyze the Kerr space-time, we appeal once again to our old friend the Killing vector which has the same form as before. It is timelike sufficiently far away from the black hole, but becomes null on the surface

$$r = m + (m^2 - a^2 \cos^2\theta)^{1/2},$$

which is obtained by setting the coefficient of dt^2 equal to zero. Stationary observers following Killing trajectories can be defined only outside this surface and hence it is the 'stationary limit'. It is consequently the infinite redshift surface for stationary sources. But the length of the normal to this surface is non-zero and therefore it is not a null surface. Particles can cross it and fly out again. The black hole itself is identified with the time-independent null surface $r = r_+ = m + (m^2 - a^2)^{1/2}$; a particle entering it cannot come out again. The stationary limit touches the black hole at the poles and bulges out at the equator. Clearly, the splitting of the two surfaces, in contrast to the Schwarzschild metric in which they were identical, is a manifestation of the rotation inherent to the space-time.

The above phenomenon can be analyzed by invoking the so-called locally non-rotating observers (or the 'rest observers') who follow the trajectories given by the vector field $\zeta^a = \left(1, 0, 0, -\frac{g_{03}}{g_{33}}\right)$, where g_{03} and g_{33} are the coefficients of the $dtd\phi$ and $d\phi^2$ terms respectively. While the Killing vector becomes null on the stationary limit, the vector ζ^a is timelike

down to the black hole on which it becomes lightlike. The rotating black hole, like a whirlpool, sets up currents in the space-time that die out asymptotically at spatial infinity. Stationary observers will have to be moving against these in order to maintain constant spatial coordinates, approaching the speed of light as their location reaches the stationary limit. On the other hand the rest observers simply drift along the currents and can exist down to the black hole itself.

At this stage of exploration, the black hole physicists realized that almost all of their results had been anticipated in the latter half of the last century by an Oxford mathematics don who wrote under the name of Lewis Carroll. Alice's fall is symbolic of gravitational collapse. The white rabbit's watch cannot maintain regular time as the rabbit flits in and out of the strong gravitational field. Alice opens up like a telescope because of tidal distortion. The Cheshire cat's grin represents the gradual disappearance of a collapsing star as viewed by an outside observer. As the red queen says, an observer has to do all the running he can in order to stay at the same place as the stationary limit is reached.

The picture of the black hole that has emerged so far makes it merely the lingering ghost of a star that has long been interred in the singularity of its own making. Far away the gravitation has nothing unusual to offer; too close, the field is strong enough to pull everything in. Incapable of emitting anything and immutable, the black hole seems to have no dynamics associated with it at all. This situation changed with the coming of the Penrose process and the black hole thermodynamics.

The Penrose process offers, at least in principle, a method of extracting energy out of a rotating black hole. We have seen that along a geodesic with four-momentum p^a, the quantity $\xi_a p^a$ is a constant and is proportional to the energy measured by the stationary observers. It should necessarily be positive outside the stationary limit where ξ^a is timelike and stationary observers are possible; whereas this quantity, although a constant, need no longer be positive inside the stationary limit since ξ^a is now spacelike. Suppose a particle (1) falling in along a geodesic from infinity splits into two (2 and 3, both following geodesics) in the region between the stationary limit and the black hole, the so-called ergosphere. Energy conservation during this process reads, $\xi_a p_1^a = \xi_a p_2^a + \xi_a p_3^a$ more or equivalently $E_1 = E_2 + E_3$. If particle 2 escapes out, then E_2 has to be positive. On the other hand if particle 3 falls into the black hole, then E_3 could very well be negative in which case particle 2 would come out with enhanced energy ($E_2 > E_1$). At the same time the black hole's rotation slows down as a result of this energy extraction. The operation can be continued until the angular momentum of the black hole is reduced to zero. However, the geodesics involved in the process have to be stringently arranged so as to make the energy extraction feasible. Investigations have shown that there is no 'natural' way to accomplish this that would make the process astrophysically significant. Needless to say, the day will probably never arrive when superpowers will compete to cast the first stone at the black hole in order to reap the outcoming energy. Incidentally, in all our discussions, we have found that the Killing vector plays a paramount role. As the old saying goes, the spoils belong to the vector!

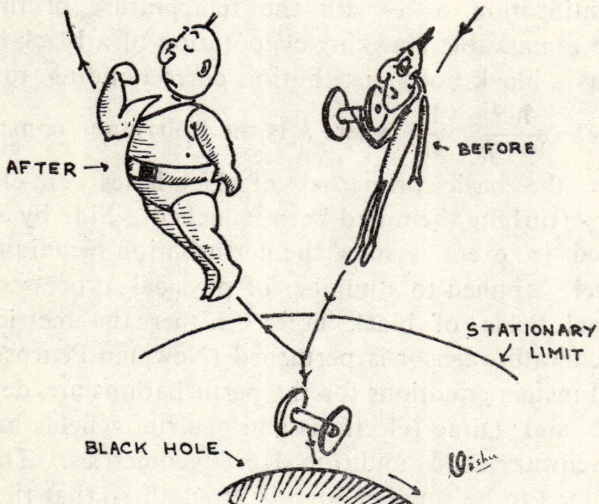

The code of conduct of black holes, stated in terms of very general laws, bears striking resemblance to ordinary thermodynamics. The analogy was initiated by the observation that the area of a black hole plays a role similar to entropy in that both of them are nondecreasing quantities in physical processes. In the simplest example of a static black hole with radially infalling particles, the area $4\pi(2m)^2$ steadily increases since particles cannot escape from within so as to decrease m.

The first law of black hole dynamics asserts that in any physical process involving black holes the energy, momentum, angular momentum and electric charge are conserved. Likewise the second law states that in any physical process the sum of the surface areas of the black holes involved can never decrease. Similarity of these two laws to the corresponding ones in thermodynamics is obvious. Better understanding of this new thermodynamics can be gained by considering the area of a Kerr black hole, $A = 4\pi(r_+^2 + a^2)$. Varying this area infinitesimally with respect to m and a and rearranging terms, it is easy to obtain the formula,

$$dm = \kappa \frac{dA}{8\pi} + \frac{a}{r_+^2 + a^2}\, da,$$

where $\kappa = (m^2 - a^2)^{1/2}/(r_+^2 + a^2)$ is the surface gravity. For a Schwarzschild black hole $\kappa = m/(2m)^2$, the acceleration due to gravity at the Schwarzschild radius. Comparing the above result with the formula for the change in energy in a thermodynamic transformation, $dE = Tds + \Omega dJ$ with Ω and J denoting angular velocity and angular momentum respectively, the correspondence between (κ, A) and (T, S) becomes evident. The analogy is further accentuated by the 'zeroth law' which showed that κ is constant over the horizon, just as a system in thermodynamic equilibrium has a constant temperature, and the third law which states that it is impossible to reduce κ to zero by any finite sequence of operations. The direct identification of κ with the temperature of the black hole results from the remarkable Hawking evaporation of a black hole in which the radiation has a black body distribution corresponding to a temperature $T = \frac{\hbar}{2\pi k}\,\kappa = \frac{\hbar c^3}{8\pi G k}\cdot\frac{1}{M}$, where k is the Boltzmann constant.

Long before the basic properties of black holes were elucidated, the programme of perturbing them had been launched. Side by side with the advances related to exact results, the perturbation techniques developed rapidly and were applied to a number of physical processes occurring in the gravitational fields of black holes. Either the metric is changed slightly or the curvature tensor is perturbed (Newman-Penrose formalism) and linearized Einstein equations for the perturbations are derived. Scalar, electromagnetic and Dirac (electrons and neutrinos) fields have also been studied in the Schwarzschild and the Kerr geometries. The perturbing fields are assumed to be small enough in strength so that the background geometries are considered to be unaltered. In the case of the Schwarzschild metric, the stability of the black hole against perturbations has been

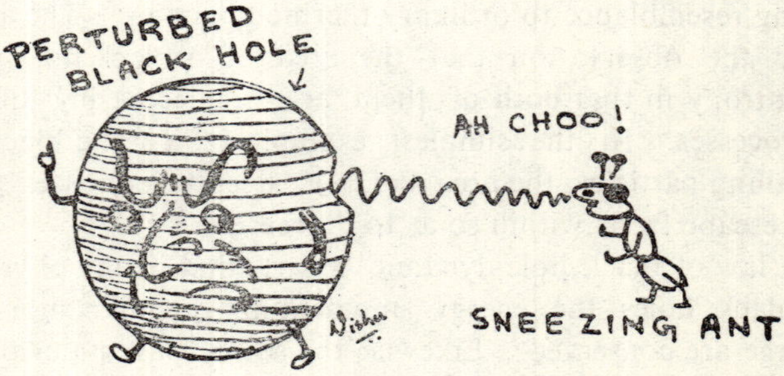

rigorously proved. Scattering of waves and production of radiation by particles in motion have been studied in detail. An analysis of spherical collapse with superimposed perturbations shows that all higher order multipoles are radiated away leaving a black hole with only mass, angular momentum and charge. As for the Kerr space-time, a completely rigorous proof of stability is lacking, but both analytic and numerical calculations indicate that the black hole is in fact stable. A new feature appearing in scattering is the superradiance which is the wave analogue of the Penrose process. Waves of certain modes coming in from infinity travel out with enhanced energy extracted at the expense of the rotation of the black hole. Production of radiation from particles has also been examined.

Perturbation calculations showed that a Schwarzschild black hole could not support static perturbations: they blew up either at infinity or at the horizon. In other words, it seemed that black holes distorted in shape could not exist. Similarly, as we have seen, non-spherically perturbed collapse showed that only mass, charge and angular momentum survived. These were the vague premonitions that preceded the startling Israel theorem which proved that the Schwarzschild black hole was the only static one possible in nature. Now we know that black holes come in only four varieties. By adding electric charge to the Schwarzschild and the Kerr black holes, say by rubbing them with fur, we get the Reissner-Nordstrom and the Kerr-Newman black holes respectively. Thus the black hole is characterized entirely by mass, angular momentum and charge and no other parameter. This uniqueness theorem is often referred to colloquially as the 'no-hair theorem'. The assertion that the black hole has no hair is probably the ultimate application of Ockham's razor!

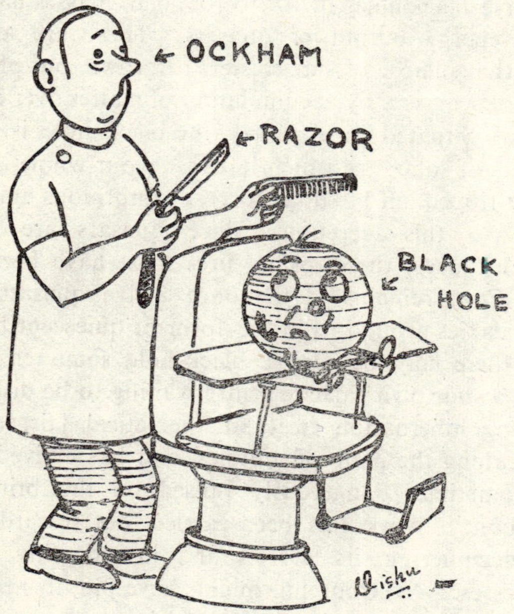

How can one detect a black hole? The philosophy of detection is that of Sherlock Holmes: When you have excluded the impossible, whatever remains, however improbable, must be the truth. If the existence of an invisible object has been established by inference and if the possibility of its being a collapsed object like a black dwarf or a neutron star is ruled out, then it must be a black hole. Of course, a black hole in isolation does not betray its existence; it has to be spotted through the company it keeps. This is precisely the procedure followed in the case of Cygnus X-1, the X-ray source in Cygnus, which is strongly suspected to comprise a black hole. This is a binary star system with a visible supergiant of mass M_1 accompanied by an invisible companion of mass M_2. From the varying Doppler shift exhibited by the spectrum of the visible star, the orbital period and the velocity can at once be determined. These two quantities when fed into Kepler's law yield the mass function $f(M) = M_2^3 \sin^3 i/(M_1 + M_2)^2$, where i is the angle between the line of sight and the normal to the orbital plane which is undetermined. However, the minimum value of M_2 is obtained for $i = 90°$ with fixed M_1, which is inferred from the spectrum of the visible star. In the case of Cygnus X-1, the mass turns out to be greater than about eight times the solar mass. This is much more than the neutron star mass limit of about three solar mass. The invisible object is therefore generally believed to be a black hole. Plasma streams out from the supergiant, forms an accretion disk around the black hole and finally spirals into it. As it pours in with enormous speed, the plasma radiates by the bremsstrahlung mechanism and makes the system a source of X-rays.

Supermassive black holes of 10^6 to 10^9 solar masses have been invoked to explain the energy output of quasars. These are assumed to have resulted from the collapse of star clusters or grown out of seed black holes of about ten solar masses by accumulation of matter over a period of about a billion years. Situated in a galaxy, the black hole is fuelled by interstellar gas, stars of solar size which it can devour whole, and fragments of giants that are ripped off by tidal forces. Enormous amount of energy is liberated due to this accretion. Since quasars are observed at only cosmological distances, these cosmic fires must have been live only in the distant past. Cold remains of the once active quasars may lie buried inside many galaxies around us in the form of quiescent black holes. For all we know, there may be such a black hole some ten thousand parsecs away from us at our own galactic centre waiting to be discovered.

After a long hibernation in its mathematical lair, the black hole has travelled fast along the path of physics and has arrived in the realm of astronomy—sometimes dangerously poised on the brink of mythology. Immense amount of work has been carried out towards describing the black hole, deciphering its secrets and investigating its influence on physical processes. As Churchill might have put it, never in the field of human knowledge so much has been said by so many about something that is so close to being nothing. Still, unanswered questions and unsolved problems remain. Except under the assumption of spherical symmetry we do not have the exact scenario for gravitational collapse; we do not even possess any information on the direct formation of a rotating black hole. We do not know whether in the process of collapse a 'naked singularity', visible to outside observers, can form. Or whether Penrose's conjecture of 'cosmic censorship', which supposes that every singularity is hidden inside an event horizon, is universally valid. The singularity, naked or clothed, is the biggest problem of all. It belongs today to the

realm of conjecture and speculation, since physical laws, including general relativity, break down when confronted with the infinite space-time curvature. Will quantum effects come to the rescue, stop the collapse at the last moment and remedy the situation? No one knows. In this respect we might as well echo Faust and say, "We do not need what we know, and do not know what we need". But, one thing is certain. The black hole is by no means merely the silent finale to the symphony of the stars, nor is it the dark crystal ball in which the fate of general relativity is revealed. The unequivocal proof of black hole's existence and a proper perspective of its role in the cosmic order both lie in the domain of the future.

Plasma Discs Around Black Holes

A.R. PRASANNA

High energy astrophysics is one of the most important topics of present day research. Particularly with the advent of HEAO and the proposed space platform the study of high energy sources like X- and γ-ray sources and Quasars has become very significant. Most of the models for such sources consider accretion discs of plasma around compact objects like neutron stars and black holes. Though work in this direction started sometime back, most of the analyses have been in the Newtonian framework only. As the gravitational fields involved near compact objects are very strong it is possible that the space-time curvature would play a significant part in the dynamical processes involved here and thus one should look into possible general relativistic effects. With this in mind we started a rigorous study of this problem and as a beginning made detailed calculations concerning the trajectories of single charged particles in an electromagnetic field superposed on curved background fields. The approach in these studies was as follows: consider test electromagetic fields superposed on background geometries like Schwarzschild and Kerr. These electromagnetic fields would not disturb the background geometry but they are modified by space-time curvature and thus different from their flat space-values. In such a background one would write a Lagrangian for the motion of a charged particle and then using the properties of axisymmetry and stationarity would get two constants of motion—one for the canonical angular momentum and the other for energy. If one restricts to the orbits in the equatorial plane by considering $\theta = \pi/2$, $d\theta/ds = 0$, then one could use these two constants of motion in the metric and further solving for $dr/ds = 0$, one could obtain the effective potential for the motion of the particles in r-direction. A study of the effective potentials would give the nature of the orbits, depending upon the physical parameters like energy, canonical angular momentum, the electromagnetic field, and the mass and angular momentum of the central body.

In our analysis we[1] used the Ginzburg-Ozernoi solution for a dipole magnetic field on the Schwarzschild background and Petterson's solution for the alectromagnetic field on the Kerr background.[2] The lagrangians and the corresponding effective potentials for the two cases are as follows:

(1) Schwarzschild geometry with a dipole magnetic field.

$$\mathcal{L} = \frac{1}{2}\left\{ c^2\left(1 - \frac{2m}{r}\right)\dot{t}^2 - \left(1 - \frac{2m}{r}\right)^{-1} \dot{r}^2 - r^2\dot{\theta}^2 - r^2 \sin^2\theta\, \dot{\varphi}^2 \right.$$
$$\left. + \frac{3e\mu \sin^2\theta}{4 M_0 c^2 m^3} \left[r^2 \ln(1 - 2m/r) + 2mr + 2m^2\right] \dot{\varphi} \right\}$$

and the corresponding effective potential for r-motion in the equatorial plane is

$$V_{\text{eff.}} = (1 - 2/\rho)\left\{1 + \rho^2\left[\frac{L}{\rho^2} + \frac{3\lambda}{8}\left\{\ln\left(1 - \frac{2}{\rho}\right) + \frac{2}{\rho}\left(1 - \frac{1}{\rho}\right)\right\}\right]^2\right\}$$

$$\rho = r/m$$

(2) Kerr geometry with an electromagnetic field.

$$\mathcal{L} = \frac{1}{2}\left\{\left(1 - \frac{2mr}{\Sigma}\right)c^2\dot{t}^2 - \frac{4amr}{\Sigma}\sin^2\theta\,\dot{t}\dot{\varphi} + \frac{\Sigma}{\Delta}\dot{r}^2 + \Sigma\dot{\theta}^2 + \frac{B}{\Sigma}\sin^2\theta\,\dot{\varphi}^2 + A_t\,\dot{t} + A_\varphi\,\dot{\varphi}\right\}$$

with $$A_t = \left(-\frac{3a\mu}{2\gamma^2\Sigma}\right)\left\{[r(r-m) + (a^2 - mr)\cos^2\theta]\,\frac{1}{2\gamma}\ln\left(\frac{r-m+\gamma}{r-m-\gamma}\right) - (r - m\cos^2\theta)\right\}$$

$$A_\varphi = \left(\frac{-3\mu\sin^2\theta}{4\gamma^2\Sigma}\right)\left\{-[r(r^3 - 3ma^2 + a^2 r) + \Delta a^2\cos^2\theta] \times \frac{1}{2\gamma}\ln\left(\frac{r-m+\gamma}{r-m-\gamma}\right) + (r-m)\,a^2\cos^2\theta + r(r^2 + mr + 2a^2)\right\}$$

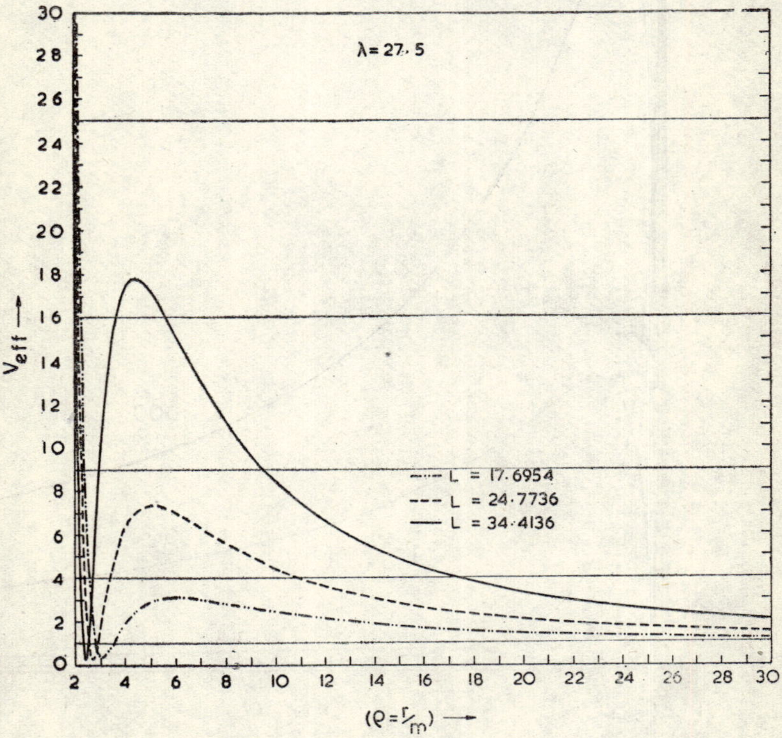

Fig. 1

and the effective potential in the equatorial plane is

$$V_{\text{eff.}} = -A_\tau + K/R$$
$$K = [2\alpha(L - \bar{A}_\varphi) + \Delta^{1/2}\{\rho^2(L - \bar{A}_\varphi)^2 + \rho R\}^{1/2}]$$
$$R = \{\rho^3 + \alpha^2\rho + 2\alpha^2\}$$
$$\rho = \frac{r}{m},\ \alpha = \frac{a}{m},\ \bar{A}_\varphi = \frac{eA_\varphi}{m},\ A_\tau = eA_t$$

Plotting $V_{\text{eff.}}$ vs. ρ in either case one could get the structure of potential wells wherein particles can be trapped in bound orbits (Figs. 1–3). In either case we have found stable orbits quite close to the event horizon, the potential well-being maintained mainly by the curvature influenced magnetic field and the centrifugal barrier. Far away from the source in both cases the potential behaves just as in their respective geometries without electromagnetic fields. Having thus found the nature of the orbits, actual orbits were obtained (Figs. 4–5) by numerically integrating the equations

$$u^i_{;j}u^j = (e/M_0)\,F^i_j u^j$$

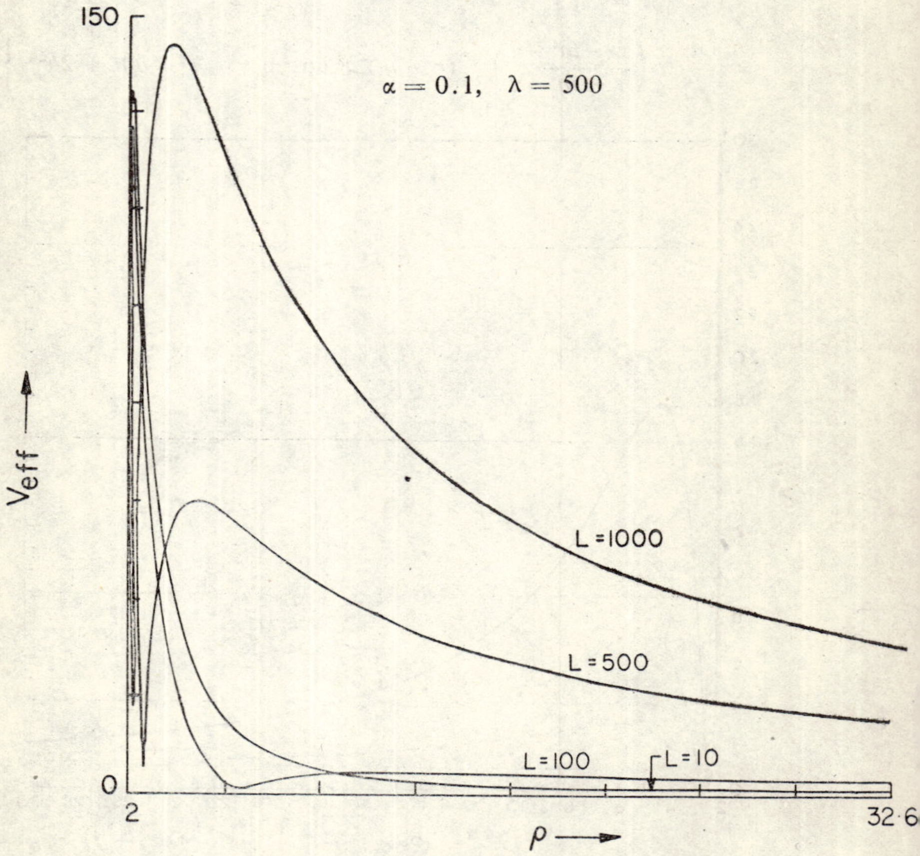

Fig. 2

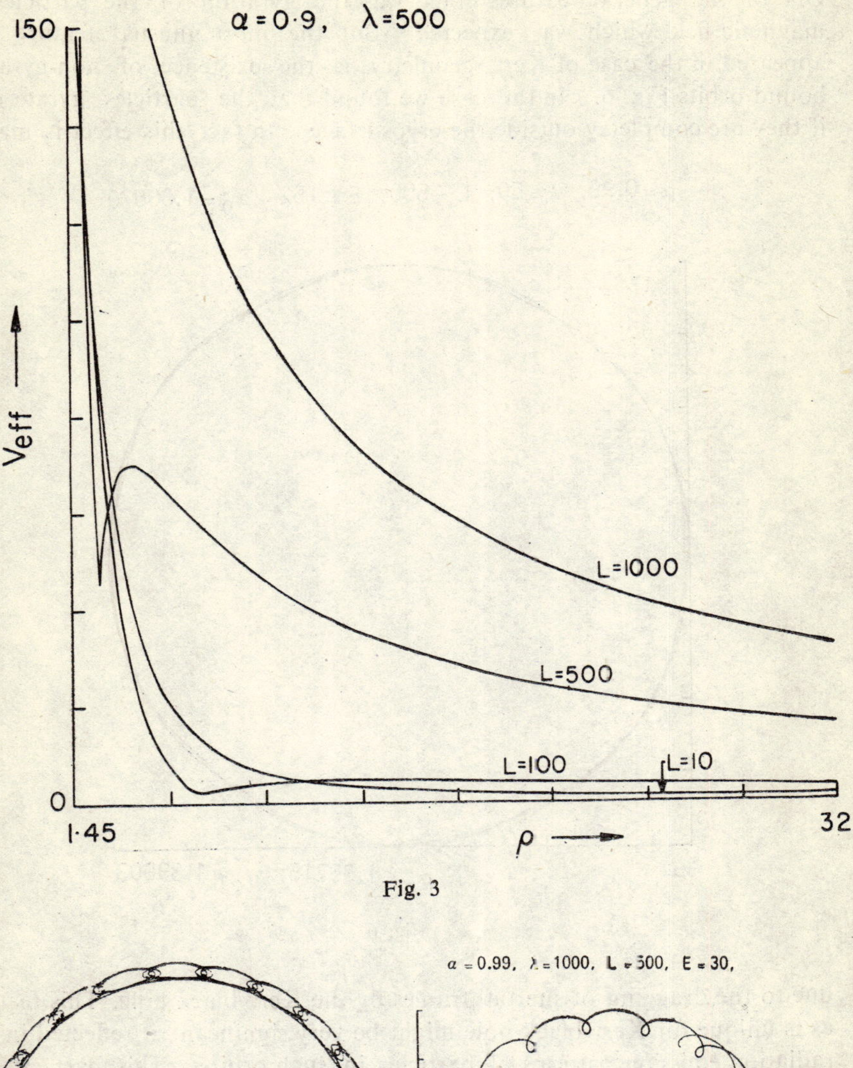

Fig. 3

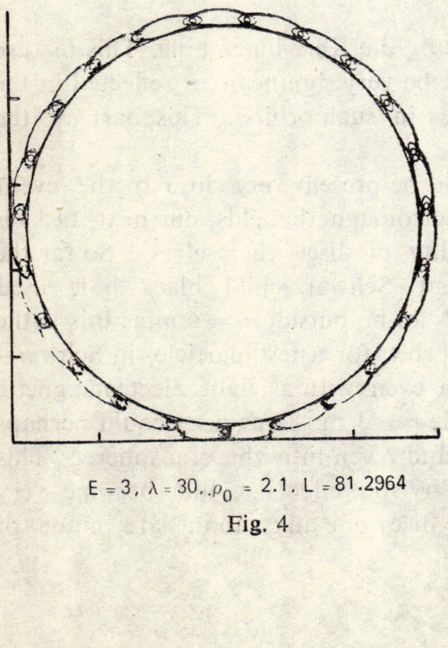

$E = 3,\ \lambda = 30,\ \rho_0 = 2.1,\ L = 81.2964$

Fig. 4

$\alpha = 0.99,\ \lambda = 1000,\ L = 500,\ E = 30,$

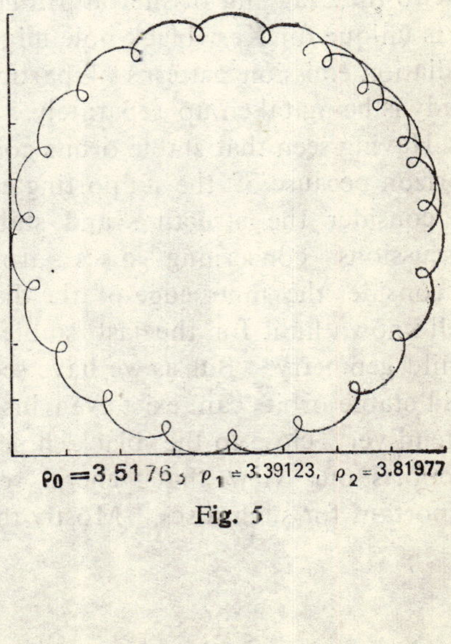

$\rho_0 = 3.5176,\ \rho_1 = 3.39123,\ \rho_2 = 3.81977$

Fig. 5

One of the general features noted was the gyration of the particles in magnetic field which was expected. But the most unusual feature that appeared in the case of Kerr geometry is the existence of non-gyrating bound orbits Fig. 6. In this case we found that the particles gyrate only if they are completey outside the ergosurface. In fact this effect is mainly

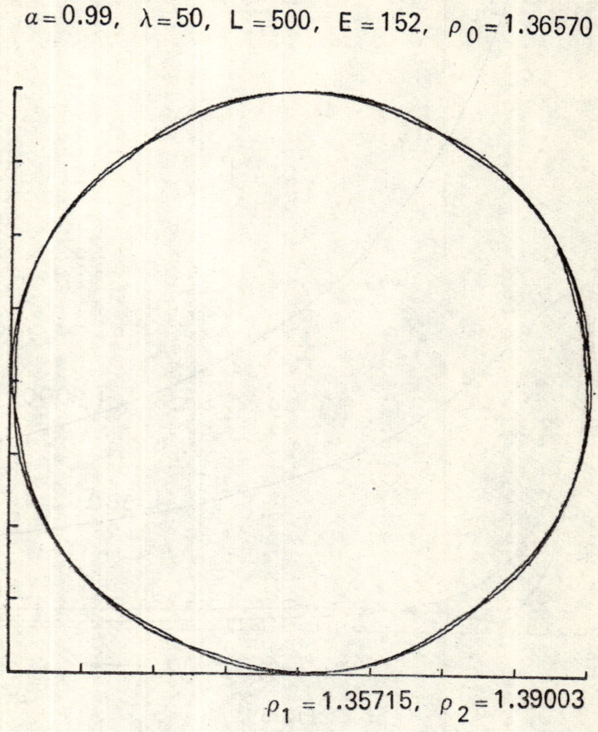

Fig. 6

due to the dragging of inertial frames by the Kerr black hole. This feature as is unique for Kerr black hole might be very significant if reflected in the radiation emission patterns of particles in such orbits. This part of the study is being taken up separately.

Having seen that stable orbits could be present very close to the event horizon because of the supporting electromagnetic fields, our next task is to consider the structure and stability of discs themselves. So far the discussions concerning discs around Schwarzschild black hole used to consider the inner edge of the discs to be outside $r = 6\ m$ as this is the well known limit for the last stable orbit for a test particle in Schwarzschild geometry. But as we have seen even with a slight electromagnetic field stable orbits can exist even inside $r = 3$ m. Thus discs could perhaps extend very close to the black hole and even into the ergosphere. This supports our view that general relativistic effects could become very important for such discs. Mostly the discs one talks about are made of quasi-neutral plasmas. But then as is known it is almost impossible to formulate and solve the plasma kinetic equations in the framework of general relativity. Hence one considers the relativistic magnetohydrodynamic equations to study these discs.

We[3] start with the Schwarzschild background geometry and consider a disc around the black hole. We assume that the total mass of the disc is very small compared to the mass of the black hole and thus the disc does not disturb the background geometry. With this we consider the fluid equations for the disc

$$T^j_{i;j} = 0 \tag{1}$$

with

$$T^j_i = \left(\rho + \frac{p}{c^2}\right) u_i u^j - \frac{p}{c^2}\delta^j_i - \frac{1}{4\pi c^2}\left(F_{ik} F^{jk} - \frac{1}{4}\delta^j_i F_{kl}F^{kl}\right) \tag{2}$$

wherein ρ is the density, p is the pressure, u^i is the four-velocity of the fluid and F_{ij} is the electromagnetic field tensor. As the electromagnetic field is generated by the motion of the charged particles of the disc the other equations governing the system are the Maxwell's equations

$$F^{ij}{}_{;j} = \frac{-4\pi}{c} J^i, \; F_{[ij,\,k]} = 0 \tag{3}$$

As the equilibrium situation sought is the one where the disc balances between the gravitational field of the central mass, its own centrifugal force and the electromagnetic field generated by the motion of the charged fluid these equations should be solved self-consistently along with the assumption $J^i = c\varepsilon u^i$ where ε is the charge density of the fluid.

In order to compare with the corresponding Newtonian problem, we first set our equations in terms of the spatial three velocity v^α given by $u^\alpha = v^\alpha u^0/c$, where $u^0 = dx^0/ds$, x^0 being the time co-ordinate, in dimensions of length.

For the Schwarzschild geometry u^0 is given by

$$u^0 = (1 - 2m/r)^{-1/2}(1 - v^2/c^2)^{-1/2} \tag{4}$$

$$v^2 = (v^{(r)})^2 + (v^{(\theta)})^2 + (v^{(\varphi)})^2 \tag{5}$$

$v^{(\alpha)}$ being components of the three-velocity in local Lorentz frame which are connected with the Schwarzschild frame components through:

$$v^{(r)} = (1 - 2m/r)^{-1} v^r, \qquad v^{(\theta)} = r(1 - 2m/r)^{-1/2} v^\theta$$

$$v^{(\varphi)} = r \sin\theta\, (1 - 2m/r)^{-1/2} v^\varphi \tag{6}$$

Similarly the components of the electromagnetic field tensor F_{ij} are given by

$$F_{(r\theta)} = \frac{1}{r}(1-2m/r)^{1/2} F_{r\theta}, \qquad F_{(rt)} = F_{rt}$$

$$F_{(r\varphi)} = \frac{1}{r\sin\theta}(1-2m/r)^{1/2} F_{r\varphi}, \qquad F_{(\theta\varphi)} = \frac{1}{r^2\sin\theta} F_{\theta\varphi}$$

$$F_{(\theta t)} = \frac{1}{r}\left(1-\frac{2m}{r}\right)^{-1/2} F_{\theta t},$$

$$F_{(\varphi t)} = \frac{1}{r\sin\theta}\left(1-\frac{2m}{r}\right)^{-1/2} F_{\varphi t}. \tag{7}$$

As we will be considering only axisymmetric fields we shall take $F_{(\varphi t)}$ and $F_{(r\theta)}$ to be zero and further have all the physical quantities independent of φ. With these the system of equations (2) and (3) becomes:

$$\left(\rho+\frac{p}{c^2}\right)\left\{\frac{Dv^{(r)}}{Dt}+\frac{mc^2}{r^2}\left(1-\frac{(v^{(r)})^2}{c^2}\right)-\left(1-\frac{2m}{r}\right)\left[\frac{(v^{(\theta)})^2+(v^{(\varphi)})^2}{r}\right]\right\}$$
$$= -\left(1-\frac{v^2}{c^2}\right)\left[\left(1-\frac{2m}{r}\right)\frac{\partial p}{\partial r}+\frac{v^{(r)}}{c^2}\frac{\partial p}{\partial t}\right]$$
$$+\frac{\varepsilon}{c^2}\left(1-\frac{2m}{r}\right)^{1/2}\left(1-\frac{v^2}{c^2}\right)^{1/2}[c^2F^{(r)}_{(t)}\ cF^{(r)}_{(\varphi)}v^{(\varphi)}-$$
$$-v^{(r)}(F^{(t)}_{(r)}\ v^{(r)}+F^{(t)}_{(\theta)}\ v^{(\theta)})] \tag{8}$$

$$\left(\rho+\frac{p}{c^2}\right)\left\{\frac{Dv^{(\theta)}}{Dt}+\left(1-\frac{3m}{r}\right)\frac{v^{(r)}v^{(\theta)}}{r}-\left(1-\frac{2m}{r}\right)^{1/2}\frac{(v^{(\varphi)})^2\cot\theta}{r}\right\}$$
$$= -\left(1-\frac{v^2}{c^2}\right)\left[\left(1-\frac{2m}{r}\right)^{1/2}\frac{1}{r}\frac{\partial p}{\partial\theta}+\frac{v^{(\theta)}}{c^2}\frac{\partial p}{\partial t}\right]$$
$$+\frac{\varepsilon}{c^2}\left(1-\frac{2m}{r}\right)^{1/2}\left(1-\frac{v^2}{c^2}\right)^{1/2}[c^2F^{(\theta)}_{(t)}+\ cF^{(\theta)}_{(\varphi)}V^{(\varphi)}$$
$$-v^{(\theta)}(F^{(t)}_{(r)}\ v^{(r)}+F^{(t)}_{(\theta)}\ v^{(\theta)})] \tag{9}$$

$$\left(\rho+\frac{p}{c^2}\right)\left\{\frac{Dv^{(\varphi)}}{Dt}+\left(1-\frac{3m}{r}\right)\frac{v^{(r)}v^{(\varphi)}}{r}+\left(1-\frac{2m}{r}\right)^{1/2}\frac{v^{(\theta)}v^{(\varphi)}}{r}\cot\theta\right\}$$
$$= -\left(1-\frac{v^2}{c^2}\right)\frac{v^{(\varphi)}}{c^2}\frac{\partial p}{\partial t}+\frac{\varepsilon}{c^2}\left(1-\frac{2m}{r}\right)^{1/2}\left(1-\frac{v^2}{c^2}\right)^{1/2}[c(F^{(\varphi)}_{(r)}v^{(r)}$$
$$+F^{(\varphi)}_{(\theta)}\ v^{(\theta)})-v^{(\varphi)}(F^{(t)}_{(r)}\ v^{(r)}+F^{(t)}_{(\theta)}v^{(\theta)})] \tag{10}$$

$$\frac{\partial}{\partial t}F_{(\varphi r)}=0, \qquad \frac{\partial}{\partial t}F_{(\varphi\theta)}=0$$

$$\frac{\partial}{\partial t}F_{(\theta t)}=4\pi\varepsilon\left(1-\frac{2m}{r}\right)^{1/2}\left(1-\frac{v^2}{c^2}\right)^{-1/2}v^{(\theta)}$$

$$\frac{\partial}{\partial t} F_{(rt)} = 4\pi\varepsilon\left(1 - \frac{2m}{r}\right)^{1/2}\left(1 - \frac{v^2}{c^2}\right)^{-1/2} v^{(\partial)}$$

$$\frac{\partial}{\partial r}\left[r\left(1 - \frac{2m}{r}\right)^{1/2} F_{(\theta t)}\right] + \frac{\partial}{\partial\theta} F_{(tr)} = 0$$

$$\frac{\partial}{\partial r}[r^2 \sin\theta\, F_{(\theta\varphi)}] + \frac{\partial}{\partial\theta}\left[r \sin\theta\left(1 - \frac{2m}{r}\right)^{-1/2} F_{(\varphi r)}\right] = 0$$

$$\frac{\partial}{\partial r}\left[r\left(1 - \frac{2m}{r}\right)^{1/2} F_{(\varphi r)}\right] + \frac{\partial}{\partial\theta} F_{(\varphi\theta)} = \frac{-4\pi\varepsilon r}{c}\left(1 - \frac{v^2}{c^2}\right)^{-1/2} v^{(\varphi)}$$

$$\left(1 - \frac{2m}{r}\right)^{1/2} \frac{\partial}{\partial r}[r^2 F_{(tr)}] + \frac{1}{\sin\theta}\frac{\partial}{\partial\theta}[r \sin\theta\, F_{(t\theta)}] = 4\pi\varepsilon r^2\left(1 - \frac{v^2}{c^2}\right)^{-1/2} \quad (11)$$

and the continuity equation

$$\left(\rho + \frac{p}{c^2}\right)\frac{D\varepsilon}{Dt} = \varepsilon\,\frac{D\rho}{Dt} \quad (12)$$

with $\dfrac{D}{Dt} = \dfrac{\partial}{\partial t} + \left(1 - \dfrac{2m}{r}\right)^{1/2}\left\{\left(1 - \dfrac{2m}{r}\right)^{1/2} v^{(r)}\dfrac{\partial}{\partial r} + \dfrac{v^{(\theta)}}{r}\dfrac{\partial}{\partial\theta} + \dfrac{v^{(\varphi)}}{r\sin\theta}\dfrac{\partial}{\partial\varphi}\right\}$

As the general equations are too complicated we make some specializations and then solve for equilibrium, then perturb the system with different kinds of perturbations and consider the stability of the equilibrium:

To begin with we consider a rigidly rotating disc so that we have $u^r = 0$, $u^\theta = 0$, $u^\phi = \Omega/c$, Ω being a constant. This gives

$$v_0^{(n)} = 0,\ v_0^{(\theta)} = 0, \text{ and } v_0^{(\phi)} = c\omega(1 + \omega^2)^{-1/2},\ \omega = \frac{r\Omega}{c}\sin\theta$$

The system in equilibrium under gravitational, centrifugal and electromagnetic forces is governed by the system of equations:

$$\left(\rho_0 + \frac{p_0}{c^2}\right)\left\{\frac{mc^2}{r^2} - \frac{\omega^2 c^2}{r(1 + \omega^2)}\left(1 - \frac{2m}{r}\right)\right\} = -(1 + \omega^2)^{-1}\left(1 - \frac{2m}{r}\right)\frac{\partial p_0}{\partial r}$$
$$+ \varepsilon_0\left(1 - \frac{2m}{r}\right)(1 + \omega^2)^{-1/2}\{F^{(r)}_{0(t)} + \omega(1 + \omega^2)^{-1/2} F^{(r)}_{0(\phi)}\} \quad (13)$$

$$\left(\rho_0 + \frac{p_0}{c^2}\right)\left\{-\left(1 - \frac{2m}{r}\right)^{1/2}\frac{1}{r}\left(\frac{\omega^2 c^2}{1 + \omega^2}\right)\cot\theta\right\}$$
$$= -(1 + \omega^2)^{-1}\left(1 - \frac{2m}{r}\right)^{1/2}\frac{1}{r}\frac{\partial p_0}{\partial\theta}$$
$$+ \varepsilon_0(1 + \omega)^{-1/2}\left(1 - \frac{2m}{r}\right)^{1/2}\left\{F^{(\theta)}_{0(t)} + F^{(\theta)}_{0(\varphi)}\frac{\omega}{(1 + \omega^2)^{1/2}}\right\} \quad (14)$$

$$\frac{\partial}{\partial r}[r^2 \sin\theta)\, F_{0(\theta\phi)}] + \frac{\partial}{\partial\theta}\left[r \sin\theta\left(1 - \frac{2m}{r}\right)^{-1/2} F_{0(\phi r)}\right] = 0 \quad (15)$$

$$\frac{\partial}{\partial r}\left[r\left(1-\frac{2m}{r}\right)^{1/2} F_{0(\phi r)}\right]+\frac{\partial}{\partial \theta} F_{0(\phi\theta)}=-4\pi\varepsilon_0 r\omega \tag{16}$$

$$\frac{\partial}{\partial r}\left[r\left(\frac{1-2m}{r}\right)^{1/2} F_{0(\theta t)}\right]+\frac{\partial}{\partial \theta} F_{0(tr)}=0 \tag{17}$$

$$\left(\frac{2}{r}+\frac{\partial}{\partial r}\right)(F_{0(tr)}+\frac{1}{r}\left(1-\frac{2m}{r}\right)^{-1/2}\left(\cot\theta+\frac{\partial}{\partial\theta}\right)(F_{0(t\theta)})$$
$$=4\pi\,\varepsilon_0\,(1+\omega^2)^{1/2}\left(1-\frac{2m}{r}\right)^{-1/2} \tag{18}$$

We first restrict to the case where we take the magnetic field configuration to be dipolar at infinity and hence the components would look like

$$F_{0(\phi r)}=\frac{-\mu\sin\theta}{r^3}\,g(r);\; F_{0(\phi\theta)}=\frac{2\mu\cos\theta}{r^3}\,f(r) \tag{19}$$

where $f(r)$ and $g(r)$ are arbitrary functions of r to be determined. Using these in eqns. (15) and (16) we get the equations for f and g as

$$\frac{d}{dr}\left(\frac{f}{r}\right)+\frac{g}{r^2}\left(\frac{1-2m}{r}\right)^{-1/2}=0$$

$$\frac{d}{dr}\left[\frac{g}{r^2}\left(1-\frac{2m}{r}\right)^{1/2}\right]+\frac{2f}{r^3}=\frac{4\pi\,\epsilon_0\Omega r^2}{\mu c} \tag{20}$$

At this stage we introduce the assumption that the charge density ε_0 is constant, and solve these equations to get

$$f=-\frac{3r^3}{8m^3}\left[\ln\left(1-\frac{2m}{r}\right)+\frac{2m}{r}\left(1+\frac{m}{r}\right)\right]$$
$$-\frac{4\pi\,\varepsilon_0\Omega}{5\mu c}\,r^5\left[\frac{1}{2}+\frac{2m}{r}+\frac{4m^2}{r^2}\ln\left(\frac{r}{2m}-1\right)\right]$$

$$g=\left\{\frac{3r^3}{4m^3}\left[\frac{m}{r}+\ln\left(1-\frac{2m}{r}\right)+\frac{m}{r}\left(1-\frac{2m}{r}\right)^{-1}\right]\right.$$
$$\left.+\frac{8\pi\,\varepsilon_0\Omega}{5\mu c}\,r^5\left[1+\frac{3m}{r}+\frac{4m^2}{r^2}\ln\left(\frac{r}{2m}-1\right)+\frac{m}{r}\left(\frac{r}{2m}-1\right)^{-1}\right]\right\}\left(1-\frac{2m}{r}\right)^{1/2} \tag{21}$$

As we need to solve the system simultaneously, we now solve for the electric field algebraically from the fluid equation and then substituting those solutions in the other two Maxwell's equations obtain constraint equations on p_0 and ρ_0. Thus we have

$$F_{0(t\theta)}\equiv F=\frac{1}{\varepsilon_0 r\,(1+\omega^2)^{1/2}}\left[\frac{\partial p_0}{\partial\theta}-c^2\omega^2\left(\rho_0+\frac{p_0}{c^2}\right)\cot\theta\right]$$
$$-\frac{2\mu\cos\theta}{r^3}\,\frac{\omega}{(1+\omega^2)^{1/2}}\,f \tag{22}$$

$$F_{0(tr)} \equiv G = \frac{1}{\varepsilon_0}\Big[(1+\omega^2)^{1/2}\Big(1-\frac{2m}{r}\Big)^{-1/2}\Big(\rho_0+\frac{p_0}{c^2}\Big)\frac{mc^2}{r^2}$$
$$+\Big(1-\frac{2m}{r}\Big)^{1/2}(1+\omega^2)^{-1/2}\Big\{\frac{\partial p_0}{\partial r}-\frac{\omega^2c^2}{r}\Big(\rho_0+\frac{p_0}{c^2}\Big)\Big\}\Big]$$
$$+\frac{\mu\sin\theta}{r^2}\;\frac{\omega g}{(1+\omega^2)^{1/2}} \tag{23}$$

with ρ_a and p_0 satisfying

$$\frac{2m}{\varepsilon_0 r^2}\Big(\rho_0+\frac{p_0}{c^2}\Big)\cot\theta+\frac{1}{\varepsilon_0 r}\Big\{\frac{1}{\omega^2}\frac{m}{r}+\frac{3m}{r}-1\Big\}\Big\{\frac{\partial\rho_0}{\partial\theta}+\frac{1}{c^2}\frac{\partial p_0}{\partial\theta}\frac{\omega^2}{1+\omega^2}\Big\}$$
$$+\frac{1}{\varepsilon_0}\Big(1-\frac{2m}{r}\Big)\Big\{\frac{\partial\rho_0}{\partial r}+\frac{1}{c^2}\frac{\partial p_0}{\partial r}\frac{\omega^2}{1+\omega^2}\Big\}\cot\theta+\frac{\mu\cos\theta}{c^2r^3}\Big\{\frac{2m}{r}\frac{f}{\omega}$$
$$-\Big(1-\frac{2m}{r}\Big)\frac{\omega}{1+\omega^2}\Big(3f-r\frac{df}{dr}\Big)\Big\}=0 \tag{24}$$

and
$$\frac{c^2\omega^2}{\varepsilon_0 r^2}\Big(\rho_0+\frac{p_0}{c^2}\Big)^2\Big[\Big(1-\frac{2m}{r}\Big)^{-1}\Big\{-3+\frac{12m}{r}-\frac{13m^2}{r^2}-\frac{1}{\omega^2}\frac{m^2}{r^2}$$
$$+\frac{\omega^2}{1+\omega^2}\Big(1-\frac{2m}{r}\Big)^2\Big\}+1-\cot^2\theta\,\frac{(2+\omega^2)}{(1+\omega^2)}\Big]$$
$$+\frac{c^2\omega^2}{\varepsilon_0 r}\frac{\partial\rho_0}{\partial r}\Big\{\frac{m}{r\omega^2}+\frac{3m}{r}-1\Big\}+\frac{1}{\varepsilon_0 r}\frac{\partial p_0}{\partial r}\Big[\frac{(2+\omega^2)\omega^2}{(1+\omega^2)}\Big\{\frac{m}{r\omega^2}+\frac{3m}{r}-1\Big\}$$
$$+2\Big(1-\frac{2m}{r}\Big)\Big]+\frac{1}{\varepsilon_0}\Big(1-\frac{2m}{r}\Big)^{3/2}\frac{\partial^2 p_0}{\partial r^2}+\frac{\cot\theta}{\varepsilon_0 r^2}\frac{\partial p_0}{\partial r}\Big(\frac{1-\omega^2-\omega^4}{1+\omega^2}\Big)$$
$$-\frac{c^2\omega^2\cot\theta}{\varepsilon_0 r^2}\frac{\partial\rho_0}{\partial\theta}+\frac{1}{\varepsilon_0 r^2}\frac{\partial^2 p_0}{\partial\theta^2}+\frac{\mu\omega}{r^3}\Big[\frac{2f\cos\theta}{r}\Big\{\tan\theta$$
$$-\frac{(2+\omega^2)}{(1+\omega^2)}\cot\theta\Big\}+\Big(1-\frac{2m}{r}\Big)^{1/2}\sin\theta.\Big\{\frac{dg}{dr}-\frac{\omega^2}{1+\omega^2}\Big(\frac{g}{r}\Big)\Big\}\Big]$$
$$=4\pi\varepsilon_0(1+\omega^2) \tag{25}$$

We now consider the case of a pressureless thin disc (thickness about 1 cm) centered along the equatorial plane $\theta=\pi/2$, and further the balancing forces acting on the plane $\theta=\pi/2$. These would reduce the equations to:

$$F=0,$$

$$G=\frac{\mu g}{r^3}\frac{\omega}{(1+\omega^2)^{1/2}}+\frac{\rho_0 c^2}{\varepsilon_0 r}\Big(1-\frac{2m}{r}\Big)^{-1/2}(1+\omega^2)^{-1/2}\Big[\frac{m}{r}-\omega^2\Big(1-\frac{3m}{r}\Big)\Big], \tag{26}$$

$$\Big(\frac{m}{\omega^2 r}+\frac{3m}{r}-1\Big)\frac{\partial\rho_0}{\partial\theta}=0, \tag{27}$$

and

$$\Big(\frac{m}{r\omega^2}+\frac{3m}{r}-1\Big)\frac{\partial\rho_0}{\partial r}+\frac{\rho_0}{r}\Big[1+\Big(1-\frac{2m}{r}\Big)^{-1}\Big\{-3+\frac{12m}{r}-\frac{13m^2}{r^2}$$

$$-\frac{m^2}{\omega^2 r^2}+\frac{\omega^2}{1+\omega^2}\left(1-\frac{2m}{r}\right)^2\bigg\}\bigg]+\frac{\mu\varepsilon_0}{\omega r^2c^2}\bigg[\frac{2f}{r}+\left(1-\frac{2m}{r}\right)^{1/2}\cdot\bigg(\frac{dg}{dr}$$

$$-\frac{\omega^2}{1+\omega^2}\frac{g}{r}\bigg)\bigg]=\frac{4\pi\varepsilon_0^2 r}{c^2\omega^2}(1+\omega^2). \tag{28}$$

f and g being given as in (21). Solving this set of differential equations for ρ_0 one can obtain the equilibrium solution for the disc. Eqn. (27) says that ρ_0 is independent of θ and thus the differential equation for ρ_0 after simplification is given by

$$\frac{d\rho_0}{dr}\bigg[1-\frac{2m}{r}+\frac{\Omega^2r^3}{mc^2}\left(-1+\frac{6m}{r}-\frac{8m^2}{r^2}\right)-\frac{\Omega^4r^5}{mc^4}\left(1-\frac{2m}{r}\right)\left(1-\frac{3m}{r}\right)\bigg]$$

$$+\frac{\rho_0 m}{r^2}\bigg[-1+\frac{\Omega^2r^4}{m^2c^2}\left(-2+\frac{10m}{r}-\frac{14m^2}{r^2}\right)-\frac{\Omega^4r^6}{m^2c^4}\left(1-\frac{3m}{r}\right)^2\bigg]$$

$$+\frac{3}{4}\frac{\mu\varepsilon_0 r\Omega}{m^3c^3}\bigg[4-\frac{14m}{r}+\frac{10m^2}{r^2}+\frac{2r^2\Omega^2}{c^2}\left(1-\frac{m}{r}\right)\left(1-\frac{3m}{r}\right)$$

$$+\frac{r}{m}\bigg\{\left(2-\frac{5m}{r}\right)\left(1-\frac{2m}{r}\right)+\frac{r^2\Omega^2}{c^2}\left(1-\frac{3m}{r}\right)\left(1-\frac{2m}{r}\right)\bigg\}\ln\left(1-\frac{2m}{r}\right)\bigg]$$

$$+\frac{8\pi\varepsilon_0^2r^4\Omega^2}{5mc^4}\bigg[\frac{9}{2}-\frac{8m}{r}-\frac{13m^2}{r^2}+\frac{20m^3}{r^3}+\frac{\Omega^2r^2}{c^2}\left(\frac{7}{2}-\frac{7m}{r}-\frac{7m^2}{r^2}+\frac{12m^3}{r^3}\right)$$

$$+\frac{4m^2}{r^2}\bigg\{\left(2-\frac{5m}{r}\right)\left(1-\frac{2m}{r}\right)+\frac{r^2\Omega^2}{c^2}\left(1-\frac{3m}{r}\right)\left(1-\frac{2m}{r}\right)\bigg\}\ln\bigg(\frac{r}{2m}$$

$$-1\bigg)\bigg]-\frac{4\pi\varepsilon_0^2r^2}{mc^2}\left(1-\frac{2m}{r}\right)\cdot\left(1+\frac{r^2\Omega^2}{c^2}\right)^2=0 \tag{29}$$

As may be seen though it is a first order linear equation it is formidable to solve this analytically. We therefore resort to two ways of solving this: (1) numerical integration of this exact equation and (2) analytical integration of the approximate equation. The approximation that we use are as follows: We retain terms up to m^2/r^2 and Ω^2r^2/c^2 and neglect all higher powers. This gives us the equation of the form:

$$\frac{d\rho_0}{dr}-\frac{m}{r^2}\rho_0\left\{1+\frac{2m}{r}+\frac{4m^2}{r^2}+\frac{2r^4\Omega^2}{m^2c^2}-\frac{5r^3\Omega^2}{mc^2}\right\}$$

$$+\frac{4\pi\varepsilon_0\Omega}{mc^3r}\left\{2+\frac{m}{r}-\frac{3m^2}{5r^2}+\frac{2r^3\Omega^2}{mc^2}-\frac{6r^2\Omega^2}{c^2}-\frac{23mr^2\Omega^2}{5rc^2}\right\}$$

$$+\frac{4\pi\varepsilon_0^2r^2}{5mc^2}\left\{-5-\frac{5r^3\Omega^2}{mc^2}+\frac{19r^2\Omega^2}{c^2}+\frac{2mr\Omega^2}{c^2}\right\}=0 \tag{30}$$

the solution of which is,

$$\rho_0=\frac{2\mu\varepsilon_0\Omega}{mc^3}\left\{1-\frac{m}{r}-\frac{m^2}{2r^2}+\frac{2r^3\Omega^2}{3mc^2}-\frac{19r^2\Omega^2}{6c^2}+\frac{13mr\Omega^2}{6c^2}\right\}\ln\frac{2m}{r}$$

$$+\frac{\mu\varepsilon_0\Omega}{rc^3}\left\{3-\frac{13m}{10r}-\frac{16m^2}{5r^2}-\frac{2r^4\Omega^2}{9m^2c^2}+\frac{49r^3\Omega^2}{18mc^2}-\frac{115r^2\Omega^2}{18c^2}-\frac{3mr\Omega^2}{2c^2}\right\}$$

$$+\frac{4\pi\varepsilon_0^2 r^3}{5mc^2}\left\{\frac{5}{3}+\frac{5m}{6r}+\frac{25m^2}{6r^2}+\frac{25r^3\Omega^2}{18mc^2}-\frac{437r^2\Omega^2}{90c^2}+\frac{mr\Omega^2}{45c^2}\right\}$$
$$+K\left\{1-\frac{m}{r}-\frac{m^3}{2r^2}+\frac{2r^3\Omega^2}{3mc^2}-\frac{19r^2\Omega^2}{6c^2}+\frac{13mr\Omega^2}{6c^2}\right\}. \tag{31}$$

K is evaluated by taking $\rho_0 = \rho_b$ at $r = r_b$. Before expansion the differential equation for ρ_0 had the coefficient $\left(\frac{m}{r\omega^2}+\frac{3m}{r}-1\right)$ for $d\rho_0/dr$ which was divided throughout. In order to expand we have used the condition

$$\left(\frac{r^2\Omega^2}{c^2}\right)\left(\frac{r}{m}-3\right) \ll 1$$

which fixes the outer edge in terms of Ω the angular velocity, m the mass and c the velocity of light. Fig. (7) gives a solution for which the various physical parameters are as follows:

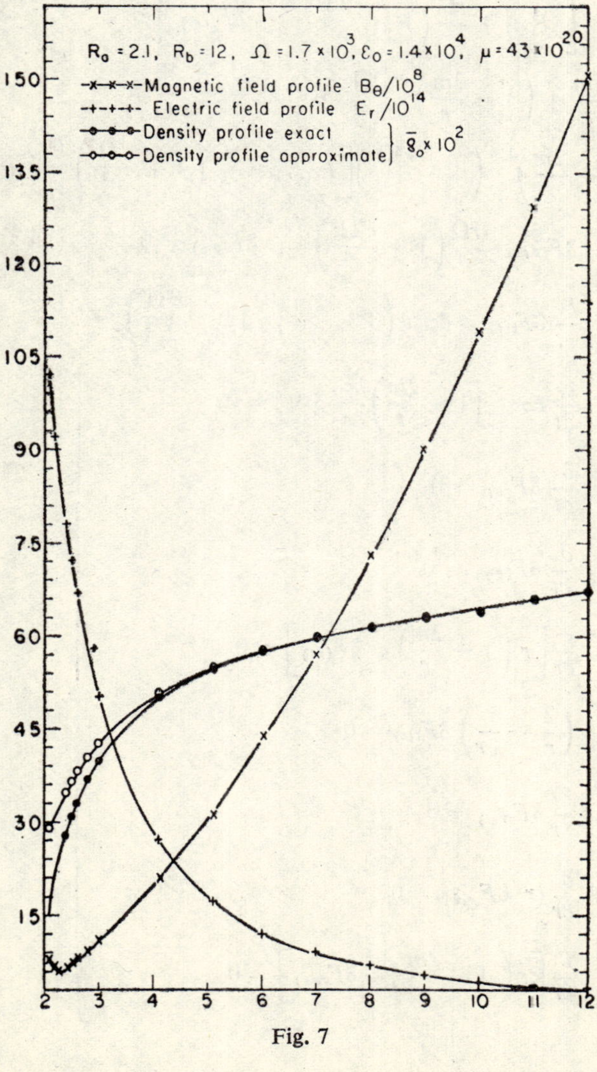

Fig. 7

$$M_{\rm disc}/M_{\odot.h.} \approx 10^{-15},\ r_b = 1.8 \times 10^6$$
$$r_a = 3.15 \times 10^5,\ \Omega = 1.7 \times 10^3,\ \rho_b = 1.01 \times 10^5$$
$$\varepsilon_0 = 1.4 \times 10^4,\ \mu = 4.3 \times 10^{21}$$

Having gotten a solution for equilibrium we now consider the stability of such a disc under radial perturbations: we restrict ourselves to the case wherein we have only $\delta v^{(r)}$, δp, $\delta \rho$, to be non-zero and function of r only. As we have assumed the charge density to be constant we take $\delta\varepsilon = 0$. With this the linearized equations governing the perturbations are:

$$\rho_0 \frac{\partial}{\partial t}\delta v^{(r)} + \frac{c^2}{r}\left(\delta\rho + \frac{\delta p}{c^2}\right)\left\{\frac{m}{r} - \frac{r^2\Omega^2}{c^2}\left(1 - \frac{r^2\Omega^2}{c^2}\right)\left(1 - \frac{2m}{r}\right)\right\}$$
$$= -\left(1 - \frac{2m}{r}\right)\left(1 - \frac{r^2\Omega^2}{c^2}\right)\frac{\partial}{\partial r}\delta p - \varepsilon_0\left(1 - \frac{2m}{r}\right)^{1/2}\left(1 - \frac{r^2\Omega^2}{2c^2}\right).$$
$$\left[\delta F_{(rt)} + \frac{r\Omega}{c}\left(1 + \frac{r^2\Omega^2}{c^2}\right)^{-1/2}\delta F_{(r\varphi)}\right]. \tag{32}$$

$$\rho_0\left\{2 - \frac{5m}{r} + \left(1 - \frac{3m}{r}\right)\frac{r^2\Omega^2}{c^2}\right\}\delta v^{(r)} = -\frac{r}{c^2}\frac{\partial}{\partial t}\delta p$$
$$-\frac{\varepsilon_0}{c\Omega}\left(1 - \frac{2m}{r}\right)^{1/2}\left(1 + \frac{r^2\Omega^2}{c^2}\right)\left[F_{0(\varphi r)} + \frac{r\Omega}{c}\left(1 + \frac{r^2\Omega^2}{c^2}\right)^{1/2} F_{0(tr)}\right]\delta v^{(r)} \tag{33}$$

$$\delta F_{(\theta t)} + \frac{r\Omega}{c}\left(1 + \frac{r^2\Omega^2}{c^2}\right)^{-1/2}\delta F_{(\theta\varphi)} = 0 \tag{34}$$

$$\frac{\partial}{\partial t}\delta F_{(rt)} = 4\pi\varepsilon_0\left(1 - \frac{2m}{r}\right)^{1/2}\left(1 \quad \frac{r^2\Omega^2}{2c^2}\right)\delta v^{(r)} \tag{35}$$

$$\left.\begin{aligned}
&\frac{\partial}{\partial t}\partial\rho + \left(1 - \frac{2m}{r}\right)\frac{\partial\rho_0}{\partial r}\delta v^{(r)} = 0\\
&\frac{\partial}{\partial t}\delta F_{(\theta t)} = 0\\
&\frac{\partial}{\partial t}\delta F_{(\theta\varphi)} = 0\\
&\frac{\partial}{\partial r}\left[r\left(1 - \frac{2m}{r}\right)^{1/2}\delta F_{(\theta t)}\right] = 0\\
&\left(\frac{2}{r} + \frac{\partial}{\partial r}\right)\delta F_{(tr)} = 0\\
&\frac{\partial}{\partial r}\delta F_{(r\varphi)} = 0\\
&\frac{\partial}{\partial r}\left(r^2\,\delta F_{(\varphi\theta)}\right) = 0\\
&\frac{\partial}{\partial r}\left[r\left(1 - \frac{2m}{r}\right)^{1/2}\delta F_{(\varphi r)}\right] = 0
\end{aligned}\right\} \tag{36}$$

Eliminating $\delta\rho$, δp, $\delta F_{(tr)}$ from (32) using (33), (35) and (36) we get the characteristic equation for perturbation.

$$\rho_0 \frac{\partial^2}{\partial t^2}\,\delta v^{(r)} = \Bigg[\frac{d\rho_0}{dr}\frac{c^2}{r}\left\{2-\frac{7m}{r}+\frac{5m^2}{r^2}-\frac{3r^2\Omega^2}{c^2}+\frac{12mr\Omega^2}{c^2}\right\}$$
$$+\frac{\rho_0 c^2}{r^2}\left\{-6+\frac{28m}{r}-\frac{32m^2}{r^2}+\frac{2r^2\Omega^2}{c^2}-\frac{10mr\Omega^2}{c^2}\right\}$$
$$+\frac{\varepsilon_0\mu c}{\Omega r^5}\left\{6-\frac{5m}{r}-\frac{22m^2}{5r^2}-\frac{4r^2\Omega^2}{c^2}+\frac{2mr\Omega^2}{c^2}\right\}$$
$$+\frac{4\pi\varepsilon_0^2}{5}\left\{-3+\frac{10m}{r}-\frac{40m^2}{r^2}+\frac{2r^2\Omega^2}{c^2}-\frac{6mr\Omega^2}{c^2}\right\}$$
$$+\frac{32\pi\varepsilon_0^2 m^2}{5r^2}\left\{-3+\frac{14m}{r}-\frac{16m^2}{r^2}+\frac{r^2\Omega^2}{c^2}-\frac{5mr\Omega^2}{c^2}\right\}\left(\ln\frac{2m}{r}\right)\Bigg]\,\delta v^{(r)} \tag{37}$$

After substituting for ρ_0 and $d\rho_0/dr$ and taking the perturbation $\delta v^{(r)} = a(r)e^{i\sigma x^0}$ one gets the equation for σ^2 which after integrating for the whole disc r_a to r_b gives the expression:

$$c^2\sigma^2\int_{R_a}^{R_b}\rho_0 dR = \int_{R_a}^{R_b}\Bigg[2\alpha\omega\Bigg[\left\{-\frac{6}{R}+\frac{18}{R^2}-\frac{62}{3R^3}-13\omega^3 R+\frac{232}{3}\omega^2\ln R\right\}\left(\ln\frac{2}{R}\right)$$
$$+\frac{6}{R}-\frac{9}{R^2}+\frac{62}{9R^3}-13\omega^2 R+\frac{116}{3}\omega^2\ln^2 R\Bigg]$$
$$+\frac{32\pi\varepsilon_0^2}{5}\Bigg[\left\{-\frac{3}{R}+\frac{7}{R^2}-\frac{16}{3R^3}-\omega^2 R+5\omega^2\ln R\right\}\left(\ln\frac{2}{R}\right)$$
$$+\frac{3}{R}-\frac{7}{2R^2}+\frac{16}{9R^3}-\omega^2 R+\frac{5\omega^2}{2}\ln^2 R\Bigg]$$
$$+\frac{\alpha}{\omega}\left\{\frac{3}{2R^4}-\frac{1}{R^5}-\frac{11}{15R^6}-\frac{4\omega^2}{R}-\frac{5\omega^2}{R^2}+\frac{32\omega^2}{R^3}\right\}$$
$$+\frac{4\pi\varepsilon_0^2}{5}\left\{-7R+\frac{55}{3}\ln R+\frac{5}{3R}-\frac{5}{3}\omega^2R^5+\frac{379}{30}\omega^2R^4-\frac{4343}{270}\omega^2R^3\right.$$
$$\left.+\frac{731}{9}\omega^2R^2\right\}+\bar{K}\left\{-\frac{6}{R}+\frac{18}{R^2}-\frac{62}{3R^3}-13\omega^2R+\frac{116}{3}\omega^2\ln R\right\}\Bigg] \tag{38}$$

wherein $\alpha = \mu\varepsilon_0/m^4, \quad \bar{k} = kc^2/m^2$

From the normal mode analysis it is known that the criterion for stability of the disc is $\sigma^2 > 0$. In the various cases that we have considered numerical estimation of the r.h.s of (38) has shown that $\sigma^2 > 0$ for all the cases and thus the disc is found to be stable under radial perturbations.

The discs considered in the above analysis were essentially very slowly rotating discs with $\omega = m\Omega/c < 0.1$. More general exact analysis of the discs using numerical integration of the exact equation has revealed that if the discs are far away then they are stable but if they are near and fast rotating they are unstable (Prasanna and Chakraborty, 1979). However, the entire analysis of the disc structure and stability has been done with the restriction that the balancing forces are confined to the equatorial plane; a restriction which is not bad considering the fact that the disc is very thin and as such the ratio of its transverse extension to radial extension is extremely small.

References

1. Prasanna, A.R. and Varma, R.K., *Pramana*, **8**, 229 (1977).
2. Prasanna, A.R. and Vishveshwara, C.V., (*Pramana*, **11**, 359 1978).
3. Prasanna, A.R. and Chakraborty, D.K., "General relativistic analysis of structure and stability of plasma discs around compact objects" (To be published) (1979).

Discussions

N Nagesha Rao: Do you have adiabatic invariants for the system you have considered? In orbit theory in plasma physics, the magnetic moment is an adiabatic invariant when the magnetic field is a slowly varying w.r.t. space variable. How far is this assumption valid in the case of a magnetic field near a black hole?

A R Prasanna: I am not yet sure about adiabatic invariants on curved background geometry. As we take the field components to be modified by the space-time curvature it is yet to be seen what adiabaticity would mean here. If you look at the modified field components you would notice that the contributions from the space-time curvature may be considered as though the dipole moment has been modified. Near a black hole as the curvature components change very fast a careful analysis of the concept of adiabaticity in the curved background is necessary.

S M Chitre: Do you include any viscosity while investigating the stability of accretion discs, as it is believed that the disc stability depends crucially on the assumption about viscosity?

A R Prasanna: No, not in the present analysis. It is true that viscosity has been found to contribute significantly in stabilizing the disc in the Newtonian analysis. As we are here trying to consider the general relativistic framework, it will be very hard to consider the fluid equations along with viscous forces. As the present analysis is just the beginning of a detailed formulation, we intend taking up viscosity terms in the subsequent analysis.

S M Chitre: Is the radiative reaction included in the calculation of the orbits of charged particles?

A R Prasanna: No, again the answer is similar. As the orbit analysis was made in Lagrangian formulation, and as there is no way of writing a Lagrangian including radiation reaction terms in a covariant way this analysis has not been possible. However, a very crude order of magnitude estimate of the Abraham-Lorentz radiative reaction terms seems to indicate that though this is cumulative it will start showing up only after a very very large number of revolutions by the particle.

Avinash Khare: To be more realistic should one not use Vlasov treatment?

A R Prasanna: Yes, one should in principle. But unfortunately there is no general relativistic treatment of kinetic theory available, which may be applicable to such systems as we are presently discussing. However, for the analysis of plasma discs the relativistic M.H.D. treatment is quite satisfactory.

Naresh Dadhich: In the Kerr case, you say there exist non-gyrating orbits, can we define the 'generalized ergosphere' through these orbits? Is the non-gyration limit same as the 'static limit'? How about the energy extraction processes with magnetic field?

A R Prasanna: The non-gyration of the particles is only when they are within the ergo surface or also called the static limit surface. As the nature of the orbits change very much within the ergosphere it would be interesting to see how this would affect the Penrose energy extraction process.

Quantized Fields, Gauge Theories
and
Foundations of Quantum Mechanics

On the Quantum Dynamics of Black Holes

N. D. HARI DASS

General relativity since its conception by Einstein nearly six decades ago has virtually exhausted all investigations into the classical aspects of gravitational phenomena. Under weak field circumstances a number of its predictions have been borne out experimentally, to higher and higher accuracies, instilling confidence in general relativity as perhaps the correct classical description of gravitation. The most remarkable prediction of general relativity is that of black holes. Although unambiguous evidence as to their existence is yet to come, there are many strong astrophysical candidates. Physics since 1915 has also witnessed the phenomenal and revolutionary developments in quantum theory culminating in quantum electrodynamics which enjoys an unprecedented degree of experimental support. Quantum theoretical concepts have proved indispensable to a proper understanding of subatomic phenomena. Despite several attempts[1] no complete quantum theory of gravitation has emerged yet. However, an area that has proved successful is the study of quantised fields in external gravitational fields.[2] The most remarkable result that has emerged from these studies is the Hawking radiation[3] of black holes. In this talk I would like to bring out the physical basis of this phenomenon as well as certain general features of quantised fields in external gravitational fields. I shall begin by utilising the formalism developed by Schwinger[4] to discuss an analogous problem for external electromagnetic fields wherein he demonstrated the instability of a static external electric field against spontaneous electron-positron pair production. I shall restrict myself to discussing the quantised scalar field only.

The scalar field ϕ in a background metric $g_{\mu\nu}(x)$ obeys

$$(g_{\mu\nu} \nabla^{\mu} \nabla^{\nu} - m^2)\, \phi = 0 \tag{1}$$

The relevant object of interest in discussing the quantum behaviour of ϕ is the propagation function $K(x, x')$ which obeys the differential equation

$$(g_{\mu\nu} \nabla^{\mu} \nabla^{\nu} - m^2)\, K(x, x') = (\sqrt{-g})\, \delta^4 (x - x') \tag{2}$$

which in an operator notation reads

$$(g_{\mu\nu} P^{\mu} P^{\nu} + m^2)\, K = 1 \tag{3}$$

where the x, x' matrix element of 1 is understood to be

$$(\sqrt{-g})\, \delta^4 (x - x')$$

The soution to (3) can be written symbolically as

$$K = \frac{1}{P^2 + m^2} \quad P^2 = g_{\mu\nu} P^{\mu} P^{\nu}$$

$$= i \int_0^{\infty} ds \; e^{-is(P^2+m^2)} \tag{4}$$

calling $P^2 + K^2 = \mathcal{H}$ we have

$$K = i \int_0^{\infty} ds \; e^{-is(m^2 - i\varepsilon)} \exp(-i\mathcal{H}s) \tag{5}$$

where the $- i\varepsilon$ in m^2 ensures convergence. We further define

$$\langle x' \,|\, \exp(-i\mathcal{H}s) \,|\, x \rangle = K(x, x', s) \tag{6}$$

whence
$$K(x, x') = i \int_0^{\infty} ds \; e^{-im^2 s} K(x, x', s) \tag{7}$$

This is essentially the representation for the propagator found by Hawking and Hartle.[5] The object of interest is obviously $K(x, x', s)$. While an exact evaluation is rather difficult and unwarranted as demonstrated by Hartle and Hawking, some important properties of $K(x, x', s)$ are worth mentioning. The boundary condition on $K(x,x', s)$ is

$$K(x, x', s) \underset{s \to 0}{\longrightarrow} (\sqrt{-g})\, \delta^4 (x - x') \tag{8}$$

(see eqn. 2.40 in Ref. 4 and 2.10 in Ref. 5.)

In flat space an explicit evaluation of $K(x, x', s)$ is available in Ref. 4 (eqns. 2.41 to 2.53): (mass less fields)

$$K(x, x', s) = \frac{-i}{(4\pi)^2} \frac{e^{i/4S\,(x - x')^2}}{s^2} \tag{9}$$

On recognising $(x - x')^2$ as the square of the Minkowski interval between x and x', we see that (9) is identical with eqn. (3.3) of Hartle and Hawking. The small and large s expansions of $K(x, x', s)$ are:[5]

$$K(x, x', s) \underset{s \sim 0}{\longrightarrow} \sum_c \frac{e^{iS_c(x, x')/4s}}{s^2} \left[D_c(x, x') + O(s) + \ldots \right] \tag{10}$$

$$K(x, x', s) \underset{s \to \infty}{\longrightarrow} s^{-2} [K_0(x, x') + s^{-1} K(x, x') + \ldots]$$

Here c stands for the class of geodesics that connect x to x'; functions D_c, K_0, K_1 are all independent of s.

Using (7), (8), (9) and (10) Hartle and Hawking derived the analyticity properties of $K(x, x')$. In order to do so they considered Kruskal coordinates and complexify the Horizon defined by either U or $V = 0$. The main result that emerged from their analysis

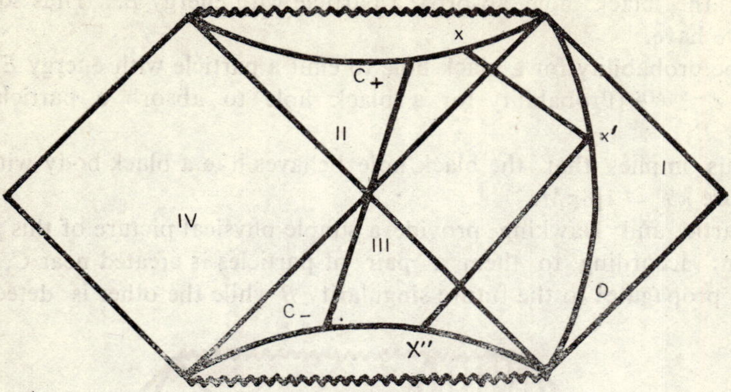

Fig. 1

is that the propagator connecting x on C_+ in region *II* to x' on 0 in region I is periodic in Imt with periodicity $8\pi M$, The periodicity is a consequence of the analyticity of $K(x, x')$ in Kruskal co-ordinates, except at the singularities dictated by the null geodesic structure and since t as a function of u, v is defined only modulo $8\pi iM$. Such a periodicity also occurs for $K(x, x')$ when both x and x' are in region I but this case is not of direct interest in determining the question of black hole radiance. The crucial information regarding this radiance is contained the amplitude defined by

$$a_E(\mathbf{R}', \mathbf{R}) = \int_{-\infty}^{+\infty} dt\; e^{-iEt} K(\mathrm{O}, \mathbf{R}'; t, \mathbf{R}) \tag{11}$$

$$= \int_{-\infty}^{+\infty} dt\; e^{-iEt} K(t, \mathbf{R}; \mathrm{O}, \mathbf{R}')$$

Since $K(x, x')$ is analytic in a strip of width $4\pi M$ (beyond this strip below the real axis there are singularities corresponding to the real null geodesic connecting a point of constant r in region III to x') it is legitimate to distort the contour of t-integration by $-4\pi Mi$.[5] Then

$$a_E(\mathbf{R}', \mathbf{R}) = e^{-4\pi EM} \int_{-\infty}^{+\infty} dt\; e^{-iEt} K(t - 4\pi iM, \mathbf{R}; 0, \mathbf{R}') \tag{12}$$

Since translating t by $-4\pi iM$ is equivalent to a reflection of the Kruskal co-ordinate about the origin, the integral

$$\int_{-\infty}^{+\infty} dt\; e^{-iEt} K(t - 4\pi Mi, \mathbf{R}; 0, \mathbf{R}')$$

may be interpreted as the amplitude to propagate to (0, $\mathbf{R}'$) from the surface C- in region III. This is just the time reversal of the process in which the black hole absorbs a particle with energy E. Thus squaring (12) we have,

{the probability for a black hole to emit a particle with energy E}

$= e^{-8\pi ME}$ {Probability for a black hole to absorb a particle with energy E} (13)

This implies that the black hole behaves like a black body with temperature $kT = 1/8\pi M$.

Hartle and Hawking provide a simple physical picture of this phenomenon: according to them a pair of particles is created near C, one of which propagates to the future singularity B while the other is detected by

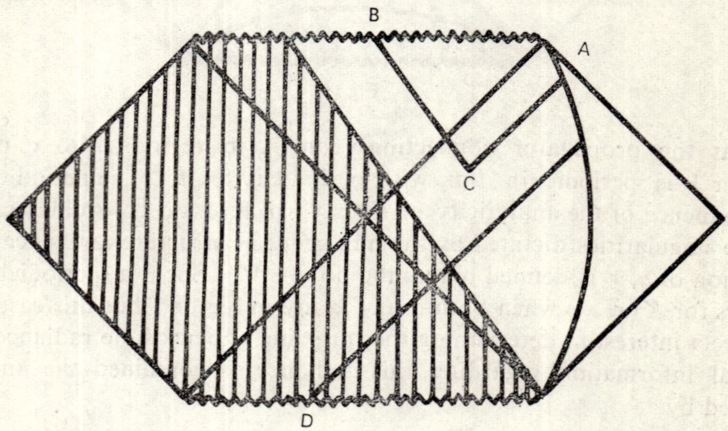

Fig. 2

the observer at A. Since t-component of the four momentum of the particle escaping to A must be positive, the t-component of the particle falling to B must be negative. In flat space this is not permitted but since in Schwarzschild geometry the role of t and r are interchanged within the horizon, these authors concludes that there is nothing wrong with a negative t-component.

There are several features of this argument that need a critical examination. Even though there seems to be nothing wrong with a negative t-component as long as it vanishes to the future singularity, it appears to me that nowhere outside the horizon should the trajectory of such a particle intersect any observer's trajectory. This seems rather hard to fulfil unless *a priori*. If such a mechanism can be justified we have introduced a new variant of the energy-momentum conservation laws in flat space and this should manifest in a number of interesting physical phenomena, like scattering near event horizons, photon pair creation, etc. It would indeed be a valuable exercise to work out general quantum phenomena around black holes. Such an exercise may even evolve means of detecting black holes hitherto unexplored. Due to limitations of time I shall not explore this topic any further.

Hartle and Hawking's analysis seems to indicate that quantities like

Fermion number, charge, etc., which are believed to be strictly conserved in flat space are conserved in a black hole environment only in a statistical sense. The precise implications of this rather drastic departure from flat space physics should be explored fully.

Another question that comes to my mind has to do with the choice of co-ordinates. All derivations of Hawking effect have relied on the use of Kruskal co-ordinates. Since the effect is so profound and fundamental it is important to explicitly verify its co-ordinate independence (this is analogous to establishing gauge invariance in quantum electrodynamics—a topic which vexed the pioneers of this field in its early phase of development). Related to the question of co-ordinate independence is the precise role played by the event horizon. Posing in a strightforward manner, can pair creation take place even if no event horizon has formed as long as the gravitational fields are sufficiently strong? An elucidation of this issue could have crucial importance in astrophysics.

DeWitt's[6] analysis of Hawking radiance provides some clarifications on this question. I shall repeat his arguments here. If t is the standard Schwarzschild time co-ordinate a vacuum state may be defined w.r.t. to the timelike killing vector $\partial/\partial t$ everywhere outside the event horizon. Even though at large distances this vacuum state is indistinguishable from Minkowski vacuum, the re-normalised t^{00} becomes negatively infinite in a local Lorentz frame near the event horizon. A particle detector (*a la* Unruh's analysis) at rest with respect to d/dt remains in its ground state. On the other hand, if one demands that a freely falling detector near the event horizon remain in its ground state, one has a vacuum state with a thermal character at infinity. The thermal nature follows from the fact that the Wightman function for the case of uniformly accelerated motion in Minkowski vacuum is identical to the mixed state Wightman function describing a bath of Rindler photons at 'temperature' $k\mathrm{T} = a/2\pi$. In the new vacuum state, therefore, a detector at rest with respect to d/dt would detect particles as if it were in a thermal bath of real photons at a temperature T given by the absolute acceleration 'a'$/2\pi$. Now, if the detector is at a position $r = 2M + \varepsilon$, in order to stay at rest w.r.t. to d/dt it must experience an acceleration $a = (2M/\varepsilon)^{1/2}/4M$ and the detector would experience a heat bath at $kT = (2M/\varepsilon)^{1/2}/8M$. The photons being real (as can be understood from the fact that t^{00} is raised to zero from $-\infty$ near the horizon) they escape to infinity where they experience the redshift factor $(\varepsilon/2M)^{1/2}$ resulting in a temperature at infinity equal to $kT = 1/8\pi M$. From this analysis it seems likely that Hawking radiance can exist even without the event horizon.

Even though Hawking's original derivation relied on linear quantised fields the results have been shown to be valid even for interacting fields.[6] The proof hinges on the fact that the green's functions if they are to be well-behaved on the horizon and at infinity must be perindic in *Imt* Since the boundory conditions on the event horizen play such an important role in Hawking phenomena it is instructive to consider the relationship of this to the well known Casimir effect. Balian and Duplantier[7] have shown that for a conductor immersed in a head bath the T^4 contribution to the internal energy is cancelled by a part of the Casimir free energy in the high temperature limit. I wonder if any such mechanism exists for the gravitational case which would inhibit the Hawking evaporation at high temperatures!

Coming back to the question of the behaviour at the event horizon I would like to bring to your attention an intriguing possibility raised by Sawyer.[8]

The well-known 'no hair' theorms of gravitational collapse are based on the requirement that fields be well-behaved at infinity and at the event horizon. Sawyer examines the consequences of relaxing the assumption of regular behaviour at the event horizon. The basic idea is that it is admissible to let the field develop singular behaviour as the event horizon is approached as the classical description would breakdown at such large field strengths and non-linear couplings have to be taken into account properly. However weak these non-linear couplings may be, they cannot be neglected when the fields approach singular values. If one considers a scalar field φ, without non-linear coupling the equation of motion is

$$\varphi^{\mu}{}_{;\mu} = 0 \tag{14}$$

whose static solutions behave as $\ln (r - r_0)$ near the event horizon. If one demands regularity both at $r = r_0$ and at infinity, one finds there are no such solutions to (14) this being the essential physical content of the 'no hair' theorems. It is conceivable that non-linear coupling would alter this singular behaviour and possibly remove it; for example, with a $\lambda\varphi^5$ interaction (14) is modified to

$$\varphi^{\mu}{}_{;\mu} = - 5\lambda\varphi^4 \tag{15}$$

whose static solution is well-behaved both at infinity and $r = r_0$ is

$$\varphi_c \sim (5\lambda)^{-1/3}\, r^{-1}\, r_0^{-1/3} \tag{16}$$

But a $\lambda\varphi^5$ is an unacceptable field theory in flat space-time. With a $\lambda\varphi^4$ interaction one has

$$\varphi^{\mu}{}_{;\mu} = - 4\lambda\varphi^3 \tag{17}$$

which admits the following static solution that is regular at infinity; viz.

$$\varphi(\gamma) \sim 4^{-1}\, (\lambda\gamma_0)^{-1/2}\, (r - r_0)^{-1/2} \tag{18}$$

By supposing that the single self-coupled field description breaks down when the field energy density becomes comparable to 10^{15} gms/cm^3, Sawyer demonstrates that such events are confined to a very small volume and do not perturb the metric in any significant way. He further finds that starting with an initial configuration, however small, but having a singular behaviour at $r \sim r_0$, the field grows to a limiting form given by (18). The possiblity is also raised that starting with a nonsingular field a singular or 'nearly singular' field develops in time. Perhaps the most interesting possibility considered is for the vacuum fluctuations to grow so much that there would be a spontaneous transition from $\langle\varphi\rangle = 0$ to $\langle\varphi\rangle = \varphi_c$. This idea is made more plausble by calculating the flux required to build the classical field φ_c.

If a fluctuating field which is singular near $r \sim r_0$ of the form

$$\begin{aligned} \varphi_{\text{fluct.}} &\sim \zeta (r - r_0)^{-1/2} && r_0 \gg r - r_0 > d \\ &\sim \zeta\, d^{-1/2} && r - r_0 < d \end{aligned} \tag{19}$$

is considered, the gradient contribution to the Hamiltonian is

$$E = 2\pi \int r^2\, dr \left(1 - \frac{r_0}{r}\right) \varphi'^2 \sim \zeta^2 r_0 d^{-1} \tag{20}$$

Uncertainty principle tells us that this fluctuation can last for a time $\Delta t \sim \hbar r_0^{-1} \zeta^2 d$. Choosing $\frac{\partial\varphi}{\partial t}(t = 0) = 0$ it is easy to estimate

$$\frac{\partial\varphi}{\partial t}(r, \Delta t) \sim \hbar d \zeta^{-1} (r - r_0)^{-1/2}$$

The outgoing radial energy flux is

$$\text{flux} = -\left(1 - \frac{r_0}{r}\right) \frac{d\varphi}{dr} \frac{d\varphi}{dt} \sim \frac{\hbar}{r_0^4} \tag{21}$$

Thus the order of magnitude of the flux is independent of ζ, d and is of the same order as the Hawking flux! Thus the possibility that quantum phenomena due to non-linear interactions among fields may thwart the Hawking effect for linear fields is not unlikely. Sawyer further argues that a classical field φ_c in conjugation with non-linear interactions seriously alters the effective potential for scalar field propagation from

$$\left(1 - \frac{r_0}{r}\right) \left[\frac{r_0}{r^3} + \frac{l(l+1)}{r^2}\right] \tag{22}$$

which vanishes at $r = r_0$ to (for

$$\left(1 - \frac{r_0}{r}\right) \left[\frac{r_0}{r^3} + \frac{l(l+1)}{r^2} + 12\lambda\varphi_c^2\right]$$

which at $r = r_0$ approaches $3/4r_0$ independent of λ. Since the barrier penetration governs the rate of emission, presence of φ_c will drastically reduce and possibly cancel the Hawking flux.

I have indicated these critical comments not in an effort to undermine the significance of the Hawking effect, but merely to emphasize that by virtue of the fundamental nature of Hawking eflect all aspects of it must be looked into in depth.

Extension of Hawking's result to other metrics has been carried out by a number of authors.[9] The thermal character of the radiance is a feature common to all these cases. In the case of Kerr black holes, irrespective of whether superradiance exists or not, spontaneous emission of quanta in classical superradiance modes has been established. Care should be taken in deciding whether these emissions are in addition to the Hawking flux or not. To appreciate this point let us consider the result obtained by Hartle and Hawking for a Kerr black hole.

Expected number of particles per mode is proportional to

$$\{\exp\,(E - m\omega_+)\,8\pi M \pm 1\}^{-1} \tag{23}$$

where E is the energy and m the projection of the angular momentum along the black hole rotation axis of the relevant mode. It is clear that in the absence of rotation significant emission takes place only as long as $E \lesssim (8\pi M)^{-1}$. But with rotation included there will be significant emission even when the black hole temperature is very small as long $E \lesssim |\,m\,|\,\omega_+$ which is the condition characterising superradiant modes.

A particularly fascinating consequence of eqn. (23) for neutrino-antineutrino emission has been demonstrated by Vilenkin.[10] According to (23) more particles will be emitted with their angular momentum along the black hole angular momentum $\mathbf{\Omega}$ than antiparallel to it. According to our present understanding only neutrinos with spin antiparallel to their momentum and only antineutrinos with spin parallel to their momentum participate in weak interactions. Thus it is at once clear that more antineutrinos than neutrinos are emitted in the direction parallel to the $\mathbf{\Omega}$ of the black hole and more neutrinos than antineutrinos in $-\mathbf{\Omega}$ direction; continuing this analysis to massive spinning particles, he concludes that, for example, more electrons than positrons are emitted along $\mathbf{\Omega}$ and more positrons than electrons along $-\mathbf{\Omega}$. I would like to make a few comments on this interesting work. First of all, the degree of parity violations in $e^- - e^+$ modes will be of order G_F in a curved space-time background while in the $\nu - \bar{\nu}$ modes parity violation will be of order unity in the same curved space-time background. This is because the $e^\pm$ are produced in the decay of $\mu^\pm$ emitted by the black hole. There is also a further suppression due to the temperature of the black hole. In any case the $e^\pm$ case should be looked into more carefully, otherwise there is an apparent non-uniformity between charged and uncharged leptons in that more *antineutrinos* come along while more *electrons* come out along $\vec{\mathbf{\Omega}}$. Vilenkin uses the fact that in μ^- (μ^+) decay more e^-(e^+) come out in the direction parallel (antiparallel) to the charged lepton spin. But this is true only in the rest frame of μ^- (μ^+) and appropriate Lorentz transformation effects should be taken into account before settling this question. Also, as mentioned before there are features of the decay in the Kerr geometry that need to be carefully accounted for. Coming to the $\nu - \bar{\nu}$ case, the parity violation here is of order unity because the violation is due to the *choice of states* (right-handed ν's) and not due to a parity violating Hamiltonian as in the $\mu^\pm$ decay. The particular choice of states is possible only for massless spinor fields. It should be stressed that *a priori* there is no reason to select right-handed ν's over left-handed ones, it is just that phenomenologically one finds that only this species participates in weak interactions at the energies we are familiar today. It is quite conceivable that at very high energies the wrong kind of neutrinos also start participating in weak interactions. In any case we know so little of the quantum gravitational interactions of elementary particles that we cannot rule out these neutrinos participating in gravitational interactions. It is, however, fascinating to note that if only one kind of neutrinos participate in gravity, we would have for the first time parity violations *independent of weak interactions.* The parity violations talked about by Vilenkin are CP-conserving and hence do not enable one to distinguish a matter black hole from an antimatter black hole. Direct parity violations in gravitational interactions, which I have analysed in great detail,[11] will also manifest themselves in Hawking phenomena and these effects will be able to distinguish matter and antimatter black holes but I shall not elaborate on them further.

Another interesting manifestation of Hawking phenomena is for the case of charged black holes; in quantum electrodynamics it is well known that a nucleus with a critical charge $Z_c > 137$ starts emitting positrons spontaneously. The reason for this is that for these critical charges the lowest bound state level is pushed below the top of the filled negative energy sea and thus it becomes energetically favourable for a positron to escape to infinity leaving behind a closely bound electron-nucleus system of charge $Z_c - 1$. We are investigating whether such direct positron production take place for charged black holes. This could have interesting astrophysical consequences as a new source for positrons. What we are interested in, apart from the rate of positron creation, is the value of the critical charge on black holes. For a Riessner-Nordstrom black hole of mass M, the maximum charge Z^*e allowed is such that $GM^2 = (Z^*e)^2$. But charged particle production of the type we have talked about would never allow the charge of a black hole to reach such a large value. Consider the Hartle and Hawking[5] result for a RN black hole:

Expected number of particles with charge of q per mode is proportional to:

$$\{\exp(E - qZer_+^{-1})\, 8\pi M \pm 1\}^{-1} \tag{24}$$

where $+ (-)$ refers to fermions (bosons) and r_+ satisfies

$$r_+^2 - 2Mr_+ - Z^2e^2 = 0 \tag{25}$$

i.e. $$r_+ = 2M \pm \sqrt{M^2 - Z^2e^2}$$

For $Ze \sim M$, $r_+ \sim 2M$ on the other hand for $Ze < M$, $r_+ \sim M$. For a positively charged black hole $Z > 0$ and it is clear that significant positron emission will take place even if $T \sim 0$ as long as $E \sim Z\alpha r_+^{-1}$. Thus the critical charge on the black hole for thermal radiation of positrons appears to be

$$\tilde{Z} \sim m_e r_+^{-1} \alpha^{-1} \tag{26}$$

where m_e is electron mass. For a black hole of solar mass one gets

$$\tilde{Z} \sim GM_\odot m_e \alpha^{-1} \sim 10^{17}$$

which is much less than

$$Z^{*2} = \frac{GM^2}{\alpha} \sim \frac{10^{-39} \cdot 10^{114}}{10^{-2}} \sim 10^{77} \quad \text{or} \quad Z^* \sim 3 \times 10^{38}\ !$$

The question we would like to answer is whether the q.e.d. process occurs ovər and above this or is simply a different manifestation of Hawking phenomena. If it is different, will it affect the estimate of $\tilde{Z}$? The q.e.d. Z_c can be estimated as follows: the binding energy $Z\alpha/r \sim m_e$ where r is the radius of the lowest orbit ($r \sim 1/Z\alpha me$) and hence $(Z\alpha)^2\ m_e \sim m_e$ or $Z_c \sim \alpha^{-1}$; now replacing r by r_+ if that is really what should be done, $\tilde{Z}\alpha r_+^{-1} \sim m_e$ which is the condition one obtains from Hawking phenomena also. Just as in the case of spontaneous emission in classical superradiant modes from Kerr black holes, we want to make sure there is no double counting. But the interesting situation remains that when the charge of a black hole exceeds $\tilde{Z}$, it rapidly loses it by electron (positron) emission. Thus even though complete evaporation by Hawking process may take very long for solar-mass black holes, quenching their charge may take a relatively short time. Making an electrostatic analogy one can ascribe an inductance and a capacitance to a charged black hole.

For fermion emission, both a Kerr black hole as well as a Reissner-Nordstrom blackhole provide at $T = 0$ a sort of Fermi level ($\varepsilon_F \sim |m|\,\Omega$, $Z\alpha r_+^{-1}$). In a practical sense, even for electrons (positrons) one requires very high Ω making superradiant emissions unlikely in realistic situations, The temperature of a black hole of solar mass is only 10^{-7} °K. Hence, it would be very difficult to observe the evaporation of such massive black holes. Also their lifetime against Hawking evaporation is much greater than the age of the universe. The search for the explosion of mini black holes of 10^{15} gms/cc has been negative. I find the speculations on the possible Baryon number non-conservation wherein two protons, for example, would tunnel inside their Schwarzschild radius and rapidly evaporate into Hawking radiation rather wild. Hawking was able to calculate the radiation only under the assumption that the black hole metric is purely classical. Once this constraint is relaxed, one simply does not know how to calculate anything.

Finally, I would like to end this review (by no means complete) with a few general remarks. Planck in his desire to keep thermodynamical laws intact in an attempt to satisfactorily explain the observed features of black body spectra of electromagnetic radiation was forced to overthrow classical physics and usher in the quantum world. Einstein's subsequent conception of 'quanta' has also completely revolutionised our thinking in the subatomic domain. Now once again in Hawkings work attempts to reconcile thermodynamic concepts with black hole phenomena have *inevitably* led to the inclusion of quantum behaviour in general relativity. Thus it is hard to think of a greater tribute to Einstein's genius and his impact on modern physics than Hawking phenomena which has provided a grand synthesis of special and general relativity, nature of quanta and quantum dynamics, statistical mechanics and thermodynamics, precisely the areas is which Einstein left indelible impressions for aeons to come.

References

1. Feynman, R. P., *Acta Physica Polon* **24**, 697 (1973); Schwinger J., *Particles, Sources and Fields,* Vol. 1, Addison Wesley (1970); Dewitt, B.S., *Dynamical Theory of Groups and Fields,* Blackie (1965).
2. Dewitt, B. S., *Phys. Rep.* **C 19** (1975).
3. Hawking, S. W., *Nature* **248**, 30 (1974): *Comm. Math Phys.* **43**, 199 (1975).
4. Schwinger, J., *Physical Review* **82**, 664 (1951).
5. Hartle, J. B. and Hawking, S.W. *Phys. Rev.* **D 13**, no. 8, 2188 (1976).
6. De Witt B. S., "Quantum gravity: The New Synthesis" to appear in a volume commemorating the hundredth anniversary of Einstein's birth, Cambridge University Press W. Israel and S. W. Hawking (Eds.)
7. Balian, R and Duplantier, B., *Ann. Phys* (N. Y.) **112** (1978).
8. Sawyer, R. F., *Phys. Rev.* **D 15**, no. 6, 1427 (1977).
9. Refs. (3) and (5)
   Unruh W. G., *Phys. Rev.* **D 10** 3194 (1974).
   Ford L. H., *Phys. Rev.* **D 12** 2963 (1975).
   Page D. N. *Phys. Rev.* **D 14** 3260 (1976).
   Iyer, B. R. and Kumar, A., *Pramana* **51**, no. 2, 171 (1978)
10. Vilenkin. A., *Phys. Rev. Lett.* **41**, no. 22, 1575 (1978).
11. Hari Dass, N. D., *Ann. Phys.* (N. Y.) **111** (1977).

Discussions

P Achuthan: We have some recent work on spontaneous pair creation (etc.) in intense inhomogeneous magnetic fields. You mentioned in the beginning of your talk about the relevance of pair creation. May I know your reaction about the possibility of using high inhomogeneous magnetic fields in the far-off regions you talked about?

N D Hari Dass: I know for fact that uniform magnetic fields cannot produce pairs. As far as pair production inhomogeneous magnetic fields is concerned, it obviously depends on the type and degree of inhomogeneity; therefore I don't want to venture any opinion on this matter except to add that exact results of the type known for uniform fields would be very hard to obtain for inhomogeneous fields.

E Gunzig: In considering the energy momentum tensor, do you take into account the improved Coleman-Jackiw form associated with the non-minimal coupling to gravitation?

N D Hari Dass: For the computation of Hawking flux I don't know if it makes any difference whether one chooses the new improved stress tensor but there are many other indications that that is the only proper choice for the stress tensor.

B R Iyer: When one considers accelerated observers, one should be careful in one's analysis. A 'vacuum' may contain particles of the other type but there might be no energy flux. For example, as shown by Fulling 'Radiation from Mirrors and Black Holes'.

N D Hari Dass: That is correct.

Arvind Kumar: About your comment on non-conservation of baryons, I want to point out the 'no superradiance theorem' for fermions. Mesons sent down a Kerr black hole can have their number amplified but not so with the electrons.

N D Hari Dass: My comment on the non-conservation of Baryon number referred to the speculation of complete black hole evaporation wherein all baryons constituting the original black hole would have evaporated into particle—antiparticle pair violating Baryon number conservation.

N Panchapakesan: In Hawking's theory, we consider an unquantised gravitational field and quantise a weak scalar field in this background. Under these conditions is it not reasonable to expect the fields to behave well on the horizon where the curvature tensor is finite?

N D Hari Dass: There is no holy reason to quantise only the linear scalar field. The point made by Sawyer is that all physical fields in nature wlll have non-linearities induced by their coupling to other fields and under such circumstances it may not be legitimate to require smooth behaviour of fields (in the single field approx.) near the event horizon. In other words, one may be neglecting effects of the same order or stronger than linear field quantum effeets in such an approximation.

A K Pandey: From your discussion I have made a picture that the Hawking radiation may be supposed to be due to our choice of defining vacuum state with respect to freely falling observer as the horizon which results in thermal radiation (of course redshifted) for the observer at infinity, If it is true that the accelerated frame of reference at horivon ('effectively' accelerated) is responsible for the observed emissions at infinity, I do not find any trouble in getting a thermal radiation out of any gravitating (sufficiently massive) object and there seems nothing special about black holes in this respect.

N D Hari Dass: I share your view (I have already discussed this question in the text). It is perhaps safe to say that this is an open issue.

M Sohnius: What is the physics content of the 'no-hair' theorem: Does it only claim the absence of higher multipole moments, or also the absence of other properties than $m, \psi, \overleftrightarrow{J}, \overleftrightarrow{m}$ even in an eventual covariant theory of all interactions (not only the em like in Maxwell-Einstein)?

N D Hari Dass: There is a point in your question, one cannot really claim to know the character of 'no hair' theorems till we have a complete theory of all fundamental fields in nature including gravitation as only then can we reasonably talk of the behaviour of physical fields near the event horizon. Thus the possibility cannot be ruled out that general quantum effects along with invalidating the area increase theorem of classical black hole physics may also invalidate the so-called no hair theorems. It is not at all clear what form these 'no hair' theorems may take in a unified treatment of all the basic forces of nature including gravitation.

Supergravity

MARTIN F. SOHNIUS

The goal of theoretical physics is generally to describe as many phenomnea as possible by a simple and 'natural' theory. In elementary particle physics, the hope is that we will eventually achieve a 'grand unification scheme' which combines all four known types of interactions into one consistent theory. We would like to make progress on the path which started with Maxwell's unification of magnetism and electricity, and which has led us to another more recent highlight—the unified gauge theories of weak and electromagnetic interactions. While colour gauge-theories may well prove fit to add the strong interactions to this list of unified (and understood) interactions, we are still far away from understanding gravitation on the elementary particle level.

The purpose of this paper is to review one recent step towards this goal which has considerably improved the chances for a consistent fusion of the ideas of general relativity and quantum theory: supergravity.[1]

First, I want to make a few remarks on the problems which arise in any attempt to do this fusion.

The work of Albert Einstein on gravitation has for almost six decades been decoupled from submicroscopic physics. The reason is that for the distances and masses involved, the deviations from the flat (Minkowskian) metric are so small that elementary particles, or atoms, can safely be treated as if gravitation did not exist. Any 'true', i.e. relativistically covariant theory should thus be closely approximated by the known, only Lorentz-invariant, non-gravitating theories. We must, however, demand of the 'true' theory that it be mathematically consistent, and that it predict the correct flat limit. Any quantum theory of gravitation so far fails to do so.

Let me argue briefly why gravitation ought to be quantized at all. Lacking a better working hypothesis, we believe that the laws of physics extrapolate into extreme regions: the theory of gravitation, has so far been tested and applied only in cases where atleast one of the gravitating masses is larger or equal to that of the earth; this same theory should describe the—however weak—gravitational attraction between, say, the proton and the electron in a hydrogen atom, We also believe that, in principle, gravitational waves can bear information and can serve to perform experiments. We believe that Heisenberg's uncertainty relation is true for all measurements, including those performed with the help of gravitational waves.

These beliefs are consistent only if gravitational waves are quantized like all other waves, and thus if gravity is an integral part of the quantum field theory which (we hope) will some day be found to describe the universe and the matter in it.

Once we assume quantized gravitation, the problem of renormalizability becomes very important. No matter how we attempt to quantize gravity, we end up with a field theory coupling constant of which (the gravitational constant) has the dimension of an area (for $\hbar = c = 1$). Such theories are in principle non-renormalizable, i.e. infinities occur which cannot be absorbed into a finite number of counter-terms, and the predictive power of the theory is nil.[2]

One way out of this dilemma could be "accidental" cancellations of the infinities, usually due to a larger symmetry of the theory. One example where something similar happens is the Weinberg-Salam theory of the weak and electromagnetic interactions, which has massive intermediate vector-bosons.[3] Their masses have in principle the same disastrous effect as a length in the coupling constant, but t'Hooft and Veltman[4] have shown that the additional, only spontaneously broken gauge invariance, the famous SU (2) $\times$ U (1), actually renders the theory renormalizable.

By the same authors it has been shown[5] that for 'pure gravity' (i.e. a world with no matter: $G_{\mu\nu} = 0$) no new infinities arise for physically measurable quantities in first-order perturbation theory. However, infinitely many Feynman graphs of order $\hbar$ and higher diverge: the theory is non-renormalizable. Furthermore, any possible coupling of matter particles with spins 0, 1/2, or 1 destroys this 'one-loop on-shell renormalizability'.

Three years ago, it was noticed that the addition of a spin-3/2 particle (later to be called the 'gravitino') in a supersymmetric way[6] not only maintains the one-loop renormalizability, but even makes the two-loop contributions finite on the mass-shell. Only on the three-loop level, and above, the situation stayed unchanged.[7]

The reason for the additional cancellations is that this theory has a higher invariance than just that under general co-ordinate transformations. The theory is locally supersymmetric, i.e. the graviton (spin 2) and the gravitino (spin 3/2) together form a higher symmetry group of local transformations. These transformations rotate particles of different spin into each other.[8]

In the following, I will describe briefly the simplest such supergravity theory[9], and then try to show the possible paths for improvements which may eventually lead to a finite, truly unified theory of all fundamental interactions.

Since we will have to treat Fermion fields in the context of general relativity, we must use the vierbein (or tetrad) formalism, i.e.. at every point we have a flat tangent space whose co-ordinate system e^a ($a = 0, 1, 2, 3$) is rotated against the 'natural' coordinate system dx^μ by the vierbein matrix $e^a_\mu(x)$:

$$e^a(x) = dx^\mu \, e^a_\mu(x) \tag{1}$$

The co-ordinate orientation should be of no physical relevance, so we demand invariance under local Lorentz rotations of the tangent space (gauge invariance of the second kind). With these transformations goes a connection field (or Yang-Mills potential) $\omega_{\mu ab}(x)$ and a covariant derivative

$$D_\mu e^a = e^a{}_{,\mu} + \omega_\mu{}^a{}_b \, e^b \tag{2}$$

which in turn gives rise to a curvature (Yang-Mills field strength $R_{\mu\nu ab}(x)$:

$$[D_\mu, D_\nu] \, e^a = R_{\mu\nu}{}^a{}_b \, e^b \tag{3}$$

The invariant quadratic form $ds^2 = e^a e^b \eta_{ab}$ ($\eta_{ab} = \text{diag}\,(1, -1, -1, -1)$) should be equal to the metric line element, i.e.;

$$e^a_\mu \, e^b_\nu \, \eta_{ab} = g_{\mu\nu} \tag{4}$$

and instead of the metric tensor, we can use the vierbein as the fundamental quantity describing the graviton. The Einstein Lagrangian is then

$$L_E = \frac{1}{8\varkappa^2} R_{\mu\nu}{}^{ab} \, e^c_\rho \, e^d_\sigma \, \varepsilon^{\mu\nu\rho\sigma} \, \varepsilon_{abcd}$$

or, in a more familiar form

$$L_E = \frac{1}{2\varkappa^2} \, eR \tag{6}$$

with

$$e \equiv \det e^a_\mu = \sqrt{-g} \tag{7}$$

$$R \equiv e_b^{-1\,\nu} \, e_a^{-1\,\mu} \, R_{\mu\nu}{}^{ab}$$

In the first-order formalism, which is used here, $R_{\mu\nu ab}$ is a function of the connection only:

$$R_{\mu\nu ab} = (\omega_{\nu ab,\,\mu} + \omega_{\mu ac} \, \omega_\nu{}^c{}_b) - (\mu \leftrightarrow \nu) \tag{8}$$

The Rarita-Schwinger field $\psi_\mu(x)$, which describes the gravitino, is a vector under general co-ordinate transformations and a Majorana spinor under local Lorentz transformations (Majorana spinor: the components of $\bar{\psi}$ are linear combinations of those of ψ). The Rarita-Schwinger Lagrangian is

$$L_{RS} = -\frac{i}{2} \varepsilon^{\mu\nu\rho\sigma} \, e_\mu{}^a \, \bar{\psi}_\nu \gamma_5 \gamma_a D_\rho \psi_\sigma \tag{9}$$

Here D_ρ is the covariant derivative for a spinor

$$D_\rho \psi = \psi_{,\rho} - \frac{i}{4} \omega_{\rho ab} \, \sigma^{ab} \psi; \; \sigma^{ab} = \frac{i}{2} [\gamma^a, \gamma^b] \tag{10}$$

so that $D_\rho \psi_\sigma - D_\sigma \psi_\rho$ is a covariant tensor.

L_E and L_{RS} are both scalar densities under general co-ordinate transformations and invariants under local Lorentz rotations. The total Lagrangian

$$L = L_E + L_{RS} \tag{11}$$

so far depends on three independent field quantities: the vierbein, the connection, and the Rarita-Schwinger field. The Euler equation of motion, however, for the connection is

$$D_\mu e_\nu^a - D_\nu e_\mu^a = \frac{i\varkappa^2}{2} \bar{\psi}_\mu \gamma^a \psi_\nu \tag{12}$$

which can be solved for the connection itself. $\omega_{\mu ab}$ thus becomes a well defined function of e_μ^a and ψ_μ and their derivatives:

$$\omega_{\mu ab} = \omega_{\mu ab}(e, \psi) \tag{13}$$

If we insert this expression for $\omega_{\mu ab}$ into the Lagrangian, we perform the step to second order formalism, and the Lagrangian will contain complicated fourth order terms in ψ_μ.

If we want to enforce

$$0 = \widetilde{D}_\mu e_\nu^a \equiv e^a_{\nu,\mu} - \Gamma^\rho_{\mu\nu} e^a_\rho + \omega_\mu{}^a{}_b e_\nu{}^b \tag{14}$$

for the fully covariant derivative $\widetilde{D}_\mu$, then

$$\begin{aligned} D_\mu e_\nu^a - D_\nu e_\mu^a &= (\Gamma^\rho_{\mu\nu} - \Gamma^\rho_{\nu\mu}) e^a_\rho \\ &= \frac{i\varkappa^2}{2} \bar{\psi}_\mu \gamma^a \psi_\nu \end{aligned} \tag{15}$$

and we have a space with torsion. The torsion is given by the spin density of the spin-3/2 field. Eqn. (14) is sufficient but not necessary to enforce $D_\mu g_{\rho\sigma} = 0$.

The remaining equations of motion are the Einstein equation

$$G^{a\mu} = \frac{i\varkappa^2}{2} \varepsilon^{\mu\nu\rho\sigma} \bar{\psi}_\nu \gamma_5 \gamma^a D_\rho \psi_\sigma \tag{16}$$

where $G_\mu{}^a$ is the (non-symmetric) Einstein tensor:

$$G_\mu{}^a = R^a_\mu - \tfrac{1}{2} e^a_\mu R;\ R^a_\mu \equiv e_b^{-1\nu} R_{\mu\nu}{}^{ab} \tag{17}$$

and the Rarita-Schwinger equation

$$R^\mu = 0 \tag{18}$$

with

$$R^\mu \equiv \varepsilon^{\mu\nu\rho\sigma} (\gamma_5 \gamma_a D_\rho \psi_\sigma e^a_\nu + \tfrac{1}{2} \gamma_5 \gamma_a \psi_\sigma D_\rho e^a_\nu) \tag{19}$$

The last contribution in eqn. (19) is a trilinear term in ψ_μ (due to (12)), which leads beyond the minimal covariantization of the Rarita-Schwinger field equation.

The Lagrangian described so far shows the above-mentioned extraordinary renormalizability properties. Why is that so, and what distinguishes our L from many other possible formulations of a gravitating spin-3/2 particle?

The action $\int d^4xL = \int d^4xL_E + \int d^4xL_{RS}$ (and only the sum, not the constituents) is found to be invariant under 'super-transformations', which in second-order formalism have the infinitesimal form

$$\delta e^a_\mu = i\varkappa\bar{\varepsilon}\gamma^a\psi_\mu$$

$$\delta\psi_\mu = \frac{2}{\varkappa}\, D_\mu\varepsilon \tag{20}$$

In first order formalism, we would have to add a suitable transformation law for the connection.

The parameter of the transformation, $\varepsilon(x)$, is an x-dependent Majorana spinor wito Fermi statistics.* Characteristically, the variation of the vierbein (spin 2) is a function of the Rarita-Schwinger field (spin 3/2), while the variation of ψ_μ contains a gradient term, $(\partial/\partial x^\mu)\,\varepsilon$ and terms with vierbein and ψ_μ, which are contained in the connection hidden in D_μ. The interlink of fields with different spins through transformations with parameters which carry spin themselves, is what defines a 'super'-transformation. The invariance under such a supersymmetry had been found before to generally improve renormalization properties.[10]

But, as we have heard above, the improvement is not yet good enough, and before we can guess what other supersymmetric theories of gravitation might exist, we will have to study the properties of this one in more detail.

From eqn. (20) we calculate the commutator of two super-transformations and we find

$$\begin{aligned}\delta_1\delta_2 - \delta_2\delta_1 = {} & \text{(a general co-ordinate transformation)} \\ & + \text{(a local Lorentz transformation)} \\ & + \text{(a supertransformation)} \\ & + \text{('off-shell' terms)}\end{aligned} \tag{21}$$

This looks rather messy. The messiness came about because we started with a minimal theory, and if, e.g., suitable 'auxiliary fields' are introduced, the contributions denoted as 'off-shell terms' do not appear. In any case, these terms vanish when the equations of motion are used.

The 'local-Lorentz transformation' is a physically irrelevant gauge transformation. It would not appear at all if we had introduced sufficiently many 'pure-gauge' fields.

*Note that we always assume Fermi statistics for the spinor fields, even though we are still working with classical fields.

The supertransformation' would vanish in the limit of a non-gravitating theory, since it is linear in the gravitational coupling constant $\varkappa$.

The 'general co-ordinate transformation', finally, would become just a translation if the parameters of δ_1 and δ_2 were not x-dependent.

Thus we have regained, in the constant-parameter, non-gravitating limit, the following supersymmetry algebra for the physically measurable quantities:

$$[\delta_1, \delta_2] = (\text{a translation}) \tag{22}$$

A more quantitative statement is that for Q (the Majorana spinor generator of the supersymmetry transformations) we get the algebra

$$\{Q, \bar{Q}\} = 2\gamma^a P_a \tag{23}$$

where P_a is the four-momentum (the generator of translations). Note that by dropping the anticommuting parameters which were still present in the symbol δ, the commutator for the variations, eqn. (22), has become an anticommutator for the generators, eqn. (23). It was the occurrence of anticommutators, leading beyond the usual Lie groups[11] for particle symmetries, that was the characteristic new feature of the supersymmetric theories which studied extensively from 1973 onwards.

The step from a supersymmetric Minkowskian theory to supergravity is quite natural; after the immense success of local gauge theories, one wanted to make the parameters of a supertransformation local, i.e. functions of x. With that, one automatically gets x-dependent translations on the right hand side of eqn. (22), and

$$x^\mu \rightarrow x^\mu + a^\mu(x) \tag{24}$$

is nothing else but a general co-ordinate transformation.

Let us inquire why we needed spin-3/2 to do it.

If eqn. (23) acts on the subspace of physical states of one massless paricle moving in the the z-direction with energy E_0, then the algebra (23) becomes the Clifford algebra[12]

$$\{a, a^+\} = 4E_0$$

$$\{a, a\} = \{b, b\} = \{b, b^+\} = 0 \tag{25}$$

where the four real components of a and b are suitable combinations of those of Q. By analogy with the Fermion creation operators, we see that with every boson corresponding to the ground state), there is one 'excited' state, a fermion. Its helicity is different by $\frac{1}{2}$ from the boson's (since Q carries spin $\frac{1}{2}$), so we have possible multiplets with spins

$$\begin{pmatrix} 0 \\ \frac{1}{2} \end{pmatrix}, \begin{pmatrix} \frac{1}{2} \\ 1 \end{pmatrix}, \begin{pmatrix} 1 \\ \frac{3}{2} \end{pmatrix}, \begin{pmatrix} \frac{3}{2} \\ 2 \end{pmatrix}, \begin{pmatrix} 2 \\ \frac{5}{2} \end{pmatrix}, \text{etc.}$$

This makes the gravitino natural as the graviton's companion. (If in doubt, we do prefer spin $\frac{3}{2}$ to spin $\frac{5}{2}$!)

So far, we have not come very far in the direction of our aim, the grand unification. We have coupled a massless spin-3/2 particle to gravity and improved renormalizability marginally. Any attempt, however, to add more matter into our world will end in the usual disaster; non-renormalizability on any loop-level, *except*, if we do it along a scheme known as extended supergravity.[13] Here we have several supersymmetries,[14] in general N, and an algebra

$$\{Q^i, \bar{Q}_j\} = 2\delta^i{}_j\gamma^a P_a;\; i, j = 1, \ldots, N \tag{26}$$

On the one- particle subspace, this reduces to the Clifford algebra for the generators of N different fermions. The multiplets have thus become larger, and the basic supergravity multiplets now have spin contents

$$\begin{pmatrix} & 2 & \\ \frac{3}{2} & & \frac{3}{2} \\ & 1 & \end{pmatrix} \text{ for } N=2\,; \quad \begin{pmatrix} & 2 & \\ \frac{3}{2} & \frac{3}{2} & \frac{3}{2} \\ 1 & 1 & 1 \\ & \frac{1}{2} & \end{pmatrix} \text{ for } N=3, \text{etc.}$$

In addition, they feature an $O(N)$ gauge symmetry and their renormalization properties are at least not worse than those of simple supergravity.[7] Possibly there could be even further improvement for some higher N, and we would then have a very attractive, finite quantum gravity with matter coupled to it, and even an 'internal' non-abelian gauge freedom (the $O(N)$). which could describe the other interactions.

One particular problem has not yet been surmounted: the helicity of the state is lowered by one-half for every step in the construction of the basic multiplets. Since we do not want higher spin than 2 (the graviton), we are thus limited to at most $N = 9$ (which gives a helicity range from $+2$ to -2), and a gauge group $G \subseteq O(8)$. The 'minimal' gauge group for the world, SU (3) × SU (2) × U (1) (color × (Weak × electromagnetism)), is not a subgroup of $O(8)$. Attempts are therefore being made[15] to enlarge the maximum possible group to U (8) by generalizing the algebra (24) to a structure which involves the full conformal group[11] instead of only the translations on the r.h.s. (conformal supergravity).

Another major field of research that I have not described is the search for a suitable tensor formalism.[16] The superspace-superfield technique[17] is the most promising path and indeed it is possible to give a geometrical meaning to the results described here, through general co-ordinate transformations in a space of four real and four Grassmann dimensions ('superspace'). The hope is that with the availability of better calculational techniques the cancellation of divergence can be more efficiently checked.

References

1. Some recent reviews on supergravity are:
Freedman, D.Z., Stony Brook preprint IIP–SB–78–54, Session Organizer's report at the 1978 Tokyo Conference.
Ferrare, S., CERN preprint TH–2514 (1978 Erice lectures);
Zumino, B., 'Supersymmetry and Supergravity' in Proc. Inter. Conf. on Particle Physics, Budapest, 1977 (Central Research Institute for Physics, Budapest, 1978):
Wess, J., "Supersymmetry-Supergravity", in Proc. 8th GIFT Seminar, Salamanca 1977 (Springer, 1978).
Freedman, D.Z., Nieuwenhuizen, P. van, *Sci. Am.* **238**, No. 2, 126 (1978).
2. Bogoliubov, N.N., Shirkov, N.N., *Introduction to the Theory of Quantized Fields*, Interscience, London (1959).
3. Weinberg, S., *Phys. Rev. Lett.* **19**, 1264 (1967);
Salam, A.. proc. 8th Nobel Symposium, Wiley, New York (1968).
4. t'Hooft, G., Veltman, M., *Nucl. Phys.* **B44**, 189 (1972);
Lee, B.W., Zinn-Justin, J. *Phys. Rev.* **D5**, 3137 (1972).
5. t'Hooft, G., Veltman, M., *Ann. H. Poincare* **20**, 69, (1974);
Deser, S., Nieuwenhuizen, P. van, *Phys. Rev.* **D10**, 401, 410 (1974);
Deser, S., Nieuwenhuizen, P. van, Tsao, H.-S., *Phys. Rev.* **D10**, 3337 (1974).
6. Freedman, D.Z., Nieuwenhuizen, P. van, Ferrara, S., *Phys. Rev.* **D13**, 3214 (1976);
Deser, S., Zumino, B., *Phys. Lett.* **B62**, 335 (1976);
Freedman, D., Nieuwenhuizen, P. van, *Phys. Rev.* **D14**, 912 (1976).
7. Grisaru, M.T., Nieuwenhuizen, P. van, Vermaseren, J., *Phys.Rev. Lett.* **37**, 1662 (1976);
Grisaru, M.T., *Phys. Lett.*, **B66**, 75 (1977).
Deser, S., Kay, J.H. Stelle, K.S., *Phys. Rev. Lett.* **38**, 527 (1977).
8. Golfand, Y.A., Likhteman, E.P. *JETP Lett.* **13**, 452 (1971);
Volkov, D.V., Akulov, V.P., *JETP Lett.* **16**, 438 (1972);
Wess, J., Zumino B., *Nucl. Phys.* **B70**, 39 (1974).
9. The presentation follows closely Zumino's Budapest lecture, ref. (1).
10. Wess, J., Zumino, B. *Phys. Lett.* **B49**, 52 (1974);
Lang, W. Wess, J. *Nucl. Phys.* **B81**, 249 (1974);
11. Hag, R., Loupszanski, J. Sohnius, M., *Nucl. Phys.* **B88**, 257 (1975).
12. Salam, A., Strathdee, J., *Nucl. Phys.*, **B80**, 499 (1974).
13. Ferrare, S., Nieuwenhuizen, P. van, *Phys. Rev. Lett.* **66B**, 1669 (1976).
14. Dondi, P., Sohnius, M., *Nucl. Phys.* **B81**, 317 (1974). Also ref. 11 & 12.
15. Kaku, M., Townsend, P., Nieuwenhuizen, P. van, *Phys. Rev.* **D17**, 3179 (1978);
Ferrara, S., Kaku, M., Townsend, P., Nieuwenhuizen, P. van, *Nucl. Phys.*. **B129**, 125 (1977).
16. Stelle, K., West, P., *Phys. Lett.* **B74**, 330 (1978);
Ferrara, S., Nieuwenhuizen, P. van, *Phys. Lett.*, **B74**, 333 (1978).
17. Salam, A., Strathdee, J., *Nucl. Phys.* **B76**, 477 (1974);
Ferrara, S., Wess, J., Zumino, B., *Phys. Lett.* **B51**, 239 (1974);
Wess, J., Zumino, B., *Phys. Lett.*, **B66**, 361, (1977).

Discussions

S Kanavi: What is the state of art in local supersymmetric gravity theories in the geometry of Arnowitt and Nath?

M Sohnius: I do not know.

Gauge Theories

G. RAJASEKARAN

1. Introduction

Until recently it has been customary for any account of subatomic physics to start with the observation that there are four fundamental forces in the universe: strong, electromagnetic, weak and gravitational.

Force	Strength	Range
Strong	1	10^{-13} cm
Electromagnetic	1/137	∞
Weak	10^{-5}	~ 0
Gravitational	10^{-39}	∞

This textbook classification is breaking down as a consequence of recent developments in high energy physics. These developments have led to a breakthrough in our understanding of one of the fundamental forces, namely, the weak force. It is this force which causes radioactive nuclei to disintegrate through beta-decay. There is every indication now that this mysterious weak force and the familiar electromagnetic force are facets of the same entity, which has been named the *electro-weak force*. This synthesis of electro and weak forces has been achieved through a non-abelian gauge theory.

The successful synthesis of electrodynamics and weak force prompts us to look for further unification with the other forces. There is good indication now that the strong force which is what binds the neutrons and protons to form nuclei is also a manifestation of a non-abelian gauge theory. Hence the chances of a grand synthesis of electro-weak dynamics with the strong nuclear force appear bright.

In the history of modern physics there have been many attempts to unify the various forces of nature. The most heroic were those of Einstein and Weyl who sought a unified theory of gravitation and electrodynamics. But these earlier attempts did not have the benefit of the new insight into the possibility of electro-weak synthesis which physicists have now obtained.

The new developments have opened the way for the possible ultimate unification of all the forces with gravitation—a goal which was very dear to Einstein and which still remains a major goal of physics. Perhaps supersymmetry and supergravity, which are under intensive study at present will provide the next step towards this goal.

However, the following cautionary remark may be in order. All our present activity in high energy physics is concerned with gaining an understanding of phenomena at the level of 10^{-13} to 10^{-15} cm. Gravitation will play a role in subatomic phenomena only when distances of the order of the Planck length, $(G\hbar/c^3)^{1/2} \sim 10^{-33}$ cm are approached. Thus, something like 20 orders of magnitude have to be crossed, before a real physical connection can be made between gravitation and other forces. We may encounter entirely new types of forces in this journey from 10^{-14} cm to 10^{-33} cm. It is even possible that our present concepts of space-time and quantum mechanics may undergo revolutionary changes before we reach down to the level of the gravitational length 10^{-33} cm. So, attempts at unification of the subatomic forces with gravitation may be premature at the present stage of our knowledge.

In any case this talk will be restricted to the gauge theory of electroweak and strong forces only. Since gauge theory now covers all of particle physics, our short review cannot be an exhaustive survey.[1] An attempt will be made to provide a perspective on the subject, albeit a coloured perspective.

2. Flavour and colour dynamics

All matter is made of leptons and hadrons. Leptons are point particles having no strong interactions. Neutrinos, electrons and muons are examples of leptons. Nucleons, pions, kaons and various hyperons are called hadrons. Hadrons have finite size with complex strong interactions among themselves. According to current philosophy, hadrons are not regarded as elementary particles, but made of certain more elementary objects, called quarks. Nucleon and other baryons are supposed to be composed of three quarks while the mesons are made of quark-antiquark pairs. So, we start with the quarks and leptons.

→Colour

↓ Flavour

$$\begin{pmatrix} u_1 & u_2 & u_3 \\ d_1 & d_2 & d_3 \\ s_1 & s_2 & s_3 \\ c_1 & c_2 & c_3 \\ \vdots & \vdots & \vdots \end{pmatrix} \qquad \begin{pmatrix} \nu_e \\ e \\ \nu_\mu \\ \mu \\ \vdots \end{pmatrix}$$

↓ Flavour

quarks leptons

The quark matrix is characterized by two types of quantum numbers, flavour which changes in the vertical direction and colour which changes in the horizontal direction. Flavour includes the well-known set of quantum numbers such as isospin, strangeness, etc., as well as newer quantum numbers charm, truth, beauty, etc. Colour is a new concept which first originated in the following way. The three-quark wavefunction of a baryon was found to be totally symmetric. So, to avoid contradiction with Fermi-Dirac statistics for the quarks, a new quantum number was introduced. It has now been called colour. The leptonic multiplet has flavour only.

So, we have quarks, flavour and colour. In high-energy physics, either arbitrary meaningless words are coined or words are used without their conventional meaning. Hopefully, the physics behind these words is not so meaningless.

Now construct a non-abelian gauge field theory in flavour space. That is called quantum flavour dynamics (QFD) and it describes the unified theory of weak and electromagnetic interactions. The non-abelian gauge theory in colour space is called quantum colour dynamics or quantum chromodynamics (QCD) and this is believed to be the theory of strong interactions.

The following table gives a compact description of both QFD and QCD.

<table>
<tr><th>QFD</th><th>QCD</th></tr>
<tr><td>Gauge theory in flavour space. Flavour-changing (vertical) transitions and corresponding emission or absorption of gauge bosons.
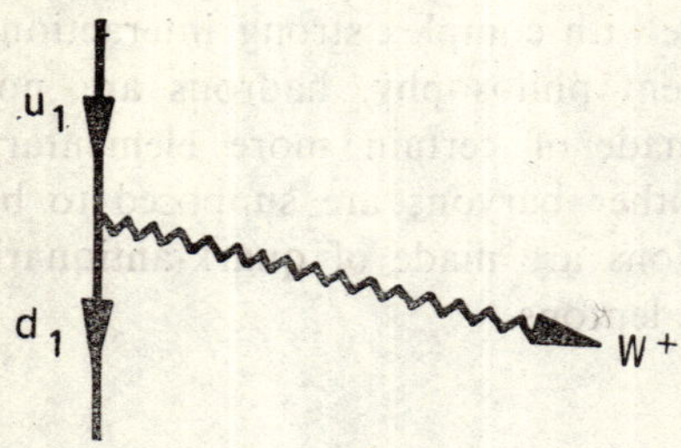
</td><td>Gauge theory in colour space. Colour-changing (horizontal) transitions with emission of colour gauge bosons (called gluons).
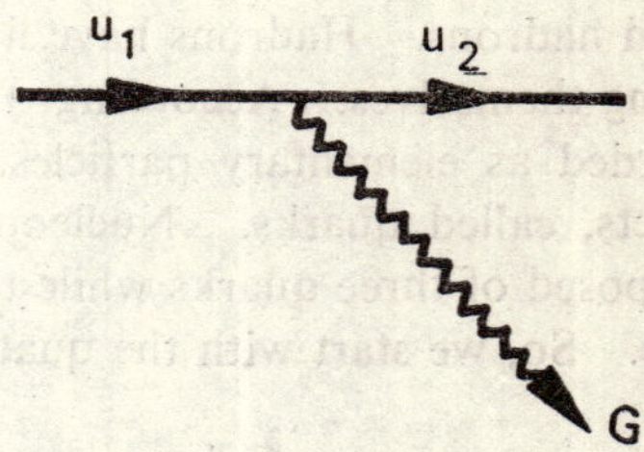
</td></tr>
<tr><td>Describes weak and electromagnetic interactions.</td><td>Strong interactions.</td></tr>
<tr><td>Based on a gauge group G_F acting on the flavour quantum numbers. What is G_F ? The most favoured candidate is the SU (2) × U (1) of Weinberg and Salam.</td><td>According to current dogma, the colour gauge group G_C is believed to be SU (3) acting on the colour quantum numbers.</td></tr>
</table>

G_F acts on both quarks and leptons.	G_C acts an quarks only and the quarks are taken to be triplets under colour SU(3).
The gauge symmetry G_F is supposed to be spontaneously broken down to the level of electromagnetic U (1) symmetry.	In contrast, colour SU(3) is regarded as an exact symmetry,
Except one gauge boson which is identified with the photon all other gauge bosons of G_F are massive. They are the various intermediate bosons of weak interaction: $W\pm$, Z, etc.	The octet of colour gauge bosons which are the gluons are all massless.

3. Quantum flavour dynamics

We shall now consider some more details about QFD. Let us take the Weinberg-Salam version for simplicity. It is based on the gauge group $SU(2)_L \times U(1)$. The left-handed weak isospin group $SU(2)_L$ is associated with the triplet of gauge fields W^i_μ ($i = 1, 2, 3$) while the weak hypercharge group U(1) is associated with an abelian gauge field B_μ. All these four vector bosons are massless, to start with.

Massive bosons are obtained through the spontaneous breakdown of symmetry. The mechanism used for this purpose is the celebrated Higgs-Kibble mechanism, namely, non-vanishing vacuum expectation values of certain chosen scalar fields ϕ:

$$\langle 0 | \phi | 0 \rangle \neq 0$$

As a result, we end up with three massive vector bosons W^+, W^- and Z and one massive boson which is the photon γ. The charged bosons $W\pm$ mediate the well-known charged-current weak interaction such as the beta decay of the neutron. the neutral boson Z mediates the neutral-current weak interaction discovered in 1973 and γ mediates the electromagnetic interaction. All these are illustrated in the following diagrams:

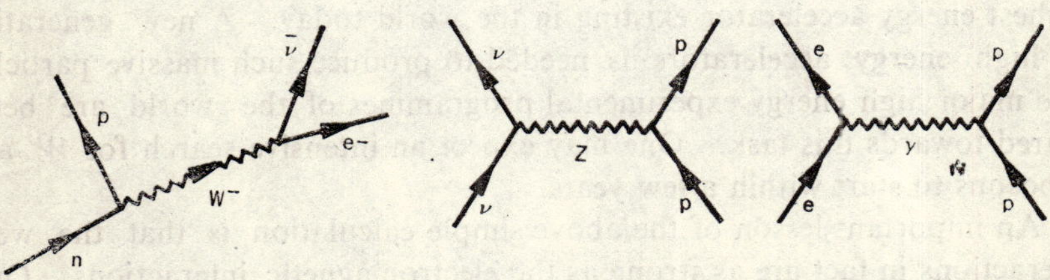

This is the unified theory of electro-weak interactions. Let us now look at some simple consequences of this unified picture.

Masses of the weak bosons W and Z

In the pre-gauge theoretical description of β-decay due originally to Fermi, the process was pictured as the four-fermion interaction and the coupling constant for this interaction was

$$G_W \simeq \frac{10^{-5}}{m_N^2}$$

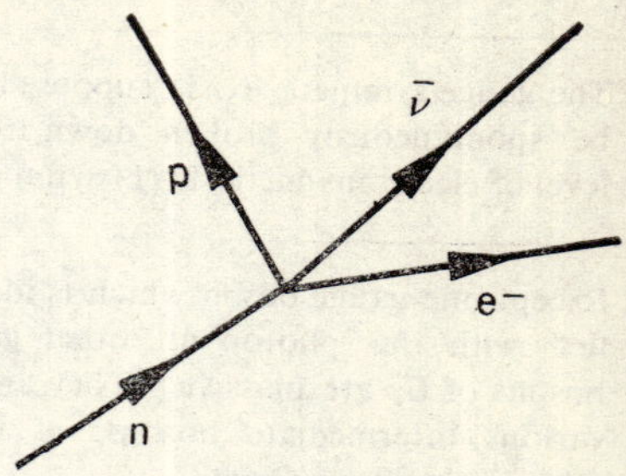

where m_N is the nucleon mass. In gauge theory, this G_W is replaced by g^2/m_W^2 where g is the gauge coupling constant and m_W is the mass of the W boson. This arises from a factor g each for emission and absorption of the W boson and a factor $1/m_W^2$ for the propagation of the W boson (for small momentum transfers).

So we have

$$G_W \simeq \frac{g^2}{m_W^2}$$

Unification of the weak and electromagnetic interactions leads to a connection between the weak coupling constant g and the electromagnetic coupling constant:

$$g \simeq e$$

Combining these relationships, we get

$$m_W \simeq \frac{e}{\sqrt{G_W}} \simeq 100 \text{ GeV}$$

where we have used $G_W = 10^{-5}/m_N^2$ and $e^2/4\pi = 1/137$.

This is only an order of magnitude calculation, but it is confirmed by a more precise analysis. One finds

$$m_W \text{ and } m_Z \simeq 80 - 100 \text{ GeV}$$

These masses are so high that W and Z cannot be produced in the highest energy accelerator existing in the world today. A new generation of high energy accelerators is needed to produce such massive particles. The major high energy experimental programmes of the world are being geared towards this task. One may expect an intensive search for W and Z bosons to start within a few years.

An important lesson of the above simple calculation is that the weak interactions in fact are as strong as the electromagnetic interactions. Only because the W boson is so massive that the weak interaction appears weak at low energies. At high energies, that is, at centre-of mass energies of the order of 100 GeV, weak interaction regains its full strength.

Neutral current

The existence of neutral-current weak interaction—reactions such as (1) with a strength comparable to that of the usual charged current weak interaction as in (2) is a natural consequence of unification with electrodynamics. In fact, because of the dynamical mixing between weak and electromagnetic interactions inherent QFD, neutral current acts as something

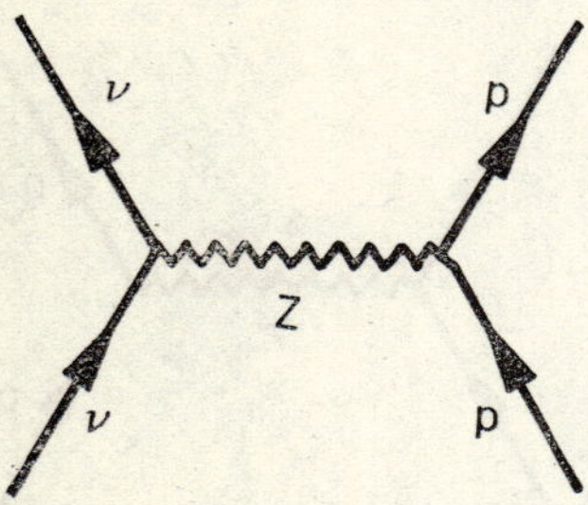

(1)

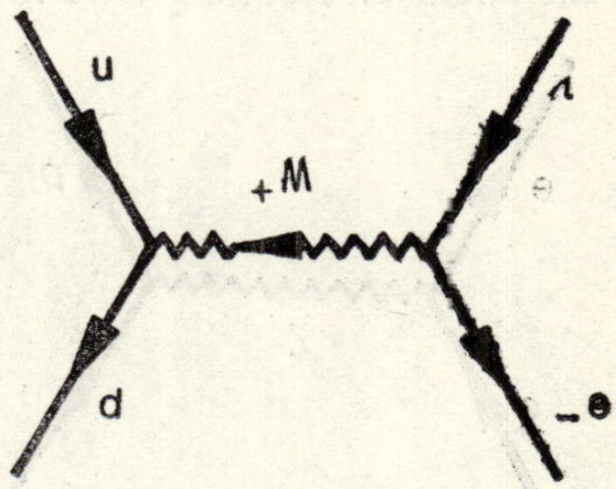

(2)

like a bridge between conventional weak and electromagnetic phenomena. Hence the discovery of the neutral current weak interaction in the neutrino reactions in 1973 and the subsequent detailed study which showed its properties to be of the type expected in QFD have helped to confirm that physicists are on the right track.

In fact, there are four parameters α, β, γ and δ characterizing the neutral current interactions in general among the neutrinos and hadrons. Their values as predicted by Weinberg-Salam (W-S) version of QFD are compared to the experimentally determined values below:

Interaction parameter	W-S ($\sin^2 \theta = 3.22$)	Experimental
α	0.56	0.58 ± 0.14
β	1	0.92 ± 0.14
γ	-0.15	-0.28 ± 0.14
δ	0	0.06 ± 0.14

Weinberg-Salam model has one free parameter $\sin^2 \theta$ which has been fixed at 0.22 for the above comparison. One can see the remarkable agreement between theory and experiment.

Since there are a number of astrophysicists in the audience, it may be relevant to point out that the neutral-current interaction of the neutrinos may be of astrophysical significance. This interaction leads to coherent scattering of neutrinos on nuclei and hence to neutrino pressure. (Without neutral currents, coherent scattering of neutrinos is not possible). Possible importance of this neutrino pressure on supernova explosion has been considered in recent literature.

Weak-electromagnetic interference

The agreement between theory and experiment noted above does not establish G_F to be $SU(2) \times U(1)$, for it concerns only one sector of the neutral current, namely the neutrino-hadron sector. Let us now consider the electron-hadron sector, where the neutral current interaction corresponding to the diagram (1) should exist. This process should exist along with the usual electromagnetic interaction via the exchange of a photon (2).

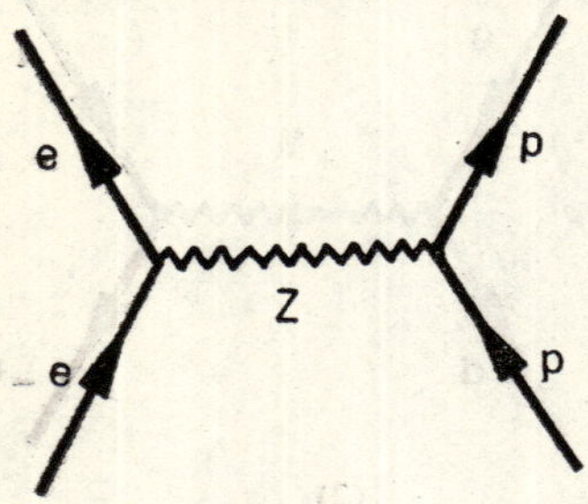

(1)

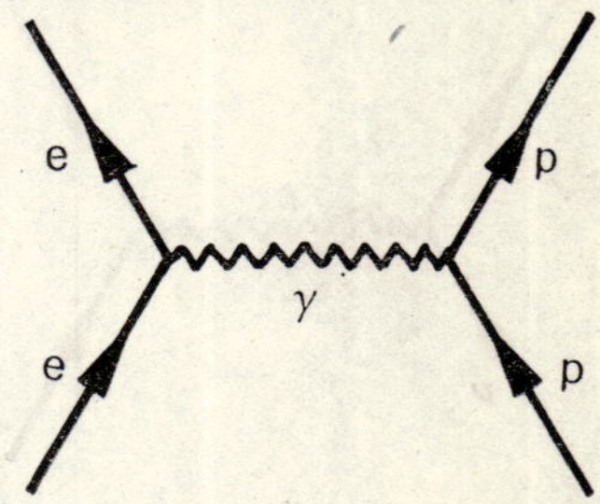

(2)

Since both the above interaction exist simultaneously, how can one disentangle the weak neutral-current effect from the dominating electromagnetic effect?

Whereas the electromagnetic interaction is parity-conserving, the weak effect is expected to be parity-violating. In fact it is definitely parity-violating in the Weinberg-Salam model. So, look for parity-violation in e-p interaction.

Such a parity-violation has been looked for, in atomic physics, however initial results were not so definitive. But, spectacular verification of this parity-violation in electron-nucleon interaction came from the high-energy electron-deuteron scattering experiment performed at the Stanford Linear Accelerator Centre (SLAC) USA last year.

It is easy to see the advantage of high energy experiment in this context, if we estimate the expected amount of parity violation. The effect arises due to the interference between the weak and electromagnetic amplitudes given by the diagrams:

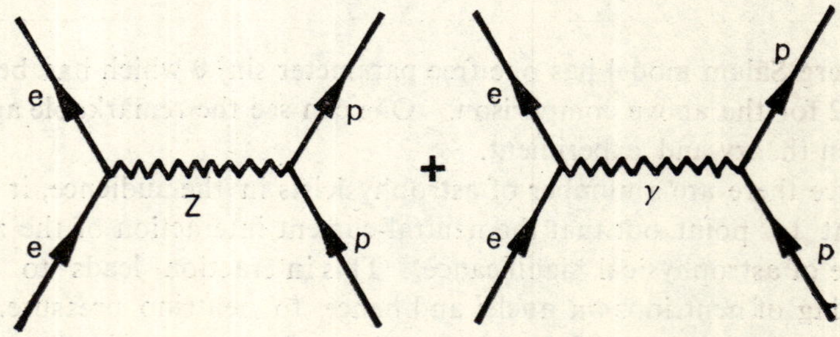

So, we have

$$A_{\text{pv}} \sim \frac{(WK)\,(EM)}{|WK|^2 + |EM|^2} \sim \frac{WK}{EM} \sim \frac{G_W}{e^2/q^2} \sim 10^{-4}\,\frac{q^2}{(\text{GeV})^2}$$

where we have again used $G_W = 10^{-5}/m_N^2$ and $e^2/4\pi = 1/137$. Thus, the parity-violating coefficient A_{pv} is seen to increase with the momentum transfer q^2 and hence the advantage of going for a high-energy experiment.

The principle of the SLAC experiment is simple. A beam of longitudinally polarized electrons of 20 GeV hit a liquid deuterium target and the inelastically scattered electrons were detected. If σ_+ and σ_- denote the cross-sections for the process with the two directions of polarization, then the parity-violating coefficient A_{pv} is simply given by

$$A_{pv} = \frac{\sigma_+ - \sigma_-}{\sigma_+ + \sigma_-}$$

The experimental result was

$$A_{pv} = -(0.95 \pm 0.16) \times 10^{-4} \frac{q^2}{(\text{GeV})^2}$$

which is of the same order of magnitude estimated above. Further it agrees very well with the prediction of the Weinberg-Salam model (with $\sin^2\theta = 0.22$): $-0.87 \times 10^{-4}\ q^2/(\text{GeV})^2$. Hence the jubilation over gauge theory!

Nevertheless, it is good to point out that there are 17 measurable parameters in all the various neutral-current sectors and only 5 have so far been measured with sufficient accuracy. In any case, until $W^\pm$ and Z are detected experimentally, gauge theory cannot be regarded as established.

Renormalizability and unification

QFD is renormalizable. The incorporation of weak interaction into a gauge-theoretic framework elevates the theory of weak interactions to the rank of a renormalizable theory—a position which it now shares with QED.

Here a remark concerning the unification of weak and electromagnetic interactions is in order. Apart from any aesthetic or other considerations, there is a logical reason for linking weak and electromagnetic interactions and that is just renormalizability. It is possible to construct a renormalizable gauge theory of weak interactions alone. But since such a theory would necessarily contain charged vector bosons (mediating the neutron beta decay for instance), one has to have a meaningful theory for these electrically charged vector bosons. It has been known for a long time that the theory of the charged vector boson interacting with the electromagnetic field does not make much sense. Not only is the theory unrenormalizable, but is, in general, not even relativistically invariant.[2] All these problems are neatly solved, once the charged vector bosons and photon are combined into a gauge multiplet, as is done in QFD. It is worthwhile to note the close similarity between this and what happens to the troubles of higher-spin fields when these are incorporated in supersymmetric theories. There is surely a moral in all this: Nature does not deal with arbitrary interactions.

Thus a meaningful theory for charged vector bosons is obtained only by dynamically combining their weak and electromagnetic interactions. This is a very satisfying thing about gauge theory.

Back to electrodynamics

To put the whole thing into proper perspective, let us now go back to electrodynamics. The basic laws of electrodynamics were first formulated by Maxwell more than 100 years ago. (The year 1979 does not only mark the centenary of the birth of Einstein, but also the centenary of the death of Maxwell. In his work Maxwell was greatly influenced by the intuitive physical pictures which Faraday had already built up for the understanding of electromagnetic phenomena.

Earlier discoveries of Oersted, Ampere and Faraday had unified electricity and magnetism into a single science of electromagnetism. Maxwell then unified optics with electrodynamics.

$$\nabla \cdot \mathbf{E} = 4\pi\rho$$

$$\nabla \times \mathbf{E} + \frac{1}{c}\frac{\partial \mathbf{B}}{\partial t} = 0$$

$$\nabla \cdot \mathbf{B} = 0$$

$$\nabla \times \mathbf{B} - \frac{1}{c}\frac{\partial \mathbf{E}}{\partial t} = \frac{4\pi}{c}\mathbf{J}$$

Laws of Electrodynamics

All this is old stuff. Maxwell's equations have stood the test of time for these 100 years. Will they remain in isolated glory for ever?

The answer has to be in the negative. Progress in physics generally consists in the discovery of more general laws of wider applicability which reduse the older laws to some restricted domain.

This is precisely what we are witnessing now. The successful development of gauge theory shows that there exist more general laws which subsume Maxwell's laws of electrodynamics. Thre exist more general fields $\mathbf{E}_i$ and $\mathbf{B}_i$ (the index i running over the set of gauge bosons) and these fields are generalizations of the usual electric and magnetic fields and describe not only electrodynamics but weak interactions also. Actually, $\mathbf{E}_i$ and $\mathbf{B}_i$ (for $i = 1, 2, 3, 4$) are related to the gauge boson fields A, W^+, W^- and Z in an analogous way as the usual electric and magnetic fields are related to the electromagnetic potentials These electro-weak fields $\mathbf{E}_i$ and $\mathbf{B}_i$ satisfy laws of motion which are similar to Maxwell's laws of electrodynamics:

$$\nabla \cdot \mathbf{E}_i + \ldots = 4\pi\rho_i$$

$$\nabla \times \mathbf{E}_i + \frac{1}{c}\frac{\partial \mathbf{B}_i}{\partial t} + \ldots = 0$$

$$\nabla \cdot \mathbf{B}_i + \ldots = 0$$

$$\nabla \times \mathbf{B}_i - \frac{1}{c}\frac{\partial \mathbf{E}_i}{\partial t} + \ldots = \frac{4\pi}{c}\mathbf{J}_i$$

Laws of Electro-weak Dynamics

There are some crucial differences between the laws of electrodynamics and those of electro-weak dynamics. The self-interaction between the gauge fields indicated by the diagrams which is a characteristic feature of non-abelian gauge theory is absent in the abelian electromagnetic theory. These self-interactions lead to non-linearities which are denoted by dots in the above equations. Further, the spontaneous breakdown of symmetry leading to the asymmetric world of massive W and Z but massless γ affects the physical interpretation of these equations.

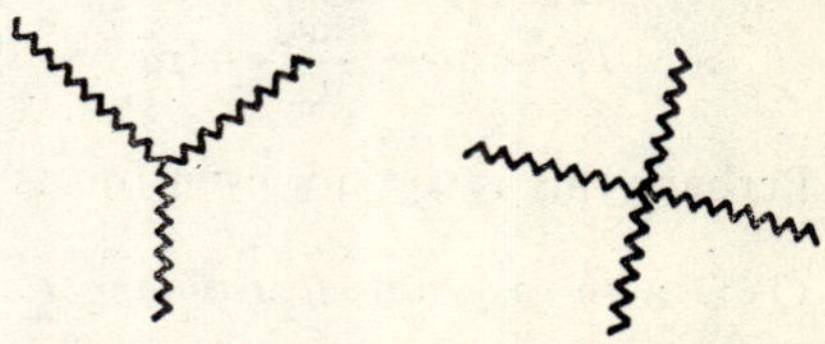

Nevertheless, it is correct to say that Maxwell's laws have been incorporated into a more general system of laws which unify electrodynamics and weak interactions. The implications of this grand unification are yet to be fully understood or realised. There is no doubt that the consequences of this unification will be as profound and as far-reaching as those of Faraday's unification of electricity and magnetism and of Maxwell's unification of electrodynamics and optics.

The truly profound consequences of the electro-weak synthesis have yet to be realised. However, the following two results are worth mentioning.

Phase transition and restoration of weak-electromagnetic symmetry

There exists a similarity between the spontaneous breakdown of symmetry and the phenomenon of phase transition. In particular Kirzhnitz and Linde[3] in 1972 pointed out the close analogy between the Higgs Lagrangian of the spontaneously broken quantum field theory and the free energy expression in Landau-Ginzburg's phenomenological theory of phase transitions. As a consequence of this analogy, there exists a critical temperature, T_c, above which the symmetry between weak and electromagnetic interactions is restored. So, a collection of leptons with conventional weak and electromagnetic interactions will behave entirely differently if their temperature is raised above T_c. The striking physical differences are as given below:

For $T < T_c$	For $T \geqslant T_c$
$\left.\begin{matrix} m_W \\ m_Z \end{matrix}\right\} \sim 80 - 140$ GeV $m_\gamma = 0$	$\left.\begin{matrix} m_W \\ m_Z \\ m_\gamma \end{matrix}\right\} = 0$
Weak-interactions weak and short-ranged Electromagnetism stronger and long-ranged	Weak and electromagnetic interactions } same strength and long-ranged

However, the critical temperature T_c is of the order

$$T_c \sim \phi \sim \frac{m_W}{e} \sim G_W^{-\frac{1}{2}} \sim 10^3 \text{ GeV} \sim 10^{16}\ ^\circ\text{K}.$$

Perhaps this is too hot even for astrophysicists!

QFD *with external electromagnetic field*

A large part of classical physics and even quantum physics, in the atomic as well as cosmological scale makes use of the concept of external classical electromagnetic field. In view of the unification, does this require any change? The following example suggests this may be so.

The electric charge of the neutrino is zero, but it can have an electromagnetic form factor induced by the weak interaction through the following diagrams:

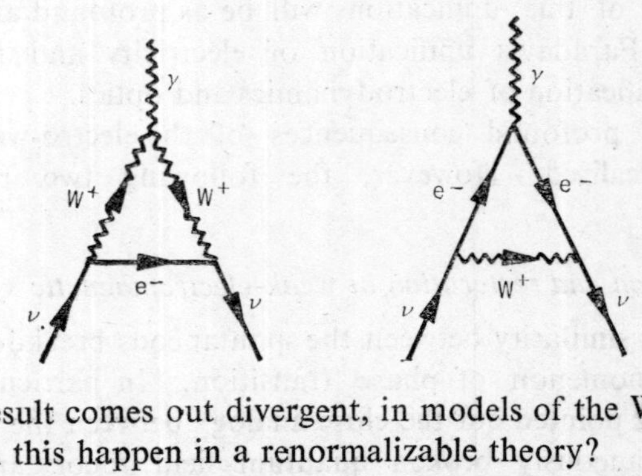

However, the result comes out divergent, in models of the Weinberg-Salam type. How can this happen in a renormalizable theory?

The answer is that only the physical S-matrix elements and decay matrix elements on the mass-shall are finite after renormalization. The form-factor, being an off-mass-shall Green's function, remains divergent.

In other words, the electromagnetic form factor is *not* a measurable quantity. To measure it, virtual photons of $(\text{mass})^2 = q^2 \neq 0$ is necessary. So, one should consider the whole process including the charged particle which emits the virtual photon. But then, in this unified model, there is the Z boson which interacts with all the charged particles. So, one should include the Z exchange diagrams also. Once we add both these classes of diagrams together, a finite result is obtained:

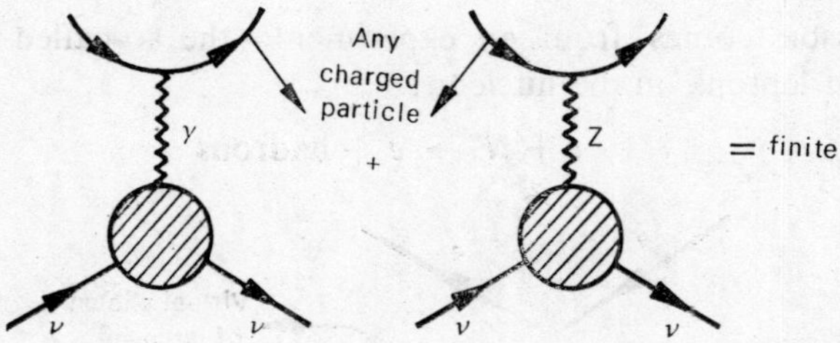

But, since electromagnetic form factors can be used to describe the interaction of a particle with an external classical electromagnetic field, this leads to the question posed above.

4. Quantum chromodynamics

We now come to QCD where the gauge group is $SU(3)_c$. The quarks are colour triplets while the gluons are colour octets and play the role of gauge bosons. The QCD Lagrangian is

$$\mathcal{L} = -\tfrac{1}{4}\left(\partial_\mu A^i_\nu - \partial_\nu A^i_\mu - g f^{ijk} A^j_\mu A^k_\nu\right)^2 + \bar{\psi}\left(i\gamma^\mu \partial_\mu - g\gamma^\mu \frac{\lambda^i}{2} A^i_\mu\right)\psi.$$

Here ψ is the quark field, A^i_μ is the gluon field, g is the gauge coupling constant, i goes over 1 to 8, f^{ijk} are the structure constants of SU (3) and $\lambda^i/2$ are the SU (3) generators in the triplet representation. The following interaction vertices are contained in $\mathcal{L}$:

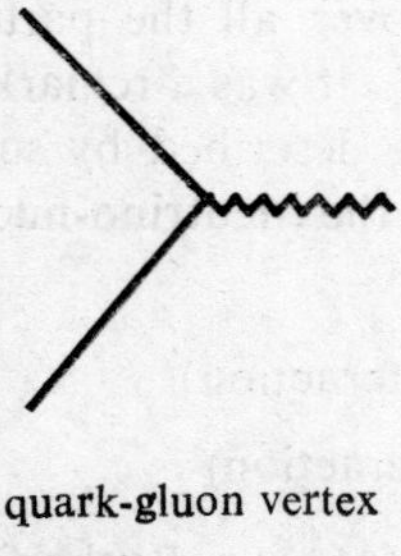

quark-gluon vertex

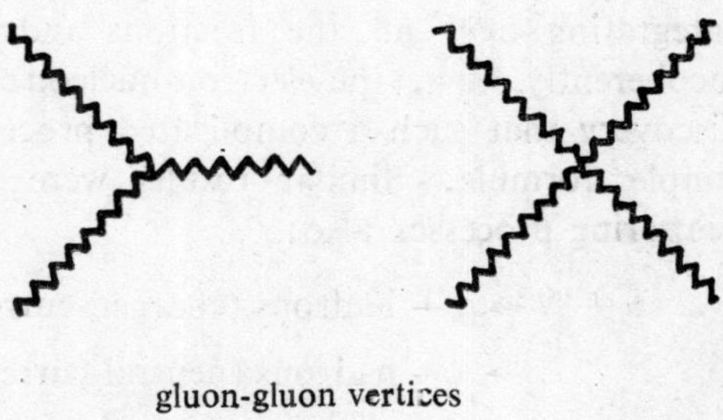

gluon-gluon vertices

The gluon-gluon vertices are characteristic of non-abelian gauge theory and distinguish QCD from the abelian QED.

The first question we have to ask is: Why QCD? What is the reason for believing QCD to be the theory of S.I.?

Why QCD?

For a long time, physicists had given up field theory as a useful approach for understanding strong interactions and taken to the S-matrix approach. So, what caused the recent resurgence of field theory in strong interaction physics and what is the reason for going for this non-abelian gauge field theory (QCD)?

The reason comes from an experiment—the so-called deep inelastic scattering of leptons on the nucleon:

$$e + N \to e + \text{hadrons}$$

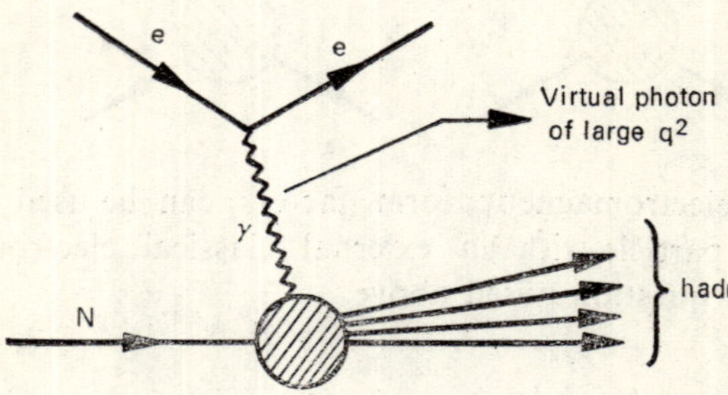

It was found that, as observed by a high q^2 probe, the nucleon behaves as if it were composed of *free*, point-like constituents (called *partons* by Feynman). The lepton scatters off each parton, elastically and incoherently. The incoherent sum of all parton cross-sections gives a very good description of the experimental results. Thus, the complete cross-section for the electron scattering off the nucleon can be written (schematically) as

$$\sigma_N \sim \sum_i \int_0^1 dx\, f_i(x)\, \sigma_i(x)$$

where $\sigma_i(x)$ is the electron-scattering cross-section on the ith parton with fractional longitudinal momentum x and $f_i(x)$ is the probability for finding the ith parton with fractional longitudinal momentum x inside the nucleon. Integrating over all the fractions and summing over all the partons i incoherently, we get the electron-nucleon cross-section. It was a remarkable discovery that such a complicated process could be described by such a simple formula. Similar results were found for the neutrino-nucleon scattering processes also:

$$\nu + N \to \mu + \text{hadrons (charged-current weak interaction)}$$
$$\to \nu + \text{hadrons (neutral-current weak interaction)}$$

This phenomenon has a rather close resemblance to Rutherford's famous α-particle scattering experiments which led to the discovery of the nucleus inside the atom. Thomson's spread-out atomic model would lead to soft scattering (i.e. small scattering angles) only. Experimentally, Rutherford and collaborators found hard scattering (i.e. large scattering angles), thus showing the presence of the point-nucleus inside the atom. In the same way, in the deep inelastic lepton-nucleon scattering, even for large q^2 (i.e. large scattering angle), scattering was observed to take place, in contrast to what would be expected for a spread out nucleon. This leads to the discovery of point-like constituents deep inside the nucleon.

More detailed study of the experimental data revealed that these partons are in fact quarks; they seemed to have the same spins and charges as expected for quarks. (We shall say more on this point later.)

Attention should now be drawn to the adjective 'free'. In addition to being point-like, the quark-partons behave as if they are free. If they are interacting, the cross-section formula would not be so simple.

Now, the quarks are bound by tremendous attractive forces to make up the nucleon. So, the interaction between quarks should really be superstrong. And yet, when observed through high q^2 probes, this super-strong interaction weakens to such an extent that the quarks behave as free particles.

For quite sometime this was a mystery. On the other hand, this provided an important clue about the nature of the strong interaction itself. We can now say that any theory of strong interactions should satisfy this property, namely, it should tend to a free particle theory or a free field theory at high q^2. Is there any such theory?

Consider non-relativistic potential scattering, i.e. non-relativistic particles interacting through well-defined smooth potentials. Since the total energy can be written as $E = T + V$, as the kinetic energy T increases, the potential energy V becomes less and less important in comparison, so that for high energies the the theory does tend to a theory of free particles, for properly defined smooth potentials.

But, of course, this is not useful for high energy physics which has to be described by relativistic quantum mechanics. Here, particle-production dominates at high energies and potential description fails.

So, we should ask the same question in the realm of relativistic quantum field theories. Here it is renormalization group which provides the required technique. By using renormalization group, one can define a momentum-dependent coupling constant $g(q^2)$, also called effective coupling constant. So, what we need is a theory in which

$$g(q^2) \rightarrow 0 \quad \text{for} \quad q^2 \rightarrow \infty.$$

Such a theory is called *asymptotically free*, i.e. the theory tends to a free field theory for asymptotic momenta.

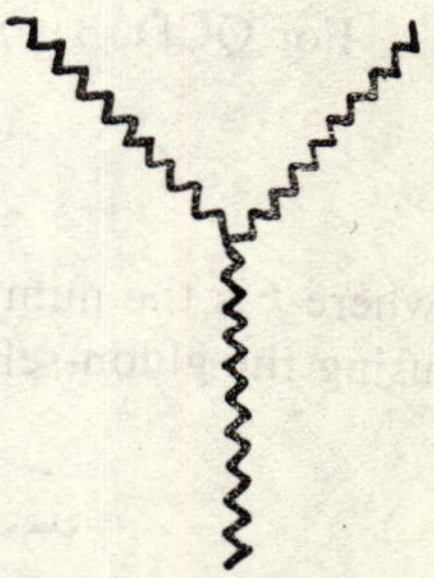

To cut the long story short, it was soon discovered that none of the conventional field theories such as ϕ^4 Yukawa interaction $\bar{\psi}\psi\phi$ or QED $\bar{\psi}\gamma_\mu \psi A^\mu$ is asymptotically free. Of all the re-normalizable quantum field theories, only non-abelian guage theory was found to possess the unique distinction of being asymptotically free. The characteristic triple gluon vertex is the essential ingradient that makes this theory asymtotically free.

So, asymtotically free non-abelian gauge theory emerged as a good choice for a theory of strong interactions. Since the colour degree of freedom with three colours was already around, the gauge group was taken as the colour SU (3) and QCD was born.

Asymptotic freedom

Some details on the theory of asymptotic freedom may be given. The re-normalization group equation for the effective coupling constant $g(t)$ can be written as

$$\frac{dg(t)}{dt} = \beta(g(t))$$

where $t = \ln q^2$, q^2 being the momentum transfer. Here $\beta(g)$ is a well-defined finite function of g which can be calculated in every re-normalizable quantum field theory at least for small g, and it characterizes the corresponding theory in a very important way. It is $\beta(g)$ which controls the asymptotic behaviour of the theory.

First consider the case where $\beta(g)$ is positive. So, dg/dt is positive and $g(t)$ increases with t or q^2. This theory is not asymptotically free. If, on

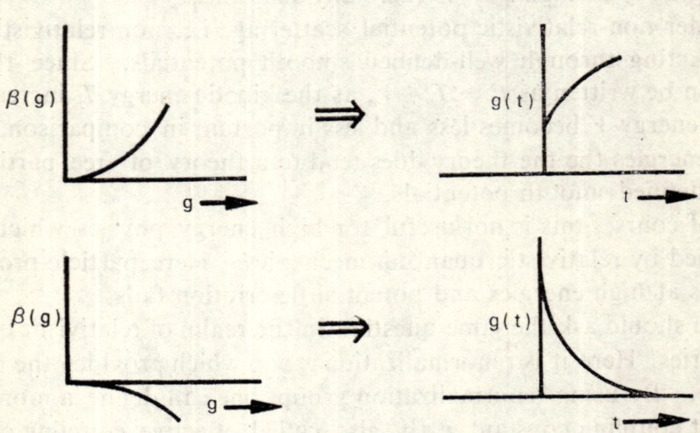

the other hand, $\beta(g)$ is negative near the origin, $g(t)$ will decrease with t and tend to zero for asymptotic values of t. So, for an asymptotically free theory $\beta(g)$ should be negative near the origin.

For QCD, $\beta(g)$ turns out to be

$$\beta(g) = \left(-11 + \frac{2}{3} f\right) \frac{g^3}{(4\pi)^2} + 0(g^5)$$

where f is the number of quark flavours. Actually, $\beta(g)$ can be calculated using the gluon-self-energy diagrams:

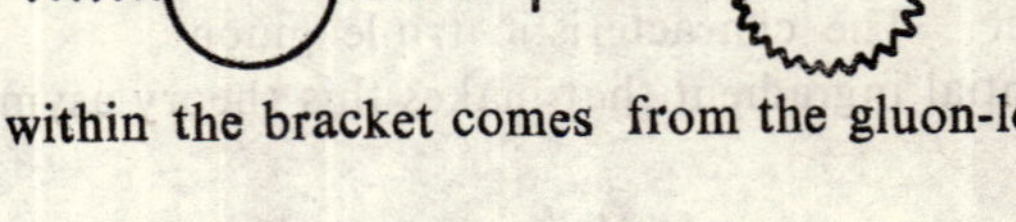

The first term within the bracket comes from the gluon-loop-diagram and hence is negative because of the three-gluon coupling while the second term proportional to f arises from the asymptotically not-free quark-gluon coupling. QCD is asymptotically free if the first term wins over the second, i.e. if $f \leqslant 16$.

So far so good. But the major problem in QCD is the following. Where are the quarks and gluons? Why are they not produced in hadronic collisions?

One possibility which is believed by many people with varying degrees of vehemence is that quarks and gluons are permanently confined within the hadrons and can never be liberated. This dogma is called *colour confinement*. Any coloured object (i.e. anything which is *not* a singlet under colour SU (3)) is confined to be within the hadrons. Quarks and gluons being colour triplet and octet respectively, they are confined. But the familiar hadrons such as nucleons and pions are colour singlets and hence can be produced in the free state. However, to this day, the hypothesis of colour confinement has not been proved in QCD.

5. QCD with a difference (a heretical view)

So far we have been discussing the orthodox version of QCD. The rest of the talk will be devoted to a heretical view.

The orthodox and unorthodox versions of QCD are based on the two types of quarks, the fractionally-charged quarks of Gell-Mann and Zweig (G-Z) and the integrally-charged quarks of Han and Nambu (H-N) respectively:

$$\underset{\text{Gell-Mann-Zweig}}{\begin{pmatrix} u^{2/3} & u^{2/3} & u^{2/3} \\ d^{-1/3} & d^{-1/3} & d^{-1/3} \\ s^{-1/3} & s^{-1/3} & s^{-1/3} \\ c^{2/3} & c^{2/3} & c^{2/3} \\ \vdots & \vdots & \vdots \end{pmatrix}} \qquad \underset{\text{Han-Nambu}}{\begin{pmatrix} u^{1} & u^{1} & u^{0} \\ d^{0} & d^{0} & d^{-1} \\ s^{0} & s^{0} & s^{-1} \\ c^{1} & c^{1} & c^{0} \\ \vdots & \vdots & \vdots \end{pmatrix}}$$

The electric charges are indicated as superscripts. For the Han-Nambu quarks, the charges vary in the colour (horizontal) direction; the colour-averaged charge is fractional and equal to that of the G-Z quarks.

It had been generally supposed that only the G-Z assignment is in agreement with experimental data on deep-inelastic scattering experiments which probe the charges of the constituents of the nucleon. But as was shown sometime ago[4], this is a false conclusion. A proper treatment of the mixing between the gluons of QCD and the gauge bosons of QFD yields the following result for the effective charge of the H-N quark as seen by a high-momentum probe:

$$Q_{\text{eff.}}^{\text{H-N}} = Q^{\text{G-Z}} + Q^{\text{C}} \frac{m_g^2}{m_g^2 - q^2}$$

For momentum-transfer $q^2 \gg m_g^2$, we see that the colour-dependent part of the charge Q^C is damped out and hence

$$Q_{\text{eff.}}^{\text{H-N}} \rightarrow Q^{\text{G-Z}}$$

In other words, although the quarks may be in fact integrally charged, they manifest only the colour-independent parts of their charges in deep-inelastic scattering experiments and hence masquerade as G-Z quarks.

How then can we distinguish between the fractionally-charged model and the integrally-charged quark model? The answer is through the gluons.

The gluons also are different in the two models. Whereas in the orthodox QCD based on G-Z quarks the gluons are electrically neutral, in the QCD based on integrally-charged quarks the gluons also are charged. This will be clear from a look at the quark matrix in the H-N case. Since the gluons cause the horizontal transitions such as

$$u^1 \rightarrow u^0 + G^+$$

they have to be charged. So, the possible electromagnetic effects of the gluons will signal integral charges for the quarks.

Recently we[5] have looked for such gluon effects in two places which can be described here briefly.

(*a*) *Two-gluon jets in e^+e^- annihilation*: The study of jets in e^+e^- annihilation is potentially a rich source of information on the hadronic constituents. The e^+e^- annihilation producing a quark-antiquark ($q - \bar{q}$) pair which in turn produces the hadronic jets is illustrated in the figure.

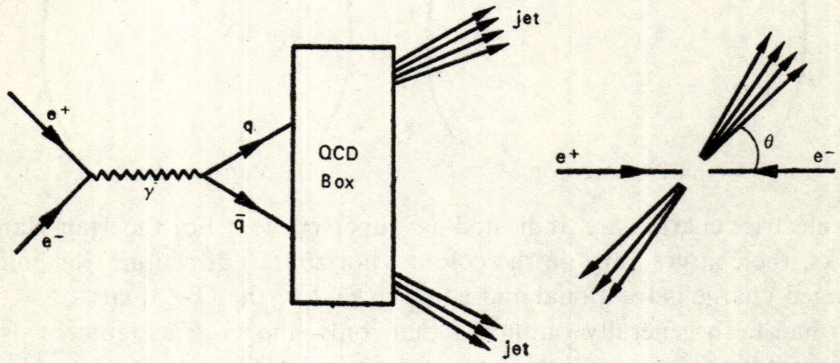

The annihilation into the quark-antiquark ($q - \bar{q}$) pair which subsequently materializes as two hadronic jets would give the angular distribution:

$$\left(\frac{d\sigma}{d\Omega}\right)_{\text{quarks}} \sim \sum_a Q_a^2 (1 + \cos^2 \theta)$$

where Q_a are the quark-charges. On the other hand, the annihilation into a charged gluon-antigluon pair which again materializes as two jets would lead to

$$\left(\frac{d\sigma}{d\Omega}\right)_{\text{gluons}} \sim \sum_i Q_i^2 (1 - \cos^2 \theta)$$

where Q_i are the gluon-charges. We find the over-all angular distribution to be

$$\left(\frac{d\sigma}{d\Omega}\right)_{\text{total}} \sim 1 + A \cos^2 \theta$$

where $A = 37/43 \approx 0.86$. This is to be compared to the experimental result: $A = 0.97 \pm 0.17$. So, the gluon contribution is not ruled out.

(*b*) *Effect of charged gluons on lepton-pair production*: The mechanism for the production of a lepton pair ($\mu^+\mu^-$ or e^+e^-) of high invariant mass, in hadronic collisions is illustrated in the following figure:

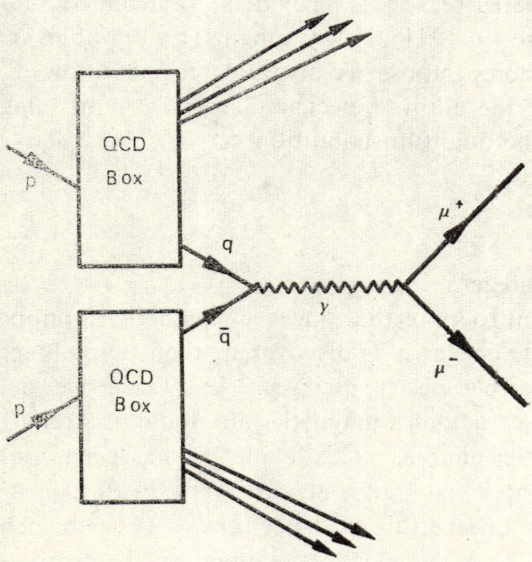

This is essentially inverse to the e^+e^- annihilation process. Again contributions from quarks as well as charged gluons are possible:

$$\left(\frac{d^3\sigma}{dq^2\, dy\, d\Omega}\right)_{\text{quarks}} \sim \sum_a Q_a^2 f_a(y_1) \bar{f}_a(y_2) (1 + \cos^2 \theta)$$

$$\left(\frac{d^3\sigma}{dq^2\, dy\, d\Omega}\right)_{\text{gluons}} \sim \sum_i Q_i^2\, g(y_1)\, g(y_2) (1 - \cos^2 \theta)$$

Here, $f_a(y_1)$ and $\bar{f}_a(y_2)$ are the probabilities for finding a quark in one proton and an antiquark in the other proton with fractional longitudinal momenta y_1 and y_2, respectively. $g(y_1)$ and $g(y_2)$ are similar probabilities for gluons.

We find that the ratio of the gluon to the quark contribution can be quite sizable in some kinematic regions. Also, the angular distribution is drastically modified from the $(1 + \cos^2\theta)$ form characteristic of quark-contribution. Hopefully, these features can be confronted with experiment soon.

To sum up, integrally-charged quarks are also consistent with experiments so far.

Preference for integral charges

Fairbank et al.[6] have claimed to have detected fractional charges in a modern version of Millikan's oil-drop experiment with superconducting balls. It is too early to say whether this will be confirmed by an independent experiment. If it is confirmed, then the heretical views expressed here would have to be abandoned.

Here, we would like to suggest that there may be some deeper reason for prefering integral charges for all particles of nature.

One such deeper reason is provided by magnetic monopoles. Dirac[7] showed long ago (in 1931) that, if magnetic monopoles exist, the laws of quantum mechanics impose a constraint on the allowed electric charges. He showed that the allowed electric charges e' and magnetic charges g' should satisfy the quantum condition:

$$e'g' = \frac{n\hbar c}{2}$$

where n is an integer.

It is imporant to note that once a magnetic monopole with a well-defined magnetic charge g' is discovered, then Dirac's condition severely restricts the possible electric charges. In 1975 Price et al.[8] claimed the observation of a track of a magnetic monopole of strength $g' = \hbar c/e$ where e is the electronic charge. This evidence has been controverted and so magnetic monopole is not yet discovered. But, just for the sake of argument, if we believe this evidence, *and* if we also believe that objects with charge $e' = e/3$ also exist (as claimed by Fairbank et al.), then there is a contradiction with Dirac's quantum condition. For, now

$$e'g' = \frac{\hbar c}{3}$$

So, *both* Fairbank and Price cannot be right!

Leptons as quarks

In this last part of the talk, all caution will be abandoned and a hypothesis will be put forward which is even further removed from conventional views. This can be stated in the form of a question. Can leptons be identified with quarks? In other words, are the hadrons made of leptons such as e, μ, ν, etc. ?

The author has had this heretical idea for quite some time, even before gauge theory became a court religion. The chief motivation for it was the mysterious principle or working hypothesis called *lepton-hadron Universality* which has guided the development of weak interaction theory for quite some time—through current algebra, Cabibbo universality and now even in the construction of gauge models.

The total weak current is always written as $l_\mu + h_\mu$ and then lepton-hadron universality is invoked to demand that the leptonic current l_μ and the hadronic current h_μ should have the same algebraic properties. Why should this be so?

Is not the simplest and the most natural explanation that l_μ and h_μ are identical? This would be the case indeed, if the hadronic constituents are simply the leptons. There is then no need to write a hadronic current separately, the leptonic current would describe the hadrons as well. So, the only current is l_μ.

Of course, such an identification of quarks with leptons would make sense only if quarks are integrally charged. And we have already argued that integral charged quarks are consistent with high-energy experiments so far.

So, let us boldly trypothesise:

$$\text{quarks} \equiv \text{leptons}$$

The problem one faces is: How to explain the absence of strong interactions for the leptons e, μ, etc? This problem, although a difficult one, may not be insurmountable. A definitive solution cannot be given immediately, but one can envisage the following possibility.

We can exploit the idea that vacuum is not simple. Vacuum can undergo a phase transition, perhaps through acquiring a non-vanishing expectation value of a Higgs field $\langle \phi \rangle$ which, however, is space-dependent. A crude model follows:

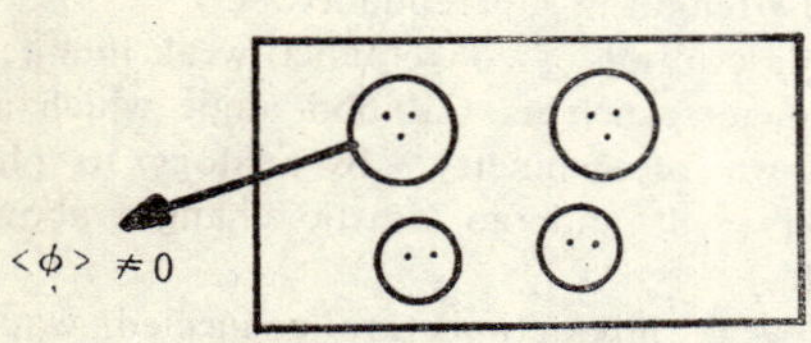

Hardonic phase

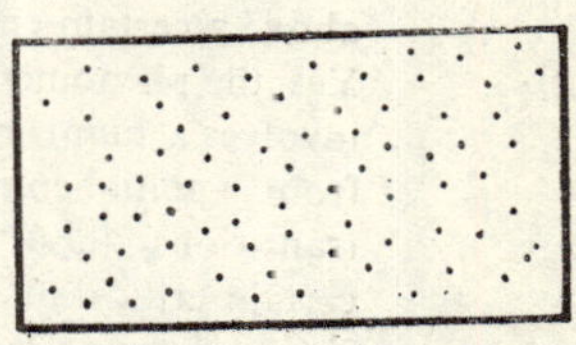

Leptonic phase

In the leptonic phase, the leptons have only weak and electromagnetic interactions. There is a phase transition. Some bubbles develop, which are regions corresponding to non-vanishing $\langle \phi \rangle$. If three leptons or a lepton-antilepton pair gets trapped in a bubble then it behaves like a hadron.

We shall stop here, by concluding that if the above idea turns out to be right, it would provide a very natural unification of the leptonic and hadronic worlds.

References

1. For more details and original references, the extensive reviews in the following conferences and symposia may be consulted:—
Tata Institute Winter School in High Energy Physics, Panchgani (December 1977); High Energy Physics Symposium, Jaipur (Dec. 1978); Proceedings of Summer Institute on Particle Physics, Stanford (July 1978) SLAC Report No. 215; International Conference on High Energy Physics, Tokyo (Aug. 1978).
2. Lee, T.D., and Yang, C.N., *Phys. Rev.* **128**, 885 (1962).
Nakamura, N., *Prog. Theor. Phys.* **33**, 279 (1965).
Tzou, K.H., *Nuovo Cimento*, **33** 286 (1964).
3. Kirzhnitz, D.A. and Linde, A.D., *Phys. Lett.* **42B**, 471 (1972).
4. Rajasekaran, G. and Roy, P , *Pramana* **5**, 302 (1975).
Pati, J.C., and Salam, A., *Phys. Rev. Lett.* **36**, 11 (1976).
5. Rajasekaran, G. and Rindani, S.D., Madras University preprint MUTP–79/4, Phys. Lett. preprint.
Jayaraman, T. and G., Rajasekaran, Madras University preprint MUTP–78/13 (to be published).
6. La Rue, G.S. et al., *Phys. Rev. Lett.*, **42**, 142 (1979).
7. Dirac, P.A.M., *Proc. Roy. Soc.* (London), **A133**. 60 (1931).
8. Price, P.B., et al., *Phys. Rev. Lett.* **35**, 487 (1975).

Discussions

C Sivaram: What is the strength of the neutral current as compared to the charged current? Does it also have a V–A structure?

G. R.: The strength is comparable to that of the charged current. The structure of the neutral current is

$$(V - A) - \sin^2 \theta\ V$$

where $\sin^2 \theta$ is Weinberg's mixing angle.

C Sivaram: The Weinberg-Salam theory seems to Suggest a suppression of the Cabibbo angle above a certain critical magnetic field strength so that the Λ particle may have a much enhanced lifetime in certain hypernuclear excited states of Co^{57} and Nb. This is analogous to phase transition above a certain critical field strength in superconductors.

G. R.: Yes, the phenomenological description of conventional weak interaction involves a number of parameters such as Cabibbo angle which arise from spontaneous breakdown of symmetry. By analogy to phase-transitions these parameters will undergo drastic changes above a certain critical magnetic field.

M Mohan: Since electromagnetic and weak interactions are connected, would it mean that the particle decay processes like radioactivity will be affected by strong electric or magnetic fields?

G. R.: The effect of the magnetic field above a critical strength in changing weak interaction phenomena was noted by Dr. Sivaram in the above comment. However, this has nothing to do with unification of weak and electromagnetic interactions. Direct manifestations of the unification are hard to find.

S C Tiwari: Is axial photon a necessary outcome of unified weak and electromagnetic theories? In four-fermion interacting type pseudoscalr Lagrangian approach, it is found that photon should appear.

G. R.: This is not generally true, although it may be so in some models. In particular, in Weinberg-Salam model there is no axial photon.

Yash Pal: Are there any deep attempts being made to understand the value of the Weinberg-Salam mixing parameter $\sin^2\theta$.

G. R.: Yes. Weinberg-Salam model does not provide a genuine unification of weak and electromagnetic interactions. For, the relative strength of the weak and electromagnetic interactions is not completely fixed, but is related to a parameter $\sin^2 \theta$ which is not determined within the framework of this model. However, if the product group $SU(2) \times U(1)$ is embedded within a simple group such as SU(5), then true unification occurs and the value of $\sin^2\theta$ is then determined theoretically. Such attempts are being made.

A K Pandey: Suppose theoreticians find by some quantization scheme, a value for fine-structnre constant consistent with electron charge, then this will contradict the experiments giving fractional charges of quarks. Does the experiment responsible for the notion of fractional charge of quark contain sufficient loopholes to be reconciled with above theory or should the theoreticians try to get a value of fine-structure constant nine times smaller?

G. R.: The theory you are referring to, is hypothetical and the experiment is not established. So, the question can be faced, if and when it arises.

P Achuthan: Identification of leptons and hadrons is radical indeed. But then many structures built with so much effort may have to be broken. Your comments please!

G. R.: Newer ideas need not break the older structures completely. Usually, the essence of the older ideas do find a niche in the more complete newer structures. It may be that the newer theory (yet to be constructed) will show under what Physical conditions the separation between leptons and hadrons become approximately valid.

P Achuthan: Hadrons as bubbles in ideal liquid (ether) have been pictured earlier by V. Weisskopf. Your way of looking at hadrons as bubbles seems to be a fruitful idea.

S Ramadurai: Phase transition in the early universe has been examined by Ruderman and he finds that the early universe does not affect this.

Foundations of Quantum Mechanics

VIRENDRA SINGH

1. Introduction

In order to deal with the observed spectral distribution of black-body radiation, for which classical physics had an unambiguous prediction (Rayleigh-Jeans law) which was in complete disagreement with observation except at large wavelengths, Planck was led in 1900 to introduce quantum of action, *h*, in physics and thus initiated quantum theory. After about a quarter century of development the present form of the mathematical structure of (non-relativistic) quautum mechanics was arrived at in two different versions, viz., matrix mechanics by Heisenberg and wave mechanics by Schrödinger.[1] These two versions were soon shown to be equivalent.

In the last fifty years or so, this mathematical formalism has been repeatedly checked for its predictions and has been found in agreement with experimental results for all known microsystems and their interactions with radiation. Not only that quantum mechanics has been found indispensable in its applications to such large macrosystems as superfluids, superconductors and lasers. For more conventional macrosystems also, such as solids, it is the quantum mechanics which has to be used to calculate their various phenomenological parameters, like conductivities and thermal expansion coefficient, required in their 'classical' description. At present there is no known disagreement with any observed phenomenon. There may be, of course, a number of unexplained facts, such as quark confinement in the theory of elementary particles, but nothing is known which contradicts quantum mechanics.

In view of the above situation there is practically no disagreement about the formalism of quantum mechanics among practising physicists. In contrast, however, the conceptual structure of quantum mechanics is still subject to violent debates. The orthodoxy is represented by the views of Copenhagen School and is associated with the views of Bohr and Heisenberg. The Copenhagen interpretation, however has never had absolute adherence. Some of the founding fathers of the quantum theory such as Einstein, de Broglie and Schrödinger were never convinced of the correctness of this interpretation. Even within Copenhagen interpretation the divergences of views have been found. Thus, for example, the views of Bohr[2] and Heisenberg[3] are not identical. Many further interpretations continue to be suggested. This is a strange state of affair for a theory which is so successful and so old.

Einstein and foundations of quantum mechanics

It is of course appropriate that we discuss this subject in the Einstein centenary symposium as Einstein did contribute a good deal to this subject. The ensemble interpretation of quantum mechanics, which is generally speaking the operative part of it in testing its probabilistic predictions, goes back really to Einstein.[4] At fifth Solvay Congress held at Brussels in October (1927), while discussing the single slit diffraction of electrons, he contrasted the following two viewpoints: (i) the ensemble according to which the modulus square of the wave function $|\psi(r)|^2$ gives information only on the probability of finding some electron in the ensemble at a particular position but does not give any information on an individual process, and (ii) the one which regards quantum mechanics as a complete theory of individual processes, and according to which it is one and the same electron whose probability of being present at a particular position is given by $|\psi(r)|^2$; needless to add that the second position is that advocated by Bohr.

In fact the evolution of Copenhagen interpretation and the views of Bohr, owe a lot to the Einstein's searching questions and analysis of the foundations of quantum mechanics. It was a fortunate circumstance that Bohr recorded his version of these discussions.[5] One of the most penetrating discussions of the foundations was given by Einstein together with Podolsky and Rosen.[6] This particular discussion has been commented on variously and has turned out to be very fruitful specially through Bell's work,[7]

Choice of matters to be discussed

The enormity of literature does call for a choice of topics to be covered here.[8] Besides my own limitations to appreciate and/or understand some of this literature is also responsible for the choice.

Basically I discuss only two aspects of the subject. There are (a) qunatum theory of measurement in sec. 2 and (b) the hidden variable approach to quantum mechanics in sec. 3.

The importance of a theory of measurement process in quantum mechanics cannot be overemphasized. After briefly discussing why classical physics is not burdened with a theory of measurement (sec. 2.1) we present the view, nearest to Bohr's views, according to which the measuring apparatus is regarded as 'classical' (sec. 2.2). Most of the discussion of the subject, however, have been based on influential Von-Neumann's theory of measurement. This is discussed in sec. 2.3 where the central concept of 'reduction of wavepacket' or 'the collapse of the wave vector' is introduced. The importance of this concept gives the reason for subtitling the whole sec.2 by referring to it. The philosophical difficulties presented by this concept are given in the next 'optional' sec. 2.4 entitled 'Schrödinger's cat and Wigner's friend'. The sec. 2 is concluded by a discussion of the 'many world interpretation' of quantum mechanics in which one discards the 'reduction of the wave packet'.

A reason for discussing 'hidden variable approach to quantum mechanics' (sec. 3) is that important progress of both mathematical and experimental nature has been made during the last 10 to 15 years. This is quite in contrast to matters discussed in sec. 2. A stumbling block to a rational discussion was the Von-Neumann's proof of impossibility of hidden variable interpretation (sec. 3.1). This was finally laid to rest by Bell's analysis (sec. 3.2). The mathematically profound work of Gleason, however, does rule out the possibility of non- contextualistic hidden variable theories (sec 3.3). A strong motivation to look for a hidden variable interpretation has been, apart from a unitary world view and/or determinism, the presence in quantum mechanics of an unintuitive quantum nonseparability. The theorem of Einstein-Podolsky and Rosen brings this out in a clear fashion (sec. 3.4). Bell's work on this system has led to the exciting possibility of experimentally testing 'Einstein-locality' or 'the principle of local causes', i.e. away from ethereal realms of metaphysics to the laboratory bench (sec. 3.5). We close this section by an attempt by Bohm and Bub to describe the process of 'reduction of the wavepacket' itself in terms of a hidden variable approach and an experimental test of this idea (sec. 3.6).

2. Theory of measurement in quantum mechanics (reduction of the wavepacket)

2.1 Theory of Measurement in Classical Physics

Classical physics forms a consistent and coherent system within which one can give an account of world as *it is*, provided, of course, nature obeys the laws of classical physics. The measurement is not problematical, unlike in quantum physics, since in principle one can use gentler and gentler probes, depending on the degree of accuracy desired in the description so as not to disturb the system unduly. Both the system under observation, and the apparatus used to make measurements on the system, and their mutual interaction can be described within classical physics. There is no limit to accuracy with which one could control the mutual interaction and as such no limit to the accuracy of the measurement. There are thus no worthwhile epistemological problems in classical physics.

2.2 Is Measuring Apparatus Classical?

According to Bohr, we describe the system quantum mechanically while the measuring apparatus is to be regarded as a macroscopic classical object. We can view the quantum-phenomenon only through experimental arrangements which can be specified by classical physics. The formalism of the quantum mechanics is only a mathematical device to predict the results of experiments on quantum systems specified by settings of measuring devices acting within classical physics.[9]

The logical priority, which has nothing to do with chronological priority, of classical physics over quantum physics in this view is thus assumed. We must have classical physics necessarily to describe the apparatus in addition to having quantum mechanics to describe the physical systems under observations.

This split of the world into two has been found disturbing by many persons. The same measuring device is to be described classically while, when we are studying functioning of the same device, or a part of it, it may have to be viewed quantum mechanically. In these days of electronic instrumentation this situation is rather the rule than exception. Besides, now we have macroscopic quantum systems.

Another difficulty faced by this view is that if measuring interaction is to be regarded as an interaction between system and apparatus like any other interaction (and one does not see why not) then we really do not have an articulated theory of interaction between a classical and quantum system.

Of course in practice all this may not cause much practical problem since one always puts enough into 'system' so that the predictions do not vary much. Here one depends on a property, the so-called 'correspondence principle', of quantum mechanics that for some large systems, and in certain limits, the description given by 'classical physics' is adequate even for quantum systems. Despite practical adequacy this whole approach is logically a little unsatisfactory.

2.3 Von-Neumann's Theory of Measurement

Von-Neumann not only axiomatized quantum mechanics using Hilbert-space framework, he also gave a theory of measurement in quantum mechanics.[11] Presumably in view of the difficulties, mentioned in sec. 2.2, it regarded both the system and apparatus as subject to quantum laws. Von-Neumann's theory of measurement has been the canonical approach to the subject and has played an important part in further development of conceptual framework of quantum mechanics.

Consider the object Σ and suppose we are interested in measuring the physical quantity Ω referring to this system. Further let the complete set of eigenfunctions of Ω be ψ_m ($m = 1, 2, \ldots$) with eigenvalues ω_m. (For simplicity we only consider the case of discrete spectrum). we denote the complete set eigenfunctions of a suitable operator referring to the apparatus M by $f(a)$ with eigenvalue a. Here a refers to the 'pointer' readings of the apparatus.

Let the object be in the pure state $|\psi_n\rangle$ while the apparatus be in the pure state $|f(a)\rangle$, i.e. initially the state of 'object-apparatus' is described by

$$|\psi_n\rangle|f(a)\rangle$$

The measurement brings an energy coupling of the object and the apparatus and, suppose, as a result of this interaction, the object-apparatus goes into the state

$$|\psi\rangle|f(a_n)\rangle$$

i.e.

$$|\psi_n\rangle|f(a)\rangle \rightarrow |\psi\rangle|f(a_n)\rangle$$

where a_n is a pointer reading of the apparatus correlated with the state ψ_n of the object. We then conclude that 'object' was in the state ψ_n of the operator Ω and this constitutes the measurement. Here the final state of object $|\psi\rangle$ may or may not be the same as $|\psi_n\rangle$. If $|\psi\rangle = |\psi_n\rangle$ we speak of the measurements of the first kind while if $|\psi\rangle \neq |\psi_n\rangle$ then the measurement is of the second kind in the terminology of Pauli. For the measurement of the first kind if the quantity Ω is measured again on the system soon after the first measurement, so that the object has not had a chance to change, then we obtain the same eigenvalue ω_n.

Since the time development which brought about the change denoted by the arrow above is unitary we must have

$$\langle\psi_m, \psi_n\rangle\langle f(a), f(a)\rangle = \langle\psi', \psi\rangle\langle f(a_m), f(a_n)\rangle$$

where

$$|\psi_m\rangle|f(a)\rangle \rightarrow |\psi'\rangle|f(a_m)\rangle$$

Since $\langle f(a_m), f(a_n)\rangle = 0\ (m \neq n)$ as the "pointer readings" a_m and a_n are macroscopically distinct it follows that we must have

$$\langle\psi_m, \psi_n\rangle = 0 \text{ for } (m \neq n)$$

This is an amusing consequence of the above measurement theory. From now on we shall consider only the measurement process of the first kind, i.e.

$$|\psi_n\rangle|f(a)\rangle \rightarrow |\psi_n\rangle|f(a_n)\rangle$$

We now consider the more general case when the object is not in a pure state, it is not an eigenfunction of the operator Ω which is to be measured. Let the initial state of $\Sigma\, C_n|\psi_n\rangle$ and the state of the apparatus be given by $|f(a)\rangle$. By the linearity of the laws of quantum mechanics we will have

$$(\Sigma\, C_n|\psi_n\rangle)\,|f(a)\rangle \rightarrow \Sigma\, C_n|\psi_n\rangle|f(a_n)\rangle$$

due to the measuring interaction. We now have the strange situation that the apparatus is in a state which is a coherent mixture of states representing macroscopically different pointer position. Of course, the probability of the different pointer readings is proportional to $|C_n|^2$ for reading a_n. If the measurement is completed then state vector

$$\Sigma\, C_n|\psi_n\rangle|f(a_n)\rangle$$

must collapse to one of the terms in the linear combination, say

$$|\psi_k\rangle|f(a_k)\rangle$$

where a_k is the *observed* pointer reading. This is referred to as collapse of of the state vector or reduction of the wavepacket.

In summary, according to Von-Neumann, the time evolution of the state vector in quantum mechanics is of two kinds:

(*a*) *Processes of the first kind*

These are acausal, discontinuous and take place when a measurement takes place. If in our discussion we had a chain of successive apparatus where we use apparatus A_1 to make measurements on the object Σ and use apparatus A_2 to make measurements on apparatus A_1 and so on, then this collapse of state vector takes place at that last point in the chain where the measurement is completed and the result of measurement is known. The final result is independent of where the chain ends.

(*b*) *Processes of the second kind*

These are causal and continuous and are governed by Schrödinger equation. The state vector evolve according to these processes of the second kind between measurements.

It was pointed out by Wigner (1952) that there cannot exist a unitary transformation which will effect the transition involved in the definition of measurements of the first kind unless dynamical observable Ω, which is being measured, commutes with all the linearly additive conserved quantities of the 'object plus apparatus' system. This would seem to be a serious technical difficulty at least as far such operators as the third component of the angular momentum is concerned, Araki and Yanase showed that the difficulty persists also for measurements of the second kind. However, later it was shown that possibility of approximate measurements remain. If we denote by $\mathcal{L}$ the size of the apparatus in the sense of the expectation in apparatus state of the conserved quantity then it turns out that an accuracy of $0\,(1/\mathcal{L}^2)$ can be achieved. We shall not go into this technicality any further.[12]

Another possible source of difficulty was pointed out by Zeh.[13] In the above discussion one has assumed the system plus apparatus to be an isolated system. As the apparatus is large its energy eigenvalues are very closely spaced, smallest perturbation is sufficient to cause transitions. As such the assumption made is not very realistic.

A number of attempts have tried to connect the irreversibility and measurement process. We refer the reader to important investigations by Ludwig,[14] by Daneri, Prosperi and Loinger[15] and by Rosenfeld[16] and others.[17]

2.4 Schrödinger's Cat and Wigner's Friend

The reduction of the wavepacket is perhaps the most mysterious thing in quantum mechanics. How unintuitive is all this is brought home by the case of Schrödinger's cat ![18,19]

Let there be a steel chamber in which are placed a cat and the following Hollenmaschine. There is a minute amount of a radioactive isotope inside a Geiger counter—so tiny that in one hour there is an equal chance that either one atom decays or none decays. If, however, one atom decays the Geiger counter triggers off and activates a hammer which breaks off the bottle of cyanide poisoning the cat. At the end of one hour the wavefunction of the system thus consists of

$$\psi = \frac{1}{\sqrt{2}}\,\psi\,(\text{cat alive}) + \frac{1}{\sqrt{2}}\,\psi\,(\text{cat dead})$$

If now we open the box and make the observation we will find the cat either alive or dead, i.e. the above wavefunction suddenly, acausally, collapses to either ψ (cat alive) or ψ (cat dead).

Was cat in a state of suspended animation between living and dead before we opened the box and forced the wavefunction to collapse to that a living or dead cat? What Schrödinger has done here is to transfer the indeterminacy from the atomic to macroscopic levels. Now there is no logical paradox here but a certain amount of uneasiness is generally felt.

Let us also comment here on the role of the consciousness of the observer. London and Bauer,[20] in their elaboration of Von-Neumann's work, noted that an observer has a quite dlfferent relationship with himself as compared to external world. By virtue of his 'immanent cognition' he knows that he is in pure state ψ_m (say) and not in some linear combination of states. The reduction of the wave packet takes place in the consciousness of the observer. According to this point of view, which also has found support in Wigner,[21] the linear laws of quantum dynamics are only transcended in the interaction of observer's consciousness with the external world. In view of the fact that one has not come across any violation of the linear laws of quantum mechanics in its application to natural phenomenon and all measurement chains terminate in consciousness of the observer this may be regarded as 'minimal' violation of the linear laws of the quantum mechnics.

Wigner inclines to the view that quantum mechanics is inapplicable when life and consciousness are involved. In this connection he considered the following situation, now known as the example of Wigner's friend. This example, dramatised even further by Everett, is as follows.[21,22]

Imagine an observer A in a room observing a system S which is not in an eigenstate of the measurement A will perform. A, being a firm believer in orthodox quantum mechanics, believes that the result of measurement is uncertain until he makes the measurement and records the result in his notebook (Process 1). Meanwhile there is another observer B. who is in possession of the wavefunction of $S + A$, and is only interested in knowing the entry in A's notebook after the lapse of a week. He therefore calculates the time evolution of the state vector of $S + A$ by using Schrödinger equation (Process 2). Ar the end of a week he opens the door to the room and examines the content of A's notebook. Being orthodox he also believes that it is his observation now which has made the contents of notebook entry definite (Process 1) and until now $S + A$ were in a linear superposition of states with various notebook entries. B therefore tells A

that it is his intervention which has given A, his notebook entry and his memory of a week ago their objective existence. A, however, does not like it at all and tells B not to feel so superior since his own objective existence may depend on a future observer.

These examples bring out the unintuitive nature of the reduction of wavepacket. If quantum mechanics is a true description of the nature then they should not appear paradoxical. It may be mentioned that these examples depend for their appeal on certain concepts, such as living-dead, consciousness, which we all feel we know but which, at least so far, cannot be given a purely physical definition. As such a number of working physicist would prefer to ignore them. While this attitude has its good points one may also note that the physics itself is impossible without some notions, which we all know but are not able to define physically.

I would like to end this section with a comment by Bell in a paper in which he analysed Coleman-Hepp model of wavefunction reduction within purely quantum mechanical framework. "So long as the wavepacket reduction is an essential component, as so long as we do not know exactly when and how it takes over from the Schrödinger equation, we do not have an exact and unambiguous formulation of our most fundamental theory".[23]

2.5 Relative State Formulation of Quantum Mechanics

Even though Krishna was sufficient unto himself, He sported individually with each Gopi by appearing in as many forms as their number—Bhāgvat Purāna **X**, 33, 20.

In view of the various difficulties of philosophical nature with the reduction of the wavepacket postulate an extreme reaction has been to discard this postulate altogether and to try to see whether we can or not proceed to build up a theory entirely based only on processes of second kind i.e. only on the basis of purely causal Schrödinger equation. This is the approach taken by Everett with support from Wheeler and further elaboration by DeWitt. A similar suggestion was put forth by Cooper and van Vlechten also.[24]

Since we now have a description in which the object and the observer are all described by quantum mechanics and evolve according to Schrödinger equation it will be expected that the observer, in general, would be in a state which is a superposition of various eigenstates. If as a result of a measurement interaction we are in a state which is, say,

$$C_1 \psi \text{ (walking east)} + C_2 \psi \text{ (walking west)}$$

then we in fact simultaneously walk east as well as walk west. We split in two editions of ourselves and one of these walks east and other one walks west. It will, of course, be pointed out that we are not aware of any such split. We may contrast this with the situation when confronted with the

Copernican theory that the earth goes round the sun, one would point out that one is not aware of any such motion. Copernican theory, of course also explained the absence of any feeling of motion of earth-bound observers. It is similarly shown that within the framework of this theory we cannot feel the split. As there is no reduction of the wave-packet all parts of the wavefunction of the observer come to pass but in different 'worlds'. Difficulties of the kind of Wigner's friend do not present themselves, as consistency of a chain of observers observing each other is established.

The usual probability interpretation of the coefficients of the amplitudes in a superposition is derived by considering an ensemble of systems which interact with an apparatus A.

We thus see that the 'relative state formulation' of the quantum mechanics gives a description which is 'realistic' in the sense that every element of the mathematical formalism has a counterpart in nature. This is achieved, however, at the cost of uncontrolled schizophrenia. At every observation the universe splits. For this reason this interpretation has also been termed 'many world interpretation'. Some people find that this interpretation is extravagant and wasteful of worlds. On the other hand, it is economical in principles in that it dispenses with reduction of the wavepacket. Besides, this is the only formalism in which one can discuss wavefunction of the whole universe as this formalism does not require an external observer. It is fair to say that despite its attractiveness in many ways this interpretation has not yet found much acceptance and some people regard the cure as worse than the disease.

3. Hidden variables and non-separability

'Our aspirations after a uniform theory of nature, on a mechanical basis or otherwise...can never be permanently repressed'—Max Planck.[25]

Quantum mechanics offers us an inherently statistical description of the phenomenon. Previous to quantum mechanics the statistical description had been used in physics, and oftentimes with great success, as in classical statistical mechanics, but it was not inherent in classical physics. It was used to deal with states of the system, such as thermodynamic ones, which were a complicated average of other better defined microscopic states, in terms of position and moments of the particles of the system. It is not unnatural to inquire whether there is another stratum to reality where things are definite and statistical predictions of quantum mechanics are only due to our averaging over the extra variables, called 'hidden variables', involved. If only we knew about these hidden variables and control their values we may expect to get definite predictions. Of course, a confirmed positivists would consider such a quest as worse than useless, but then one may remember Mach's misgiving about the existence of atoms, and need not regard positivism as a sure guide to fruitful physics.

(a) *Von-Neumann's impossibility proof*

For a long time however no such attempt was made as the 'supposed proof' of the absence of hidden variables in quantum mechanics given by Von-Neumann effectively deterred everybody. Von-Neumann's proof was based on very general and plausible looking assumptions.[26] Assuming (i) if two observables are represented by operators A_1 and A_2 then the observable represented by the sum of these two observables is represented by $A_1 + A_2$ and (ii) if A_1 and A_2 are any two observables, whether commuting or non-commuting, and a_1 and a_2 two real numbers then

$$\langle a_1A_1 + a_2A_2\rangle = a_1\langle A_1\rangle + a_2\langle A_2\rangle$$

where $\langle \ldots \rangle$ represents the expectation value, Von-Neumann first thowed that there exists a statistical operator ρ. such that for any operator A, we have

$$\langle A\rangle = \mathrm{Tr}\,(\rho A)$$

Further ρ is shown to be independent of the operator A and depends only on the ensemble. If we require that observables which are non-negative have non-negative expectation values then ρ can be shown to be a non-negative operator.

If we now further assume, as is reasonable, that if an observable is represented by an operator A then the square of the same observable is represented by the operator A^2, we have for dispersion-free ensembles

$$\langle A^2\rangle = \langle A\rangle^2$$

i.e.

$$\mathrm{Tr}\,(\rho A^2) = [\mathrm{Tr}\,(\rho A)]^2$$

Now let A be a projection operator $P = |\,\psi > < \psi\,|$ where $|\,\psi>$ is an arbitrary normalised vector. Since $P^2 = P$ we obtain

$$[\mathrm{Tr}\,(\rho P)[\,[\mathrm{Tr}\,(\rho P) - 1] = 0$$

This implies that ρ is either a null or unit operator. Since either of these is unacceptable we cannot have dispersion-free states. As Von-Neumann concluded "It is therefore not, as is often assumed, a question of re-interpretation of quantum mechanics, the present system of quantum mechanics would have to be objectively false, in order that another description of the elementary processes than the statistical one be possible".[27]

(b) *Bell's analysis of Von-Neumann's proof*

The first person to pinpoint the fallacy in, or rather the irrelevance of, Von-Neumann's proof for the hidden variable question was Bell[28] even though this was also indicated by Bohm's 1952 hidden variable theory.[29]

Consider a spin 1/2 particle. without space-degree of freedom, in quantum mechanics. It will have a two-dimensional state space. The most general observable would be represented by

$$M(\alpha, \boldsymbol{\beta}) = \alpha 1 + \boldsymbol{\beta}\cdot\boldsymbol{\sigma}$$

where α is a real number and $\boldsymbol{\beta}$ a real vector. If one measures the operator $M(\alpha, \boldsymbol{\beta})$ the result would be one of its eigenvalues and these are $\alpha \pm |\boldsymbol{\beta}|$. On the other hand the expectation value in the state ψ is given by

$$\langle M(\alpha, \boldsymbol{\beta})\rangle = \langle \psi, M(\alpha, \boldsymbol{\beta})\, \psi\rangle$$

Bell then showed that all these results of quantum mechanics can be reproduced by the following hidden variable approach: The dispersion-free states are specified by giving ψ and a variable λ, $(1/2 \geqslant \lambda - 1/2)$. Let the value of the observable, having the $M(\alpha, \boldsymbol{\beta})$ as the operator, be given in the dispersion free state (ψ, λ) by

$$M(\alpha, \boldsymbol{\beta}; \psi, \lambda) = \alpha + |\boldsymbol{\beta}| \operatorname{sign} (\lambda |\boldsymbol{\beta}| \langle \psi, \psi\rangle + \tfrac{1}{2}\boldsymbol{\beta}\cdot\langle \psi | \boldsymbol{\sigma} | \psi\rangle \operatorname{sign} (\boldsymbol{\beta}\cdot\langle \psi | \boldsymbol{\sigma} | \psi\rangle)$$

The quantum-mechanical state ψ will be taken to be the state obtained by uniform averaging over λ. Notice, that we have

$$\langle \psi | \alpha + \boldsymbol{\beta}\cdot\boldsymbol{\sigma} | \psi\rangle = \int_{-1/2}^{+1/2} d\lambda \,\{M(\alpha, \boldsymbol{\beta}; \psi, \lambda)\}$$

as required. We thus have a perfectly reasonable hidden variable theory of a spin 1/2 system.

If we examine Von-Neumann's assumptions in the light of this example we find the culprit to be the assumption, even where A_1 and A_2 do not commute

$$\langle aA_1 + a_2A_2\rangle = a_1 \langle A_1\rangle + a_2 \langle A_2\rangle$$

which, though true for quantum-mechanical states, is not true for dispersion free (ψ, λ) states. Von-Neumann had required it to be true for even the dispersion-free states. After all the eigenvalues $\pm \sqrt{2}$ of the operator

$$\sigma_x + \sigma_y$$

are not linear combination of the eigenvalues of σ_x and σ_y, i.e. $(\pm 1 \pm 1)$. The physical explanation of this phenomenon is that the measurement of σ_x, σ_y, and $\sigma_x + \sigma_y$ requires three different orientations of Stern-Gerlach magnets.

We shall not discuss further descendants of Von-Neumann's impossibility proof such as those of Jauch-Piron, Misra and others.[30] These have been commented on sufficiently in the literature. We must, however, mention some implications of Gleason's work for the present problem.

(*c*) *Gleason trouble*

Gleason, in 1957, proved a remarkable mathematical theorem about measures on a Hilbert space.[31] Let us recall that an essential ingredient in Von-Neumann's proof was establishing the existence of a unique semi-definite density matrix operator ρ such for any observable

$$\langle A \rangle = \mathrm{Tr}\,(\rho A).$$

In proving this Von-Neumann had used the assumption that sum of expectation values of two operators A_1 and A_2 even when they do not commute, is equal to the expectation value of $A_1 + A_2$ and which Bell criticized. Gleason was able to obtain the same result by making Von-Neumann's assumption only for commuting operators for separable real or complex Hilbert spaces of dimension greater or equal to 3. More precisely to each projection operator P we associate a non-negative number $\langle P \rangle$ such that:

(i) $\langle I \rangle = 1$

(ii) $\langle \sum_i P_i \rangle = \sum_i \langle P_i \rangle$ where P_i are mutually orthogonal,

then there exists a unique self-adjoint semidefinite density operator ρ such that

$$\langle P \rangle = \mathrm{Tr}\,(\rho P)$$

Since Von-Neumann's assumption is made here only for commuting operators our previous objections do not apply. The eigenvalues of $A_1 + A_2$ are equal to the sum of eigenvalues of A_1 and A_2, if A_1 and A_2 commute. Even physically the same experimental set up can be used to measure all three, i.e. A_1, A_2 and $A_1 + A_2$. Thus for Hilbert spaces of dimension > 3 Gleason's result again allows us to prove the impossibility of hidden variable theories of the following type.

If we measure the observable A on a system, which is described in quantum mechanics of dimension > 3, and if we demand that result of measurement of A dyes not depend on what other compatible observables are measured along with A, then the objective predictions of quantum mechanics are not reproducible by a hidden variable theory. Thus Gleason's work rules out non-contextualistic hidden variable theories.

We thus cannot associate, in general, hidden variables only with the system. The results of observation thus dependent not only on the system but also on the complete experimental set up. As expressed by Belinfante[32] "hidden variables cannot predict what 'is' but only what 'would be found'.

Kochen and Specker (1967) work on hidden variables leads to similar conclusions.[33]

(*d*) *Quantum non-separability and Einstein-Podolsky-Rosen theorem*

Another strong motivation for studying a possible hidden variable interpretation quantum mechanics is the presence of a kind of inseparability of two systems which have once interacted together. Once they have done so one cannot assign them separate state vectors any more. As Schrödinger put it "I would not call that one but rather the characteristic of quantum mechanics".[34] Since these two systems might have separated arbitrarily far, after the interaction between them is over, a number of persons have found this situation very unintuitive.

The unintuitive, and as some would say paradoxical, nature of quantum non-separability was brought out forcefully by Einstein, Podolsky and Rosen in their famous paper of 1935 'Can Quantum-mechanical Description of Reality be Considered Complete?'[6] Since the title of the paper may appear somewhat imprecise let us reproduce two of their definitions.

(i) A *necessary* condition for the completeness of a theory is that "every element of the physical reality must have a counterpart in the physical theory".

(ii) A *sufficient* condition to identify an element of physical reality: "If, without in any way disturbing a system, we can predict with certainty (i.e. with probability equal to unity) the value of a physical quantity, then there exists an element of physical reality corresponding to this physical quantity".

As a result of analysis of the thought experiment given in their paper one can prove the following theorem, as was stated by Einstein, in 1949, himself.[33]

Theorem: The following two assertions are not compatible with each other:

(1) the description by means of the ψ-function is *complete*.

(2) the real states of spatially separated objects are independent of each other.

Einstein's predilection was to believe the second assertion and thus answer the question raised in the title of their paper in the negative. He expressed this commitment unambiguously. "But on one supposition we should, in my opinion, absolutely hold fast: the real factual situation of the system S_2 is independent of what is done with the system S_1 which is spatially separated from the former".[36] We shall refer to it as the 'Einstein locality' postulate.

We now describe briefly EPR thought experiment in the Bohm version as it has some pedagogic advantages.[37] Consider a two particle system, with each particle having spin 1/2, in singlet (i.e. total spin 0) state. Let this system decay into these two particles as a result of angular momentum preserving interactions. As a result of disintegration the two particles will spatially separate and interaction between them will tend to vanish with their separation. The spin wave function would still be that corresponding to that a singlet state composed of two spin 1/2 particles.

Let us now measure a particular component of spin, say, z-component $S_z^{(1)}$ of one of the particles. Now the singlet wave function is given by

$$\psi\,(\text{singlet}) = \frac{1}{\sqrt{2}}\left(\chi_{\frac{1}{2}}^{(1)}(\hat{z})\,\chi_{-\frac{1}{2}}^{(2)}(\hat{z}) - \chi_{-\frac{1}{2}}^{(1)}(\hat{z})\,\chi_{\frac{1}{2}}^{(2)}(\hat{z})\right)$$

where we use the notion $\chi_m^{\alpha}(\hat{a})$ to denote the eigenfunction of the operator $\vec{S}_\alpha^{(\alpha)}.\hat{a}$ having the eigenvalue $m\hbar$ and $\vec{S}^{(\alpha)}$ is the spin angular momentum of the particle α and $\hat{a}$ is a unit vector in space. We can therefore assert that a measurement of the $S_z^{(1)}$, if it gives a value $\pm 1/2\hbar$, implies that the second particle has the value of S_z equal to $\mp 1/2\hbar$. Since we can thus predict without uncertainty the value of $S_z^{(2)}$ without in any way disturbing the second particle which is spatially far away, the *z*-component of spin for the second particle must be regarded as an element of reality and existing separately in the second particle alone. If so the second particle must have had the same value of S_z even before the measurement.

We now note that we are equally free to measure, instead of z component of $S^{(1)}$, some other component of $S^{(1)}$, say, *x*-component. Since the *same* singlet wave function can be reexpressed as

$$\psi\ (\text{singlet}) = \frac{1}{\sqrt{2}}\left(\chi_{\frac{1}{2}}^{(1)}(\hat{x})\,\chi_{-\frac{1}{2}}^{(2)}(\hat{x}) - \chi_{-\frac{1}{2}}^{(1)}(\hat{x})\,\chi_{\frac{1}{2}}^{(2)}(\hat{x})\right)$$

we can obtain the value $S_x^{(2)}$ without disturbing the second particle. Thus elements of reality must be existing for particle 2 which correspond to its spin component in any arbitrary direction. This, however, is impossible since different components of $\vec{S}^{(2)}$ do not commute. Hence the theorem follows.

In view of EPR theorem we must renounce at least one of two views. Most physicists would have wanted to have both.

(*e*) *Bell's inequalities and experimental tests of 'locality'*

Since 'Einstein locality' has a very appealing nature one can try to investigate whether supplementing quantum mechanics with hidden variable would allow us maintain this locality character of the physical theory. Bell's analysis of EPR thought experiment brings this question from the rarefied domains of philosophy to that experimental measurements.[1,38]

Consider the same two spin 1/2 particle system in a singlet state such that these two particles are moving in different directions towards two Stern-Gerlach magnets. Let the hidden variables of the system be collectively dented by λ. Let the first Stern-Gerlach magnet measure the spin-component along the direction $\hat{a}$, i.e. $\vec{\sigma}_1 . \hat{a}$. Clearly the result A, which would be either $+1$ or -1, can depend on $\hat{a}$ and λ, i.e. $A = A(\hat{a}, \lambda)$ and $A(\hat{a}, \lambda) = \pm 1$. Similarly for the result of measurement B of $\vec{\sigma}_2 . \hat{b}$ by the second Stern-Gerlach magnet $B = B(\hat{b}, \lambda)$ and $B(\hat{b}, \lambda) = \pm 1$. Note that locality requires that A does not depend on $\hat{b}$ and B does not depend on $\hat{a}$. We take it that measurement epochs are such that no direct light signal can reach between the two Stern-Gerlach magnets.

The average value of the product AB is then given by

$$P(\hat{a}, \hat{b}) = \int d\lambda\, \rho(\lambda)\, A(\hat{a}, \lambda)\, B(\hat{b}, \lambda)$$

where $\rho(\lambda)$ is the probability distribution of the hidden variables for the given quantum mechanical state ψ. We clearly must have $\rho(\lambda)$ positive semidefinite and normalized. To be somewhat more general the measuring magnets may also have some hidden variables and if we average over them we obtain

$$P_{\text{h.v.}}(\hat{a}, \hat{b}) = \int d\lambda \rho(\lambda) \bar{A}(\hat{a}, \lambda)\, \bar{B}(\hat{b}, \lambda)$$

where $|\bar{A}(\hat{a}, \lambda)| \leq 1$, $|\bar{B}(\hat{b}, \lambda) \leq 1$. We have put a subscript h.v. to refer to the local hidden variable nature of our expression. A consequence of this representation of $P_{\text{h.v.}}(\hat{a}, \hat{b})$ is the Bell's inequality[1,39]

$$|P_{\text{h.v}}, (\hat{a}, \hat{b}) - P_{\text{h.v.}}(\hat{a}, \hat{b}')| + |P_{\text{h.v.}}(\hat{a}', \hat{b}') + P_{\text{h.v.}}(\hat{a}', b)| \leq 2.$$

Important progress in deriving these inequalities with weaker assumptions[40] as well in getting further inequalities has been made.[41] We now note that for singlet state, quantum mechanics, leads to

$$P(\hat{a}, \hat{b}) = \langle \vec{\sigma}, \hat{a}\, \vec{\sigma}_2 . \hat{b} \rangle_{\text{singlet}} = -\hat{a}.\hat{b}.$$

Since this form of $P(\hat{a}, \hat{b})$, given by quantum mechanics cannot satisfy Bell's inequality it follows that "no local hidden variable theory can reproduce all the results of quantum mechanics".

The exciting possibility of checking the locality assumption directly by experiments on $P(\hat{a}, \hat{b})$ is opened up by Bell's inequality and its generallzation irrespective of quantum mechanics. A number of experiments have been done or are in progress.[42] Lamehi and Mittig[43] use proton spin correlation use in a S state for a two proton system. The polarisation correlation of two γ rays emitted in positronium annihilation were used in the experiment of Kasday, Wu and Ullmann[44] and in that of Faraci, Gutkowski, Notarringo and Pennisi.[45] The two photon polarisation correlation in atomic cascades were used in experiments of (i) Freedman and Clauser (1971) using

$$^1S_0(4p^2) \xrightarrow{5513°\text{A}} {}^1p_1(4S^4p) \xrightarrow{4227°\text{A}} {}^1S_0(4S^2)$$

cascade in calcium.[46]

(ii) Holt and Pipkin (1974) $9\,{}^1p_1 \xrightarrow{5676°\text{A}} 7\,{}^3S_1 \xrightarrow{4047°\text{A}} 6\,{}^3p_0$ cascade in a spin zero isotope of mercury.[47] (repeated by Clauser (1976)).[48] (iii) Fry and Thompson (1976).[49] Is the experiment of Aspect,[50] under planning, the polariser orientation will be set randomly so that at the time of emission of photons, the polariser orientation is still unknown.

Except for the experiment of Faraci and that of Holt and Pipkin there is at present no evidence in favour of local hidden variable theories. If we disregard these then nature clearly has opted for a "non-local" version which is consistent, within present experimental accuracies, with quantum mechanics.

(*f*) *Bohm and Bub's hidden variable theory of the reduction of the wavepacket*

The challenge of constructing a hidden variable deterministic theory which views the reduction of the wavepacket as a continuous and causal deterministic process—in contrast to usual quantum mechanics where it is discontinuous acausal and probabilitistic—was met by Bohm and Bub in their 1966 model.[51] It is a very unusual suggestion and works by introducing non-linear terms in Schrödinger equation which are operative only during the period of measurement.

Consider a spin 1/2 particle without translational motion (or any other quantum system with two-dimensinal Hilbert space) with wavefunction $|\psi\rangle = \psi_1|S_1\rangle + \psi_2|S_2\rangle$, where $|S_1\rangle$ and $|S_2\rangle$ are basis states in the Hilbert space in which the spin S is diagonal $|\psi\rangle$ is also normalized, i.e.

$$|\psi_1|^2 + |\psi_2|^2 = 1$$

To this we append the hidden variables ξ_1, ξ_2 in the *dual* Hilbert space

$$<\xi| = \xi_1 < S_1| + \xi_2 < S_2|$$

It is assumed that for the 'normal' quantum-mechanical ensemble these hidden variables ξ_1 and ξ_2 are distributed randomly over hypersphere

$$|\xi_1|^2 + |\xi_2|^2 = 1$$

We postulate the non linear equations of motion

$$\frac{d\psi_1}{dt} = \gamma(R_1 - R_2)\,\psi_1 J_2$$

$$\frac{d\psi_2}{dt} = \gamma(R_2 - R_1)\,\psi_2 J_1$$

where $J_1 = |\psi_1|^2 = R_1|\xi_1|^2,\ J_2 = |\psi_2|^2 = R_2|\xi_2|^2$

and γ is a quantity which is practically zero before and after the (impulsive) measurement and is a constant during the interaction period. Note that it follows from these equations

$$\frac{d}{dt}[|\psi_1|^2 + |\psi_2|^2] = 0$$

An interesting characteristic of these equations is as follows: If $R_1 > R_2$ initially then $|\psi_1|^2 \rightarrow 1$, $|\psi_2|^2 \rightarrow 0$ while if $R_2 > R_1$ initially then $|\psi_1|^2 \rightarrow 0$ and $|\psi_2|^2 \rightarrow 0$ after a short time. Thus whether the spin is found up or down on the initial values of the hidden variables ξ_1 and ξ_2. If one averages over ξ_1 and ξ_2 according to the quantum-mechanical ensemble distribution, given earlier, then one encounters the probability of finding spin up to be $|\psi_1|^2$ while spin down to be $|\psi_2|^2$ as required.

While there are many undesirable features in this suggestion, and one does not see how these equations could possibly be generalized to satify relativistic consideration, the challenge posed at the beginning of the section has clearly been met.

Not only that if some such non-linear model is really behind the reduction of wave packet phenomenon then in successive measurements, which follow each other at such short time intervals that the distribution of hidden variables has not become random, one should find violations of quantum mechanics. Papaliolios used this possibility to devise experiments to reveal hidden variables.[52] However, within the accuracy of the measurements no violations of quantum mechanics were found. His results set an upper limit of 1.9×10^{-14} for the relaxation time of hidden variable to reach random distribution. It is not impossible that future experiments of this type may reveal the variables which are now hidden from us.

4. Apologia

We had to leave out a large number of important and interesting aspects of our subject. We have, for example, said nothing about quantum logic approach and have not discussed stochastic interpretation at all. We make slight amends to it by referring to two recent review articles.[53]

A very desirable thing might have been to discuss the ontological aspets of quantum mechanics. What is the quantum mechanical nature of reality? It has, for example, been asserted that completeness of quantum mechaics is incompatible with external existence of space-time.[54] A number of attempts to work out quantum-mechanical relate to Whitehead philosophy of a world-process.[55] Another attempt is the theory of local beables.[56]

There are probably many other things of which we are not even aware of. Despite various limitations we hope that we have been able to bring a little excitement in the subject.

References

1. For historical aspects we refer to the following two excellent books: Jammer, M., *The Conceptual Development of Quantum Mechanics*, McGraw-Hill, New York (1966); *The Philosophy of Quantum Mechanics* Wiley, New York (1974).
2. Bohr, N., *Atomic Theory and the Description of Nature*, Cambridge (1934); *Atomic Physics and Human Knowledge*, Wiley, New York (1958); Essays 1958/1962 on Atomic Physics and Human Knowledge, Wiley, New York (1963).
3. Heisenberg, W., *The Physical Principles of the Quantum Theory*, Dover, New York, (1930); *Physics and Philosophy*, Harper, New York (1958).
4. Einstein, A., in '*Electrons et Photons*'—Report and Discussions of the 5th Solvay Congress, Gauthier-Villars, Paris (1928); For a recent review of this viewpoint see Ballentine, L.E., 'The Statistical Interpretation of Quantum Mechanics',

Rev. Mod. Phys. **42**, 358 (1970); Blokhintsev, D.I., *The Philosophy of Quantum Mechanics*, Reidel, Dordrecht, Holland, (1968).

5. Bohr, N., Discussions with Einstein on Epistemological Problems in Atomic Physics, in *Albert Einstein: Philosopher-Scientist*, P.A. Schilpp (Ed.), Library of Living Philosophers, Evanston, III (1949), reprinted by Harper and Brothers, New York (1959) p. 199–242. This book will be referred to as Schilpp volume (Ref. 5).
6. Einstein, A., Podolsky, B., and Rosen, N., *Phys. Rev.* **47**, 777 (1935).
7. Bell, J.S., *Physics* (N.Y.) **1**, 195 (1964).
8. Nilson, D.R., 'Bibliography on the history and philosophy of quantum physics' in '*Logic and Probability in Quantum Mechanics*', edited by P. Suppes Reidel, Dordrecht-Holland, (1976). (For a recent bibliography listing 37 items on history of quantum physics and 944 items on philosophy of quantum theory). d'Espagnat, B., *Conceptual Foundations of Quantum Mechanics*, W.A., Benjamin, Calif. (1971). We also list among useful general references for our subject: *Foundations of Quantum Mechanics*, D'Espagnat (Ed.), Academic Press, New York, (1971). This book will be referred to as D'Espagnat, 'Foundations' (ref. 8). Belinfante, F.J., *A Survey of Hidden Variable Theories*, Pergamon, Oxford (1973).
9. Landau, L.D., and Lifshitz, E.M., *Quantum Mechanics* (trans. by J.B. Sykes and J.S. Bell), Addison-Wesley, Reading, (1958); Stapp, H.P., *Am. Jour. Phys.* **40**, 1098 (1972).
10. Von-Neumann, J., Mathematische Grundlagen der Quanten Mechanik, Springer Verlag, Berlin (1932), Mathematical Foundations of Quantum Mechanics (trans. by R.T. Beyer), Princeton (1955). [We refer to english translation]
11. Wigner, E.P., *Am. Jour. Phys.*, **31**, 6 (1963).
12. ———, *Phys. Zeits.*, **133**, 101 (1952); Araki, H., and Yanase, M.M., *Phys. Rev.*, **120**. 622 (1960); Yanase, M.M., *Phys. Rev.*, **123**, 666 (1961).
13. Zeh, H.D., *Found. Phys.*, **1**, 67 (1970).
14. Ludwig, G., *The Measuring Process and an Axiomatic foundation of Quantum Mechanics* in D'Espagnat 'Foundations' (Ref. 8).
15. Daneri, A., Loinger, A.. and Prosperi, G.M., *Nucl. Phys.*, **33**, 297 (1962); *Nuovo Cimento* **44B**, 119 (1966).
16. Rosenfeld, L. in supplement of the *Prog. Theor. Phy.*, p. 222 (1965).
17. Bub, J., *Nuovo Cimento* **57 B**, 503 (1968); Jauch, J.M., Wigner, E.P., and Yanase, M.M., *Nuovo Cimento*, **48 B**, 144 (1967); Loinger, A.. *Nucl. Phys.*, **108**, 245 (1968); Prosperi, G.M., in D'Espagnat 'Foundations' (Ref. 8) Macroscopic physics and the problem of measurement in quantum mechanics.
18. Schrödinger, E., Naturwissenschaften 23, pp. 807, 823, 844 (1935).
19. An english translation of the passage describing the "Schrödinger cat" is given in Jauch, J.M., *Foundations of Quantum Mechanics*, p. 185, Addison-Wesley, Reading, (1968).
20. London, F., and Bauer, E., *La Theorie de L'observation en Mecanique Quantique*, Hermann and Cie, Paris (1939).
21. Wigner, E.P., *Symmetries and Reflections*, Indiana University Press (1967).
22. Everett, H., III, 'The Theory of the Universal Wavefunction' in Dewitt, B.S., and Graham, N. (Eds.), '*The Many-Worlds Interpretation of Quantnm Mechanics*', Princeton University, N.J., (1973).
23. Hepp, K., *Helv. Phys. Acta*, **45**, 237 (1972); Bell, J.S., *Helv. Phys. Acta*, **48**, 93 (1975).
24. Practically all the relevant work is available in the volume mentioned in Ref. 22, See also J.B. Hartle, *Am. Jour Phys.*, 36, 704 (1968); 'Quantum Mechanics debate', Physics Today, p. 36–44 (April 1971); Bell, J.S., Th 1424–CERN (1971); Ballentine, L.E., *Found. Phys.* **3**, 229 (1973).

25. Quoted in the obituary of H.S. Allen 'Biographical Memories of Fellows of the Royal Society' Vol. 1, p. 8–9 (1955).
26. Ref. 10, Ch. 4 See also Albertson. J., *Am. Jour. Phys.* **29,** 478 (1961).
27. Ref. 10, p. 325.
28. Bell, J.S., On the Problem of Hidden Variables in Quantum Mechanics, *Rev. Mod. Phys.* **38,** 447 (1966).
29. Bohm, D., *Phys. Rev.*, **85,** pp. 166, 180 (1952).
30. Jauch, J.M., and Piron, C., *Helv. Phys. Acta* **36,** 827 (1963), Misra, B., *Nuovo Cimento*, **47**A, 841 (1967).
31. Gleason, A.M., *J. Math. Mech.*, **6,** 885 (1957).
32. Belinfante, F.J., (quoted in Ref. 8) p. 35.
33. Kochen, S., and Specker, E., *J. Math. Mech.*, **17,** 59 (1967).
34. Schrödinger, E., *Proc. Cambridge Phil. Soc.* **31,** 555 (1935).
35. Einstein, A., Reply to Criticisms in Schilpp volume (Ref. 5) p. 682.
36. Einstein, A., Autobiographical Notes in Schilpp volume (Ref. 5) p. 85.
37. Bohm, D., *Quantum Theory* (Prentice-Hall, Englewood Cliff. N.J., (1951).
38. Bell, J.S., 'Introduction to the Hidden Variable Question' in D'Espagnat Foundations' (Ref. 8).
39. Clauser, J.F., Horne, M.A., Shimony, A. and Holt, R.A. *Phys. Rev. Lett.*, **26,** 880 (1969).
40. Wigner, E.P., *Am. Jour. Phys.* **38,** 1005 (1970); Stapp, H.P., *Phys. Rev.* **D3,** 1303 (1971); d'Espagnat, D., *Phys. Rev.*, **D11,** 1424 (1975); *Phys. Rev.* **D18,** 349 (1978); Gutkowski, D., and Masotto, G., *Nuovo Cimento*, 22B, 1921 (1974); Eberhard, P.H., *Nuovo Cimento*, **B38,** 75 (1977).
41. Selleri, F., *Found. Phys.*, **8,** 103 (1978); Roy, S.M., and Singh, V., *Jour. Phys.* A, **11,** L167 (1978); Roy, S.M., and Singh, V., Completeness of Tests of Local Hidden Variable Theories, *Jour. Phys.*, A **12,** 1003(1979).
42. For a general review of the experimental situation we refer to; Pipkin, F.M., 'Atomic Physics Tests of the Basic Concepts in Quantum Mechanics' (Advances in Atomic and Molecular Physics, **14,** 281 (1978).
43. Lamehi-Rachti, M., and Mittig, W., *Phys. Rev.* **D14,** 2543 (1976).
44. Kasday, L.R., Ullmann, J.D., and Wu, C.S,, *Nuavo Cimento* **25,** 633 (1975); Kasday, L.R., in d'Espagnat 'Foundations' (Ref. 8).
45. Faraci, C., Gutkowski, D., Natarrigo, S., and Pennisi, A.R., *Nuovo Cimento Letters*, **9,** 667 (1974).
46. Freedman, S.J., and Clauser, J.F., *Phys. Rev. Lett.* **28,** 938 (1972).
47. Holt, R.A., and Pipkin, F.M., Harvard University, preprint (1947).
48. Clauser, J.F., *Phys. Rev. Lett.* **36,** 1223 (1976).
49. Fry, E.S., and Thompson, R.C., *Phys. Rev. Lett.*, **37,** 465 (1976).
50. Aspect, A., *Phys. Lett.*, **54 A,** 117 (1975) and *Phys. Rev.* **D 14,** 1944 (1976).
51. Bohm, D., and Bub, J., 'A Proposed Solution of the Measurement Problem in Quantum Mechanics by a Hidden Variable Theory', *Rev. Mod. Phys.*, **38,** 453 (1966).
52. Papaliolios, C., *Phys. Rev. Lett.* **18,** 622 (1967).
53. Ghirardi, G.C., Omero, C., Rimini, A.. and Weber, T., The Stochastic Interpretation of Quantum Mechanics: critical review, *Revista del Nuovo Cimento*, **1,** no. 3 (1978); Beltrametti, E.G., and Cassinelli, G., 'Logical and Mathematical Structures of Quantum Mechanics', *Revista del Nuovo Cimento*, **6,** 321 (1976).
54. Stapp, H.P., *Am. Jour. Phys.*, **40,** 1098 (1972); Chew. G.F., *Science Progress* **51,** 529 (1963).
55. Whitehead, A.N., *Process and Reality*, Macmillan, New York (1929); Shimony, A., *Quantum Physics and the Philosophy of Whitehead*, Humanities, New York (1956); Stapp, H.P., *Phys. Rev.* D 3, 1303 (1971); *Nuovo Cimento*, **29 B,** 270 (1975); *Found. Phys.* **7,** 313 (1977).
56. Bell, J.S., Th 2053–CERN (1975) and *Epistemological Lett.* **15,** (Feb. 1977).

Predictive and Retrodictive Einstein Correlation in the *S*-Matrix Formalism

O. COSTA DE BEAUREGARD

1. Introduction

As early as 1927, over the very cradle of the 'new' quantum mechanics, Einstein[1] pointed at one paradoxical feature which, through the years, was to draw ever more talking and thinking, and finally develop as heralding an important new paradigm. This paradox of 1927 is better known as the Einstein-Podolsky-Rosen[2] 1935 paradox, then formulated in non-relativistic quantum mechanical style. Subsequent versions of the paradox have been mainly in terms of polarizations of electrons or photons as correlated through a common origin, discussed in well-known writings by Bohm[3], Furry[4], Bohm-Aharonov[5], and Shimony[6] et al.

This trend of thought, coming in touch with Bell's[7] problematic of hidden variables (which will not be discussed in this paper), has finally led to very striking experimental demonstration[8] of the *reality of the Einstein paradox*, which thus joins the historical Copernicus and Michelson paradoxes* as the promotor of an important new paradigm[9] to be adequately defined.

Briefly, the *Einstein paradox* consists in a *correlation between distant measurements at* L *and* N *upon systems that have previously interacted at* C, the point being that, in the chance game that is at stake, "*the dice, so to speak, are definitely not cast at* C, *while shaken in the cup, but at* L *and* N, *when rolling on the table*." This is exactly the point made as early as 1927 by Einstein, who believed the thing to be unacceptable[10] as also did Schrödinger,[11] Renninger,[12] de Broglie,[13] and many others, but which is today an experimental truth.

What is less known is that there also exists a *time-reversed Einstein paradox* consisting in a *correlation between distant preparations* L *and* N *that will later interact at* C, also demonstrated experimentally, as a correlation of occupation numbers of waves issuing from two independent lasers.[14] Even at the reception of one single photon in the interference region there is a phase coherence between the two waves, and the usual 2-slit discussion then shows that one cannot retrodict from which of the two lasers the photon has come. In other words, the photon has been emitted by both

*Paradox: a very surprising but perhaps true statement (meaning *n*°1 in all dictionaries).

lasers in unison, and they are coupled by their future interaction.*

So, there is a predictive and retrodictive Einstein correlation, the paradox being, in both cases, that there is no 'present' interaction, but *an interaction through, respectively, the past or the future*.

It should thus be clear that, in one of its aspects, the paradox consists in *intrinsic time symmetry at the elementary level*, very much like the celebrated Loschmidt and Zermelo paradoxes did in their days.

The new trait in the Einstein paradox, however, and the one which makes its sting so painful, is that this intrinsic time symmetry is expressed in terms no more of the classical probability calculus, obeying the principle of addition of partial probabilities, but rather of the new 'wavelike' probability calculus defined in 1926 by Born, obeying the law of addition of partial amplitudes.

As is well known, squaring the absolute value of the sum of partial amplitudes (for obtaining the probability) yields diagonal terms which, if alone, would give back the old law, plus off-diagonal terms containing the phase differences. It is because of these terms that the Einstein correlation is so paradoxial.

Of course, as Nature is both relativistic and quantal, it is in relativistic quantum mechanics that the full import of the Einstein paradox shows up. While this has been more or less stated verbally quite a few times ever since 1927, the mathematical formalism has not been produced. This I intend to do here by expanding previous papers.[15]

So, the full fledged Einstein paradox belongs to quantum mechanics in its relativistically covariant expression. Both the relativity and the quantum theory (especially in its 'new', 1924–1927, version) are born from the physics of waves; also, Born's 1926 statistical interpretation of wave mechanics truly defines a new *wavelike probability calculus*. It is thus inside the *relativistic quantum field theory* that the Einstein paradox gains its full stature, as expressing *intrinsic symmetry between statistical prediction and retrodiction* (the paradox already inherent in the classical probability calculus itself[16] and underlying the Loschmidt and the Zermelo paradoxes) together with *intrinsic symmetry between retarded and advanced waves*. Fock[17] and Watanabe[18], among others, have been adamant in emphasizing that this binding is inherent in the 'wavelike probability calculus' of quantum mechanics.

2. *n*-Uple Einstein correlation in general

We consider a state vector $|\Phi\rangle$ expandable in the form[19] (summation sign over repeated indices ommitted)

$$|\Phi\rangle = c^{ij\cdots}|\phi_I\rangle|\chi_J\rangle\cdots \tag{1}$$

*To take the two statements as equivalent to each other implies of course that probability and chance are taken as irreducible in quantum mechanics.

where the $| \phi \rangle$'s, span disjoint Hilbert spaces, and a Hermitean operator $M = m \otimes p \otimes \cdots$, where the m's act on the $| \phi |$'s. The mean value

$$\langle \Phi \mid M \mid \Phi \rangle = c^{i'j'} \cdots c^{ij} \cdots \langle \phi_{i'} \mid m \mid \phi_i \rangle \langle \chi_{j'} \mid p \mid \chi_j \rangle \cdots \tag{2}$$

is the correlated mean value of the measurement $m \otimes p \otimes \ldots$.

We write it

$$\langle \Phi \mid M \mid \Phi \rangle = \langle \Phi \mid M \mid \Phi \rangle_0 + \Delta \langle \Phi \mid M \mid \Phi \rangle \tag{3}$$

as it contains the diagonal terms

$$\langle \Phi \mid M \mid \Phi \rangle_0 = w^{ij} \cdots \langle m_i \rangle \langle p_j \rangle \cdots \tag{4}$$

with, by definition,

$$w^{ij} \cdots = c^{i'j'} \cdots c^{ij} \cdots \tag{5}$$

$$\langle m_i \rangle = \langle \phi_i \mid m \mid \phi_i \rangle, \text{ etc} \ldots \tag{6}$$

and the off-diagonal terms

$$\Delta \langle \Phi \mid M \mid \Phi \rangle = \langle \Phi \mid M \mid \Phi \rangle - \langle \Phi \mid M \mid \Phi \rangle_0 \tag{7}$$

The contribution (4) has the classical form of a sum of partial probabilities and, if alone, would represent a mixture. The contribution (7) expresses Born's interference of probabilities, and stems, of course, from the addition (1) of partial amplitudes.

Contrary to (2), neither (4) nor (7) is basis invariant, and, in particular, the neoquantal correction (7) is rendered 0 if the representation diagonalizes all the m's. But this of course yields merely the semblance of a mixture, as it is *relative* to the reference frame (i.e. the basis).

The crux of the paradox is that, due to the contribution (7), the subsystems $| \phi \rangle \ldots$ *cannot* be thought of as *possessing* the eigenvalues of the corresponding operators $m \ldots$. While this does not cause too much headache as long as the subsystems can be thought of as 'presently' interacting (as in the theories of ortho- and para-hydrogen, or of ortho- and para-helium), it *does* become *extremely* mind shaking when concerning distant preparations, or measurements, made upon subsystems that will interact, or have interacted.

This can be shown[19] quite dramatically by discussing the predictive correlation of polarizations of a pair of photons issuing from a cascade. There is absolutely no doubt that if the corresponding experiments[8] had been performed in the days of the old quantum theory, they would have caused the same sort of stupefaction as did the Michelson experiment.

3. *n*-Uple Einstein correlation in the *S*-matrix formalism

The Schwinger-Feynman transition amplitude (in the interaction picture)

$$\langle \Phi \mid \Psi \rangle = \langle \Phi^2 \mid U \Psi_1 \rangle \langle \Phi_2 U \mid \Psi_1 \rangle \tag{8}$$

between an initial $| \Psi(\sigma) \rangle$ and a final $| \Phi(\sigma) \rangle$ state, being a scalar product is intrinsically symmetrical between the two vectors $| \Phi \rangle$ and $| \Psi \rangle$. Of course, it can be thought of, or performed, asymmetrically, either by projecting $|U\Psi_1\rangle$ upon $|\Phi_2\rangle$, or by projecting $|U^{-1}\Phi_2\rangle$ upon $|\Psi_1\rangle$ (U evidently denotes that specification of the unitary evolution operator leading from σ_1 to σ_2).

In the first case we speak of *collapsing* the state vector in view of *predictions* using retarded waves with *sources* on σ_2; in the second case we speak of *anticollapsing* the state vector in view of *retrodictions* using *advanced waves* with *sinks* on σ_1.

Now we introduce the complete orthogonal set of projectors $|\Theta\rangle\langle\Theta|$ corresponding either to a measurement (predictive case) or to a preparation (retrodictive case), and rewrite (8) in the form (omitting the summation sign over the θ's)

$$\langle\Phi\,|\,\Psi\rangle = \langle\Phi\,|\,\Theta\rangle\langle\Theta\,|\,\Psi\rangle \tag{9}$$

This is none else than a covariant expression of (1). In a predictive problem we interpret the $\langle\Theta|\Phi\rangle$'s as the components of final state (in the Θ representation) and the $\langle\Psi|\Theta\rangle$'s as the coefficients of the expansion. In a retrodictive problem we interpret the $\langle\Theta|\Psi\rangle$'s as the components of the initial state and the $\langle\Theta|\Psi\rangle$'s as the coefficients of the expansion.

For example, in quantum electrodynamics, the $|\phi_\lambda\rangle$'s in (1) are the states $|A\rangle$ of the photon, $|\bar{\psi}\rangle$ of the electron and $|\psi\rangle$ of the positron. Indeed, the Feynman transition amplitude is a direct product of the form (1). A specific discussion of the Einstein correlation has been given along these lines.[20]

Logically there was one missing link in the Schwinger-Feynman theory: an explicitly covariant definition of the states $|\phi\rangle$ at σ_1 and σ_2, of their Hermitean scalar product, etc. This has been given.[21]

Once made clear that (9) is covariant expression of (1) all the rest

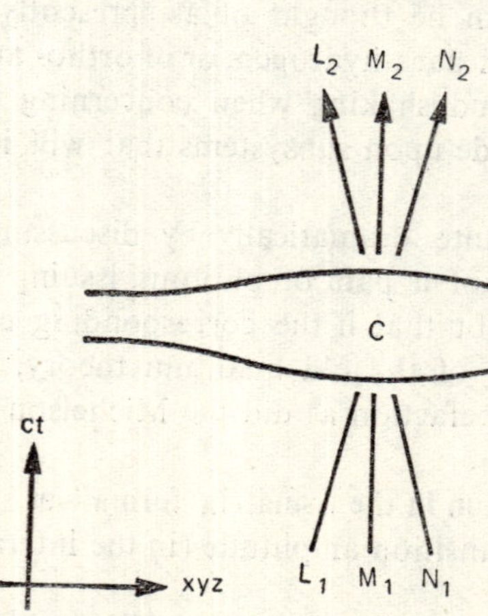

follow, as shown in the figure. Due to the interaction in the C region, there is an n-uple predictive Einstein correlation between the outgoing states L_2, M_2, N_2, ... and an n-uple retrodictive Einstein correlation between the incoming states L_1, M_1, N_1, ... It is thus obvious that the *deus exmachina* of these distant correlation is the Feynman zigzag, as *I*[22] have stated quite a few times since 1953, and is also apparent in writing by Stapp[23] and Davidon.[24]

4. Conclusions

The paradoxical Einstein[1] correlations, either predictive[8] or retrodictive[14], herald the advent of an important paradigm implicit—that is, implied—in the very heart of both the relativity and the quantum theory, which both are born from the physics of waves. This belongs to the *wavelike probability calculus* initiated by Born in 1926 and, as shown in this paper, to the very formalism of the *S*-matrix and the Feynman zigzags.

The Einstein paradox is born from the union of two (already) paradoxical genitors: *intrinsic time symmetry inside the calculus of probabilities* (that is, *intrinsic symmetry between (blind statistical) prediction and retrodiction**); and *existence of the off-diagonal, interference Born terms,* inside the expression of the probability.

As belonging to the relativistic formalism, the paradox, consisting in *spacelike connections established via Feynman zigzags* made of two timelike vectors, with a relay either in the past (Einstein paradox proper) or in the future (time-reversed Einstein paradox), definitely shows that *Einstein's prohibition to 'telegraph into the past' was of the factlike*[25], *or macroscopic nature.* It does not hold at the elementary level, where time symmetry prevails, This, in turn, entails the *possibility of telegraphing indirectly outside the light cone, via Feynman zigzags, in the specific sense of the Einstein correlations*, as definitely proved experimentally[8, 14].

So, contrary to an opinion that has been expressed[26], the existence of the Einstein correlations does not contradict relativistic covariance**, but rather the prejudice excluding advanced waves and connections *via* Feynman zigzags. It is in this sense, and in de Broglie's[13] words, that *they upset "our accepted notions concerning space and time"*.

As belonging to the wavelike probability calculus, the Einstein correlations do stress, *via* the intrinsic symmetry between (blind statistical) prediction and retrodiction, the symmetry between the *two faces of the information concept, gain in knowledge, and organizing power* (as exemplified in the reception and the emission of a *signal* endowed with a *negentropy*†)

*Macroscopically speaking, even when the transition probabilities are symmetric (as in card shuffling or in radioactive decays) blind statistical prediction is physical while blind statistical retrodiction would be anti-physical. But this is "a factlike, not lawlike"[28], statement. This is the common root of the Loschmidt and the Zermelo paradoxes. (Another instance of lawlike symmetry versus factlike asymmetry is the one of particles and anti-particles.)

**The time symmetric, 'micro' relativity is more covariant than the time asymmetric, 'macro' relativity.

†Both of these aspects of the information concept are present in the writings of Aristotle and of Thomas Aquinas.

This, togeter with the assumption that the quantal chance concept is *irreducible*, implies that there *is* a reaction of the observer upon the observed system, because the act of consciousness tied with *preparation* and *measarement* is *indissolubly cognizance and will.** *Therefore two* (*or more*) *observers of one quantal transition are not independent from each other* (as Einstein[10] and Schrödinger[11] assumed they should be). They are truely cooperating or competing for producing the *same* wave collapse—in their common past. This, again, *is* the Einstein paradox.

It does not belong to a technical paper like this one to expand the epistemological consequences of such theoretical facts. This we have attempted elsewhere.[27] Suffice it to recall here not only what de Broglie[13], but also what Einstein[10] and what Schrödinger[11] had to say. Both Einstein and Schrödinger had deemed unacceptable these distant correlations between stochastic occurrences as implying 'telepathy'[10] or 'magic'[11]. Now that the Einstein correlations are proxed as facts, it is well worth pondering the implications of the relativistic non-separability between the past and the future (as the light cone produces a trichotomy and no more a dichotomy) together with the non-separability of distant preparations or measurements (due to the interference of probabilities). It may well turn out[27] that the space-time telegraph using quantized waves at the micro-level displays features quite alien to the one obeying the irreversibility principle of macro physics.

References

1. Einstein, A., in *Repports et Discussions du 5e Conseil Solvay*, Gauthier Villars, Paris, pp. 253–256 (1928).
2. Einstein, A., Podolsky, P., and Rosen, N., *Phys. Rev.* **47**, 777 (1935).
3. Bohm, D., *Quantum Theory*, Prentice-Hall, N.J. 164, (1951).
4. Furry, N.H., *Phys, Rev.* **49**, pp. 393, 476 (1936).
5. Bohm, D., and Aharonov, Y., *Phys. Rev*, **108**, 1070 (1957).
6. Clauser, J.F., Horne, M.A., Shimony, A., and Holt, R.A., *Phys. Rev. Lett.* 23, 880 (1969).
7. Bell, J.S., *Physics* **1**, 195 (1964).
8. Freedman, S.J., and Clauser, J.F., *Phys. Rev. Lett.* **28,** 938 (1972); Clauser, J.F., *Phys. Rev. Lett.* **36,** 1223 (1976); Fry, E., and Thompson, R.C., *Phys. Rev. Lett.* **37,** 465 (1976); Kasday, L.R., Ullman, J.D., and Wu, C.S., *Nuovo Cimento*, **25 B,** 633 (1975); Wilson, A.R., Low, J., and Butt, D.K., *Jour. Phys. G,* **2,** 613 (1976); Lamehi-Rachti, M., and Mittig, W., *Phys. Rev.* **D 14,** 2543 (1976).
9. Colodmy, R G., (Ed.), *Paradox and Paradigm*, Pittsburgh, 1973.
10. Einstein, A., in *Einstein*: *philosopher scientist*, (Ed.), P.A. Schilpp, The Library of Living Philosophers, Illinois, pp. 85, 683 (1949).

*By discussing the covariant formalism of reciprocal Fourier transforms of solutions of wave equations (implying positive and negative frequencies), resolution of the Cauchy problem (towards both the past and the future), and the covariant position measurement problem, the none separability of these symmetrically related concepts is explicit in the formulas.[21]

11. Schrödinger, E., *Naturwissenschaften* **23,** 844 (1935).
12. Renninger, W., *Physik* **158,** 417 (1960); *Phys. Zeits.* **136,** 251 (1953).
13. de Broglie, L., *Une tentative d'interprétation... de la mécanique ondulatoire,* Gauthier Villars, Paris, p. 73 (1956).
14. Pflegor, R.L., and Mandel, L., *Phys. Rev.* **159,** 1084 (1967); *Jour. Opt. Soc. Amer.* **58,** 946 (1968).
15. de Beauregard, Costa, O. *Comptes Renaus* **A 286,** 535 (1978); *Phys. Lett.* **A 67,** 171 (1978).
16. de Beauregard, O. Costa, *Cah. de Phys.* **12,** 317 (1958).
17. Fock, V., *Dokl. Akad. Nauk SSSR* **60,** 1157 (1948).
18. Watanabe, S., *Rev. Mod. Phys.* **27,** 26, 40, 179 (1952).
19. de Beauregard O. Costa, *Nuovo Cimento Letters,* **19,** 113 (1977); *Nuovo Cimento.* **42 B,** 41 (1977); Garuccio, A., and Selleri, F., *Nuovo Cimento* **36 B,** 176 (1976).
20. ———, *Phys. Lett.* **60 A,** 93 (1977).
21. ———, *Précis de Mécanique Quantique Relativiste,* Dunod, Paris, 1967.
22. ———, *Comptes Rendus* **236,** 1632 (1953); *Rev. Int. Philos.* **61–62,** 1 (1962); *Dialectica* **19,** 280 (1965); *in Proceedings of International Conference on Thermodynamics,* (Ed.), P.T. Landsberg, Butterworths, London, p. 539 (1970).
23. Stapp, H.P., *Nuovo Cimento* **29 B,** 270 (1975).
24. Davidon, W.C., *Nuovo Cimento* **36 B,** 34 (1976).
25. Mehlberg, H. in *Current Issues in the Philosophy of Science,* (Ed.), H. Feigl and G. Maxwell, p. 105, Holt, Rinehart & Winston, New York, 1961. For an overall view and a guide to the literature of irreversibility, see O. Costa de Beauregard, in *Proceding International Congress for Logic, Methodology and Philosophy of Science,* (Ed.), Y. Bar Hillel North Holland p. 73 (1964); *Studium Generale* **24,** 10 (1947).
26. d'Espagnat, B., *Conceptual Foundations of Quantum Mechanics,* 2nd ed., Benjamin, New York, 1976 (passim).
27. O. Costa de Beauregard, *Found. Phys.* **6,** 539 (1976); *Synthese* **35,** 129 (1977).

Deterministic Model Equation of Motion for Quantum Mechanics and a Probability Amplitude Description in Classical Mechanics

RAM K. VARMA

1. Introduction

This paper is devoted to the question relating to the statistical character of the quantum mechanical description. This question, as is well known, had interested Einstein greatly ever since the discovery of quantum mechanics and was the subject of a heated controversy between him and Copenhagen School led by Bohr and Heisenberg.

While the Copenhagen School regarded the statistical nature of quantum mechanics as something intrinsic and irreducible, Einstein though convinced of the correctness of quantum mechanics was not willing to accept the irreducibility of its statistical character. Einstein, appeared to suggest the possibility of the existence of a deterministic substructure underlying the statistical formalism of quantum mechanics.

The investigation reported here has been motivated by the desire to find a possible deterministic basis for non-relativistic quantum mechanics and to see if the latter can be obtained as a probabilistic description of an ensemble of systems following deterministic trajectories. As in statistical mechanics, the probabilistic nature of the description would then be a consequence of the lack of information about some of the initial conditions specifying the deterministic trajectories. The investigation at the same time also seeks to clarify whether the probability in quantum mechanics is of a different nature from the classical probability or is identical with it.

In a recent work (Varma 1978) we have given a deterministic set of equations of motion in a space extended by introducing an extraangular co-ordinate. The motion in classical mechanics is then identified as the motion of the position co-ordinate in an adiabatic approximation, while the quantum behaviour of an individual system is identified as the motion of the system in the extended space in the neighbourhood of the adiabatic (that is, classical) motion.

By applying the laws of classical probability to an ensemble of systems in the extended phase space (in particular, the law of conservation of probability along trajectories—the Liouville theorem) we arrive at a probabilistic description, in terms of probability amplitudes, which turns out to be a generalization of the Schrödinger description with a built-in probability interpretation. The generalized description admits of an infinite number of wavefunctions $\Psi(n)$, $n = 1, 2, 3, \ldots$ which obey a set of coupled Schrödinger-like equations with the total probability density being $|\Psi(n)|^2$ summed over all the modes $n = 1, 2, 3, \ldots$. In the particular case that all the modes $n \neq 1$ are negligible, the system of equations reduces to the Schrödinger set. In general, the generalized set of equations predict the existence of other modes $n \neq 1$ for the propagation of probability.

It is, therefore, concluded that (i) the individual quantum events can, in principle, be described by deterministic equations of motion in an extended space, and (ii) behaviour of an ensemble of trajectries representing a quantum state in this formalism can be described by a set of probability amplitudes obeying generalized Schrödinger equations; and that (iii) the classical probability applied appropriately to individual events not only embraces probability in quantum mechanics, but allows for new modes for its propagation.

A somewhat more general question may be posed in this connection, namely, whether it is possible to describe the ensemble behaviour of at least a certain class of deterministic systems in terms of a set of probability amplitudes; in particular, whether such a description is possible for a certain situation in classical mechanics (which is a deterministic system).

We have, in fact, found such a system in classical mechanics (Varma 1971) which admits of a description in terms of a set of probability amplitudes obeying Schrödinger-like equations. Recently experiments carried out at the Physical Reseacrch Laboratory (Bora et al. 1979) have confirmed some of the rather surprising predictions of the description in terms of the probability amplitudes. We shall briefly describe the problem and the results of the experiments verifying the predictions of the theory.

We shall briefly outline here the derivation of the generalized set of Schrödinger equations starting from a postulated set of deterministic equations of motion. For details the original paper (Varma 1978) may be referred to.

2. Deterministic trajectories, state preparation and the quantum-mechanical ensemble

As set out in the Introduction, the main contention of the point of view being presented here is that an individual quantum event can, in principle, be described by deterministic trajectories in an extended space of one higher dimension obtained by the addition of an angular co-ordinate to the original three spatial dimensions. The equations of motion describing the motion in this extended space are postulated. It is assumed, however, that any procedures which are employed to prepare what is identified as a 'quantum state' are unable to specify all the initial conditions required to define a trajectory. In particular, the new angular co-ordinate cannot be fixed to any pre-assigned value for any individual quantum system, though a particular event will correspond to a particular initial value for the angular co-ordinate.

Following the terminology introduced by Khinchin (1949) in his exposition of statistical mechanics, we introduce the concept of 'controllable' and 'free' integrals of motion. All those integrals of motion of the deterministic trajectories which can be specified through a state preparation procedure are termed as 'controllable' while the others are termed 'free'. In terms of the deterministic trajectories, a state preparation procedure then prepares an ensemble which is specified by certain pre-assigned values of the controllable integrals of motion and by an unknown distribution in the values of the free integrals of motion. It is the unknown nature of the latter which imparts the probabilistic character to the description. One is thus led to identify a quantum state with an appropriately prepared ensemble of the deterministic trajectories. The dispersive nature of the quantum state is thus well described by the ensemble since the various members constituting the ensemble representing a quantum state would, in general, follow different trajectories, for though they have the same set of values for the controllabie integrals, they have a distribution in the values of the free integrals.

3. Generalized equation of motion and the classical limit

The generalized determininistic trajectories are introduced through a variational principle which gives the equations of motion for them as the Euler-Lagrange equations.

If one considers for simplicity the case of a particle of mass m, then the equations of motion are postulated to follow from the variational principle

$$\delta S = \delta \int \Lambda \, dt = 0, \tag{1}$$

with the Lagrangian Λ of the form:

$$\Lambda = \tfrac{1}{2} m \dot{\mathbf{x}}^2 + \tfrac{1}{2} g\,(\mathbf{x}, \chi)\, \dot{\chi}^2, \tag{2}$$

where χ is the extraangular co-ordinate introduced and $\mathbf{x}$ is the usual position co-ordinate of the particle. The Euler-Lagrange equations resulting from (1) and (2) are given by:

$$m\ddot{\mathbf{x}} = \frac{1}{2}\, \dot{\chi}^2 \frac{\partial g}{\partial \mathbf{x}} (\mathbf{x}\chi), \tag{3a}$$

$$\frac{d}{dt}(g\dot{\chi}) = \frac{1}{2}\, \dot{\chi}^2 \frac{\partial g}{\partial \chi} (\mathbf{x}\,\chi) \tag{3b}$$

These equations constitute a generalization of the equations of classical mechanics. That being the case, one must show under what conditions and with what identification do these equations go over into those of classical mechanics. It has been shown that if the fractional change in g due to the change in $\mathbf{x}$ co-ordinate is small per period of the χ-motion, that is, if

$$\eta \equiv (g\dot{\chi})^{-1}\,\dot{\mathbf{x}}\,.\frac{\partial g}{\partial \mathbf{x}} \ll 1, \tag{4}$$

then the momentum p_χ conjugate to χ is an adiabatic invariant.

$$p_\chi \equiv g\dot{\chi} = 2\varepsilon\,.(\text{const.}). \tag{5}$$

If this invariance (5) is used to eliminate $\dot{\chi}$ from equation of motion (3a), then one gets

$$m\ddot{\mathbf{x}} = -\frac{\partial}{\partial \mathbf{x}}\left(\frac{2\varepsilon^2}{g}\right). \tag{6}$$

Eqn. (6) can then be identified as the equation of motion for classical mechanics where now $2\varepsilon^2/g$ can be identified as the potential for the classical motion. In this picture, the potential, therefore, appears as a property of the 4-dimensional space, and in particular of the metric component g.

It is to be noted that classical motion as identified with Eqn. (6) appears only as an approximation to the exact motion in the 4-dimensional space inasmuch as it is based on the approximate (adiabatic) invariance of $p_\chi = g\dot{\chi}$. Departures from this approximate behaviour will occur under conditions which violate the adiabatic invariance. Such departures are identified with the quantum behaviour and it has been shown that the ensemble description of these departures is governed by a generalized set of Schrödinger equations.

4. Formulation in terms of classical probability and the Liouville equation for the system

Another important question that was posed was whether the nature of probability in quantum mechanics was same or different from the classical probability. We have postulated it to be the same as classical probability, and have used Liouville's equation for describing the evolution of the distribution function of the ensemble representing a quantum state.

The ensemble corresponding to a given pure quantum state is stipulated to be microcanonical in the initial value of the adiabatic invariant p_χ. We later assume it to be microcanonical in energy also. Thus if f represents the Liouville density for the ensemble in the phase space of either variables $(\mathbf{x}, \dot{\mathbf{x}}, p_\chi, \chi)$ at the time t, the Liouville equation for f is,

$$\frac{\partial f}{\partial t} + \boldsymbol{\nu}.\frac{\partial f}{\partial \mathbf{x}} + \dot{\boldsymbol{\nu}}.\frac{\partial f}{\partial \boldsymbol{\nu}} + \dot{p}_\chi\frac{\partial f}{\partial p_\chi} + \dot{\chi}\frac{\partial f}{\partial \chi} = 0 \tag{7}$$

where $\dot{\boldsymbol{\nu}}$ are $\dot{p}_\chi$ are given by Eqn. (3a) and (3b), respectively. We assume the adiabatic invariance to hold approximately and thus write $\dot{p}_\chi \cong 0$. Further, a change of variable is effected from χ to Φ

$$\Phi = x - \frac{1}{\varepsilon}\int dt\,\frac{1}{2}\,m\nu^2. \tag{8}$$

The Liouville equation is then given by:

$$\frac{\partial \hat{f}}{\partial t} + \mathbf{v} . \frac{\partial \hat{f}}{\partial \mathbf{x}} + \dot{\mathbf{v}} . \frac{\partial \hat{f}}{\partial \mathbf{v}} + \left(\dot{\chi} - \frac{\frac{1}{2} m \mathbf{v}^2}{\varepsilon} \right) \frac{\partial \hat{f}}{\partial \Phi} = 0, \tag{9}$$

where
$$f = \delta (p_\chi - 2\varepsilon) \hat{f} (\mathbf{x}, \mathbf{v}, \Phi, t). \tag{10}$$

The characteristics of Eqn. (9) are

$$\dot{\mathbf{x}} = \mathbf{v},$$

$$\dot{\mathbf{v}} = - \frac{1}{m} \frac{\partial}{\partial \mathbf{x}} \left(\frac{2\varepsilon^2}{g} \right),$$

$$\dot{\Phi} = \frac{2\varepsilon}{g} - \frac{1}{2} \frac{m v^2}{\varepsilon} = - \frac{L}{\varepsilon}, \tag{11}$$

where L is the Lagrangian describing the motion (6) identified with the classical motion.

5. Equations for the probability amplitudes

An event in an experiment in quantum mechanics is identified as the end-point of the trajectory of an individual system in this space. Since in an experiment one records the density of trajectory end-points without regard to either $\mathbf{v}$ or Φ, the total integrated density is given by

$$G(\mathbf{x}, t) = \int d\Phi \, d\mathbf{v} \, \hat{f}(\mathbf{x}, \mathbf{v}, \Phi, t)$$

$$= \int d\Phi \, \tilde{f}(\mathbf{x}, \Phi, t; \varepsilon) \tag{12}$$

A very important demand is now made, namely, that the probability functions $\hat{f}$ and $\tilde{f}$ be positive definite at all space points at all times. To ensure this we write

$$\hat{f}(\mathbf{x}, \mathbf{v}, \Phi, t) = \hat{\psi}^2, \tag{13}$$

where $\hat{\psi}$ is a real quantity.

Now from the Liouville eqn. (9) for an infinitesimal time τ one gets,

$$\hat{f}(\mathbf{x}, \mathbf{v}, \Phi_{t+\tau}, t + \tau)$$

$$= \hat{f}(\mathbf{x} - \mathbf{v}\tau, \mathbf{v} - \dot{\mathbf{v}}\tau, \Phi_{t+\tau} - \dot{\Phi}\tau, t). \tag{14}$$

Using (13) in (14) and taking the square root,

$$\hat{\psi} (\mathbf{x}, \mathbf{v}, \Phi_{t+\tau}, t + \tau)$$

$$= \pm \hat{\psi} \left(\mathbf{x} - \mathbf{v}\tau, \mathbf{v} - \dot{\mathbf{v}}\tau, \Phi_{t+\tau} + \frac{L\tau}{E}, t \right) \tag{15}$$

where $\dot{\Phi}$ has been substituted for from eqn. (11).

We now start from $\tilde{f}$ defined through eqn. (12) and written for the time $(t+\tau)$. Using (13) and introducing a δ-function $\delta(\mathbf{v}-\mathbf{v}')$ one obviously obtains,

$$\tilde{f}(\mathbf{x}, \Phi_{t+\tau}, t+\tau = \int d\mathbf{v} d\mathbf{v}' \, \hat{\psi}(\mathbf{x}, \mathbf{v}, \Phi_{t+\tau}, t+\tau)$$

$$\delta(\mathbf{v}-\mathbf{v}')\hat{\psi}(\mathbf{x}, \mathbf{v}', \Phi_{t+\tau}, t+\tau) \tag{16}$$

One can now substitute for $\hat{\psi}(\mathbf{x}, \mathbf{v}', \Phi_{t+\tau}, t+\tau)$ in the integrand in eqn. (16) from eqn. (15), taking the positive sign, and carry out integration over $\Phi_{t+\tau}$ to obtain:

$$G(\mathbf{x}, t+\tau) = \int d\Phi d\mathbf{v} d\mathbf{v}' \, \hat{\psi}(\mathbf{x}, \mathbf{v}, \Phi_{t+\tau}, t+\tau)$$

$$\delta(\mathbf{v}-\mathbf{v}')\hat{\psi}\left(\mathbf{x}-\mathbf{v}'\tau, \mathbf{v}'-\mathbf{v}'\tau, \Phi_{t+\tau}+\frac{L'\tau}{\varepsilon}, t\right) \tag{17}$$

where $L' = \frac{1}{2}mv'^2 - V(\mathbf{x}, \chi, t).$ (18)

We have so far indicated the most crucial inputs pertaining to the derivation. The rest of the algebra leading to the final result though important need not be repeated here. It consists of carrying out the various integrations after taking Fourier expansions of the $\hat{\psi}$ functions with respect to the phase and making Taylor expansions of the Fourier coefficients $\hat{\Psi}(\mathbf{x}-\mathbf{v}\tau, \mathbf{v}-\dot{\mathbf{v}}\tau, n, t)$ about the point $(\mathbf{x}, \mathbf{v}, t)$.

After carrying out the analysis given in the cited reference (Varma 1978) one obtains a set of equations for the functions $\Psi(\mathbf{x}, n, t)$ which are Schrödinger-like and are the generalizations of the Schrödinger equation,

$$\frac{i\hbar}{n}\frac{\partial\Psi}{\partial t}(\mathbf{x}, n, t) = -\left(\frac{\hbar}{n}\right)^2 \frac{1}{2m}\nabla^2\Psi(n) + V\Psi(n)$$

$$+ \sum_{n' \neq n} \frac{n'}{n} \langle n|V|n'\rangle \Psi(n'),$$

$$n = 1, 2, 3, \ldots . \tag{19}$$

while eqn. (12) furnishes an expression for the total probability density in terms of the $\Psi(n)$ functions

$$G(\mathbf{x}, t) = \sum_n \Psi^*(\mathbf{x}, n, t)\,\Psi(\mathbf{x}, n, t) \tag{20}$$

It should pointed out that $\hbar$ is obtained in eqn. (19) by identifying ε with it. The last terms in eqn. (19) are some small coupling terms.

It may be noted that if only the wave functions $\Psi(1)$ corresponding to $n=1$ is non-zero then eqn. (19) and (20) reduce to the usual Schrödinger formalism of non-relativistic quantum mechanics. The set of equations obtained then constitute a generglization of the Schrödinger theory, and thus furnish new modes $n \neq 1$ for the propagation of probability. These new modes predict new observable effects which can be looked for. It can be seen that the new modes represent an inhomogeneity of the distribution in Φ (or χ), for it is obvious that if only mode $n = 1$ is present, the resulting distribution is uniform in Φ. Any interaction processes that tend to make the distribution in Φ inhomogeneous will lead to the production of the new modes.

The new modes have many observable consequences. The $n = 2$ mode, for instance, will lead to the appearance of half-integral order maxima in electron diffraction from crystals. The $n = 3$ mode will give rise to third-integral order maxima, etc. The half-integral order maxima have actually been reported (Gervais et al. 1969) with as yet no clear explanation for their appearance in terms of the usual theory.

There are obvious consequences for the atomic spectra some of which have been discussed in the original reference and there are many others in other situations. We do not wish to detail them all here. But it is clear that the observation of any of these effects will have profound consequences for the foundations of quantum mechanics.

Summary of the above discussion

(i) we have provided a deterministic basic for non-relativistic quantum mechanics which, in our opinion, is very much along the conceptual lines that Einstein appears to have argued for, namely, that the wave function represents merely the behaviour of an ensemble of systems corresponding to a given quantum state, while an individual event should be described deterministically.

(ii) We have shown that an appropriate application of classical probability to describe the behaviour of the ensemble of systems yields not only the Schrödinger formalism but a generalization thereof, which furnishes new modes for the propagation of probability via the new equations. These lead to new observable predictions. It is thus concluded that the probability in quantum mechanics is already contained as one mode of propagation of classical probability when the latter is applied appropriately to describe an ensemble of deterministic trajectories.

(iii) We have also provided an identification for $\hbar$, as the 'initial' value for the canonical momentum conjugate to the new angular variable χ introduced.

(iv) The Schrödinger formalism of quantum mechanics and its generalization along with the prescription for the probability as the modulus squared of the wave functions have been obtained deductively.

(v) In the process of these investigations we have given what may be described as a geometrization of mechanics whereby the potential appears as a property of the extended space.

(vi) Though it has not been given here, we have also derived momentum position uncertainty relation in terms of our formalism.

(vii) An explanation for the quantum tunnelling comes out quite naturally. This again has not been presented here.

6. A probability amplitude description in classical mechanics

We now turn to the other question posed in the Introduction, namely, whether a probability amplitude description such as the one obtained is possible for at least some classical-mechanical systems. We have actually succeeded in finding one such situation where the behaviour of a certain appropriately prepared ensemble of systems belonging to classical mechanics can be described by a set of Schrödinger-like equations (Varma 1971) which are completely analogous to the set (19) and (20) obtained above.

The system, referred to above, consists of a charged particle moving in a magnetic field configuration known as the magnetic mirror trap, involving an inhomogeneous magnetic field. The exact equation of motion of a charged particle in a magnetic field is:

$$m\dot{\mathbf{v}} = \frac{e}{c}\,\mathbf{v} \times \mathbf{B} \tag{21}$$

where **B** represents the magnetic field, e, the charge and m, the mass. The motion, in general, consists of a gyration perpendicular to the magnetic field and a motion parallel to the magnetic field.

If the magnetic field, taken to be static, is a slowly varying function of the position, such that

$$\frac{v_\perp}{\Omega}\frac{d}{dx}\ln B \ll 1 \tag{22}$$

(where $v_\perp$ is the component of velocity perpendicular to the magnetic field locally, and $\Omega = eB/mc$ is the cyclotron frequency), then the quasi-periodic gyrating motion perpendicular to the magnetic field admits of an adiabatic action invariant (Northrop 1963).

$$\mu = \tfrac{1}{2}mv_\perp^2/\Omega \tag{23}$$

If one makes use of the abiabatic invariance, the equation for the motion of the particle along the magnetic field lines can be shown to be given by (Northrop 1963):

$$m\dot{v}_{11} = -\mu\nabla_{11}\Omega \tag{24}$$

where v_{11} is the component of the velocity parallel to the magnetic field locally. One sees from here that $\mu\Omega$ acts as a 'potential' for the parallel motion. A charged particle can be trapped 'adiabatically' in this potential if $\mu\Omega$ exhibits a well, that is, if a region of weaker magnetic field is bounded on either sides by regions of stronger magnetic field. The trapping will occur, if the total energy of the particle ξ, is less than the maximum of the potential hump $(\mu\Omega)_{max}$; that is, $\xi < (\mu\Omega)_{max}$.

The (adiabatic) trapping of charged particles that results in eqn. (24) is, of course, approximate inasmuch as μ is only an adiabatic invariant. Since violations of the invariance will, in general, occur during the motion, the particles traped adiabatically at an initial time (that is, with $\xi < \mu\Omega$ max) may leak out of the well at some later time. Such leakage of particles from potential wells bounded by potential humps have indeed been observed to occur in the experiments (Dubinina et al. 1969) for instance.

The leakage of particles with $\xi < \mu\Omega_{max}$, from the (adiabatic) potential well can be compared with the similar leakage of particles from (classical) potential wells in quantum mechanics, and one may ask whether such leakage from adiabatic magnetic wells can be described by Schrödinger-like equations. This problem has a structure very much analogous to the one considered in sec. 3 and 4, since the exact equation of motion (21) of the present problem corresponds to the exact equations of motion (3) of sec. 3, the adiabatic action invariant μ of eqn. (23 corresponds to the adiabatic invariant ε of eqn. (5) and the adiabatic equation of motion (24), with the adiabatic potential $\mu\Omega$ corresponds to the classical equation of motion (6) with the potential $V \equiv 2\varepsilon^2/g$. It may be, further, noted that the approximate (classical) equation of motion is obtained from the exact equation in consequence of the adiabatic invariance of eqn. (5). Likewise, the approximate (adiabatic) equation of motion (24) is obtained in consequence of the adiabatic invariance of μ (eqn. 23).

It is not surprising therefore that the behaviour of an ensemble of charged particles moving in a magnetic mirror trap, defined by the same total energy (microcanonical) and the same initial value of the adiabatic invariant is given by a set of Schrödinger-like equations completely analogous to the set (19) and (20) obtained earlier. We have indeed obtained such a set of equations (Varma 1971) which are as follows:

$$\frac{i\mu}{n}\frac{\partial\Psi(n)}{\partial t} = -\left(\frac{\mu}{n}\right)^2 \frac{1}{2m}\frac{\partial^2\Psi(n)}{\partial X^2} + (\mu\Omega)\,\Psi(n) \quad n = 1, 2, 3, \ldots \tag{25}$$

$$G(\mathbf{x}, t) = \sum_n \Psi^*(n)\,\Psi(n) \tag{26}$$

where, as expected from the correspondence pointed out aboue, μ appears in place of $\hbar$ and the adiabatic potential $\mu\Omega$ in place of the classical potential.

Using these equations one can calculate the probability of transmission or tunnelling of particles across the potential hump as in quantum mechanics and obtain lifetimes for residence of particles in the trap. If we model the potential humps bounding the well by the expression

$$\mu\Omega = (\mu\Omega)_{max} \cosh^2(dx) \tag{27}$$

then the nth Schrödinger-like equation gives for the lifetime for the residence of particles in the trap

$$\tau_n = T_n \exp\left[(2m)^{1/2}\frac{2\pi n}{\alpha\mu}\{(\mu\Omega)_m^{1/2} - \xi^{1/2}\}\right] \tag{28}$$

Different Schrödinger-like equations for $n = 1, 2, 3, \ldots,$ give different lifetimes for residence in the trap, τ_1, τ_2, τ_3, etc., as given by eqn. (28) where T_n is some effective bounce time. This implies that a system of particles injected with a given value of ξ and initial value of the invariant μ would exhibit a multiplicity of lifetimes. That is, a fraction A_1 of the total number of particles reside in the trap with the time τ_1, another fraction A_2 reside with the lifetime τ_2, etc. One expects from general considerations that $A_1 > A_2 > A_3 \ldots.$ This is somewhat of a surprising prediction, since the existence of a multiplicity of lifetimes for a system of particles defined by the same value of energy ξ and the action invariant μ is not easily understood in terms of the standard approach to this problem. It was, therefore, a big surprise (though a pleasant one) indeed when experiments performed recently here at the Physical Research Laboratory (Bora et al. 1979) confirmed the existence of at least two distinct lifetimes in the decay of particles from the magnetic mirror trap. The lifetimes τ_1 and τ_2 were, furthermore, found to vary with the various parameters in accordance with the expressions (28). Of particular interest is the dependence on the integer n.

To study this dependence more conveniently it is useful to transform μ in terms of the energy ξ and the pitch angle θ defined by $v_\perp = v \sin\theta$, so that $\frac{1}{2}mv_\perp^2 = \frac{1}{2}mv^2 \sin^2\theta = \xi \sin^2\theta$ and $\mu = \xi \sin^2\theta/\Omega$. Then expression (28) becomes

$$\tau_n = T_n \exp\left[(2m)^{1/2}\frac{2\pi}{\alpha}\, nB\left\{\frac{(B_{\max}/B)^{1/2}\sin\theta - 1}{\sqrt{\xi}\,\sin^2\theta}\right\}\right] \tag{29}$$

In a particular experiment we keep ξ, α, the ratio $(B_{\max}/B)$ and θ fixed and determine the two lifetimes τ_1, τ_2 as a function of the magnetic field B. From expression (29) if we plot $\ln \tau_n$ as a function of B, we would get straight lines

$$\ln \tau_n = \ln T_n + n\beta B \tag{30}$$

where
$$\beta = (2m)^{1/2}\frac{2\pi}{\alpha}\left[\frac{(B_m/B)^{1/2}\sin\theta - 1}{\sqrt{\xi}\,\sin^2\theta}\right] \tag{31}$$

is a constant with respect to B, because the ratio (B_m/B) does not change with B. From eqn. (30) it is clear that the slope of $\ln \tau_1$ vs. B line would be twice the slope of $\ln \tau_2$ vs. B line. Figs. 1-3 give the experimentally determined curves for $\ln \tau_1$ vs. B and $\ln \tau_2$ vs. B. As is clearly seen from these straight line curves the ratios of their slopes lie in the range 1.82 to 2.2 which numbers are in good agreement with the theoretically expected value of 2.

It may be mentioned that some earlier experiments (Dubinina et al. 1969, Ponomarenko et al. 1968) had determined the life times as a function of B, and plotted these $\ln \tau$ vs. B curve. But none had so far reported the observation of multiple lifetimes.

There was also found to be good agreement between the theoretically predicted value of the slope (for $n = 1$) and the experimentally determined slope for the only lifetime measured. This was reported in the earlier reference (Varma 1971).

7. Conclusions and observations

The prediction of the quantum-mechanical-like theory and subsequent observation of multiple lifetimes of charged particles in magnetic mirror traps shows that the description in terms of the probability amplitudes for

Plots of ln τ_1 and ln τ_2 vs. the magnetic field B.

For Fig. 1 $\left\{\begin{array}{l} E = 2.2 \text{ KeV} \\ \alpha^{-1} = 8 \text{ cm} \end{array}\right.$

For Fig. 2 $\left\{\begin{array}{l} E = 2.2 \text{ KeV} \\ \alpha^{-1} = 11 \text{ cm} \end{array}\right.$

For Fig. 3 $\left\{\begin{array}{l} E = 2.9 \text{ KeV} \\ \alpha^{-1} = 13 \text{ cm} \end{array}\right.$

($\theta = 33° \pm 3°$ for all the cases)

(m_1 and m_2 values are the slopes of the curves for the lifetimes τ_1 and τ_2. m_2/m_1 gives the ratio of the two slopes)

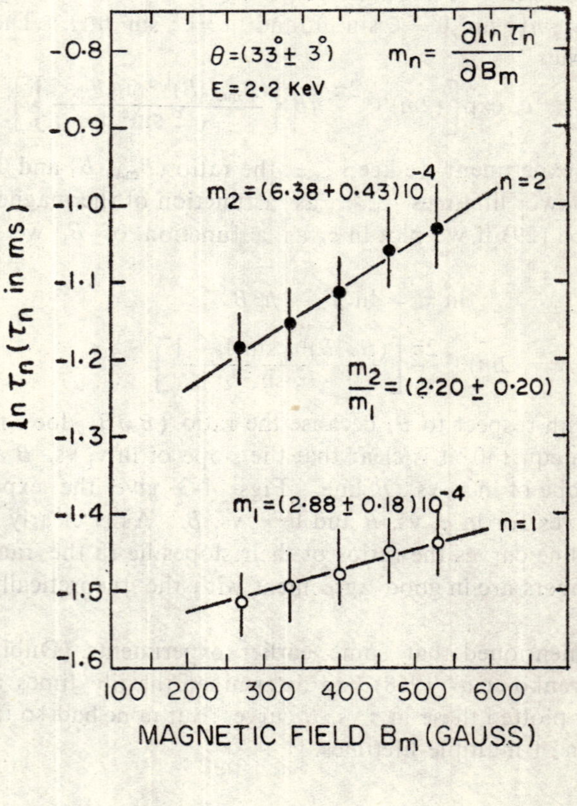

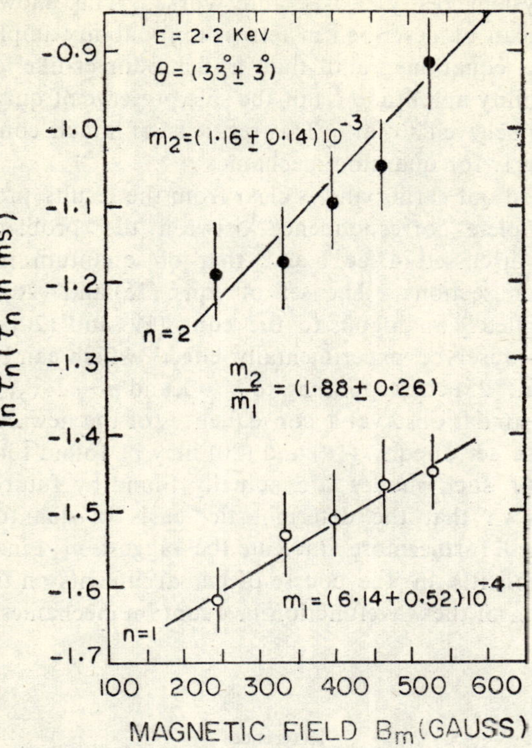
E = 2·2 KeV
θ = (33° ± 3°)
m₂ = (1·16 ± 0·14) 10⁻³
n = 2
m₂/m₁ = (1·88 ± 0·26)
m₁ = (6·14 + 0·52) 10⁻⁴
n = 1
ln τₙ (τₙ in ms)
MAGNETIC FIELD Bₘ (GAUSS)

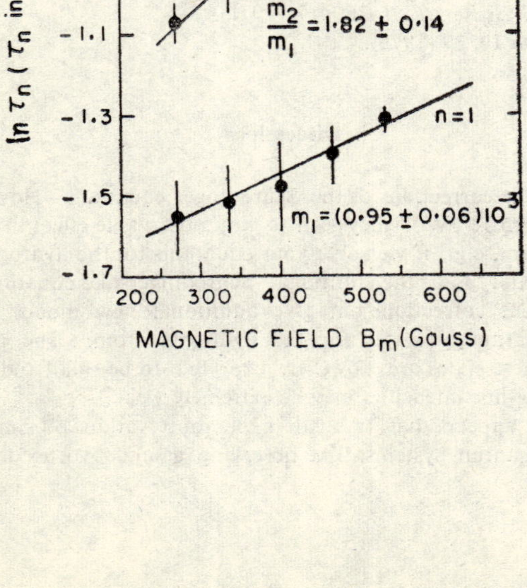
θ = (33° ± 3°)
E = 2·9 KeV
n = 2
m₂ = (1·7 ± 0·06) 10⁻³
m₂/m₁ = 1·82 ± 0·14
n = 1
m₁ = (0·95 ± 0·06) 10⁻³
ln τₙ (τₙ in ms)
MAGNETIC FIELD Bₘ (Gauss)

the classical system really exists and works. This shows that a deterministic system can be described in terms of probability amplitudes obeying Schrödinger-like equations, and that a Schrödinger-like description in terms of probability amplitude is not the sole preserve of quantum mechanics as has been believed so far. This result is of much conceptual significance particularly for quantum mechanics.

As we pointed out earlier and is clear from the results presented above there is a complete correspondence between the problem of charged particle motion discussed in sec. 6 and that of quantum mechanics discussed in earlier sections. The set of eqns. (25) and (26) governing the former are completely analogous to the eqns. (19) and (20) for the latter. Since we have observed experimentally effects which can be explained in terms of the eqn. (25) corresponding to $n = 1$ and $n = 2\ldots$, it is tempting to expect that similar observable consequences of the new modes $n = 2, 3,$ etc., given by the set of eqns. (19) and (20) may be found for the quantum domain also. If such modes are actually found by future experiments, it will have shown that the deterministic basis of quantum mechanics is real and it will furthermore vindicate the suggestion Einstein seems to have made implicitly in the course of his arguments on the question of the interpretation of the wavefunction in quantum mechanics.

References

1. Bora, D., John, P.I., Saxena, Y.C., and Varma, R.K. *Phys. Rev. Lett.* Preprint. (1979).
2. Dubinina, A.N. et al., *Plasma Phys.* **11**, 551. (1969).
3. Gervais, A., Stern R.M. and Menes M,, *Acta Cryst.* **A 24**, 191 (1969).
4. Khinchin, A.I., *Mathematical Foundations of Statistical Mechanics* p. 47, Dover, Pub. New York (1949).
5. Northrop, T.G. *Adiabatic Motion of Charged* Particles, Inter-science, New York. (1963).
6. Ponomarenko, V.G. et al, *Zh. Exsp. Teor. Fiz.* **53** 3. (1968).
7. Varma, R.K. *Phys. Rev. Lett.* **26,** 417 (1971).
8. ———, *Pramana* **10,** 89 (1978).

Discussions

T Padmanabhan: The corrections to the Schrödinger equations—How small are these terms? Will they lead to any observable effect in hydrogen spectrum, e.g., if we solve your equations for the hydrogen atom?

R K Varma: First of all, the additional Schrödinger-like equations do not represent corrections but give additional new modes. They give new stationary states for the hydrogen atom. The probabilities for these states are, however, expected to be small and the corresponding line intensities may be extremely weak.

R Cowsik: It appears that one can make observation on single macroscopic quantum systems, like observing a single vortex line in a rotating superfluid. How are such phenomena understood within the framework of quantum mechanics using ensemble interpretation of measurement?

R K Varma: I am quite aware of the situation that you have pointed out. I do not have a complete answer to this question yet, but I am sure, one should be able to find an appropriate interpretation.

C K Raju: According to the results you have stated, classical probabilities include quantum probabilities. How can this be correct in view of Wigner's theorem?

R K Varma: If by Wigner's theorem you mean the non-positive definiteness of the Wigner distributions, then I think the latter has nothing to do with classical probability. However, one should try to find out whether the Wigner distributions can be made positive definite by including the other modes found here.

Rajeev Gupta: I have a very fundamental question. Consider a system with a discrete spectrum. Quantum mechanics tells us that the result of any measurement will give us precisely one of the eigenvalues. Would your theory reproduce this fact? I think your theory has lost more determinism than what quantum mechanics requires.

R K Varma: What you have stated in your question corresponds to the measurement theory according to the Copenhagen School. In the present picture, the measurement theory is totally different. But I don't see how this theory has lost more determinism than what quantum mechanics requires. I am not sure I understand your question completely.

B G Sidharth: If the mathematical result that the amplitude of a wave function, when Fourier analysed multiplied by the width is of the order of h has to be reconciled with the purely physical result of position momentum uncertainty as deduced from the Heisenberg λ-ray experiment, the Born interpretation of the wave function has to be invoked. This points to the fact that the wave function gives a complete description of a single system, i.e. that the uncertainty is inherent and not just of statistical significance.

R K Varma: I don't think that the argument you have given shows that the uncertainty as given by the uncertainty relation does not have a statistical significance. In fact, if one were to verify the uncertainty telation one would not be able to do it without the use of an ensemble—as indeed all the proofs of the uncertainty relation require.

A Maheshwari: There is a third view of quantum mechanics due to Feynman in which the quantum amplitudes are defined as integrals over the space of paths and the trajectory of the quantum particle is completely undeterministic. However, in the path integral calculation in $\hbar$ expansion the leading term is WKB approximation which gives quantum amplitudes in terms of classical trajectories and the adiabatic invariants.

R K Varma: Yes, I am aware of the Feynman formulation, where the trajectories are completely undeterministic, but there is still only one wave function, namely, the Schrödinger function that one is dealing with. Perhaps one can ask the question whether the existence of an infinite set of wave functions in the present theory is a reflection of the trajectories being deterministic, and vice versa.

Avinash Khare: Your programme is very similar to Heinrich Hertz's efforts of generalization of Mechanics by introducing a latent variable. Otherwise he did not have quantum mechanics to cope with (comment). Will not Gleason's trouble arise in your hidden variable theory which is a general objection to all hidden variable theories?

R K Verma: You are quite right. Heinrich Hertz was probably the first to try to geometrize mechanics, and what I have done is, in some sense, similar to his attempt. However, as you can see, what is important here is not just geometrization but, if you like, an explanation or model of quantum mechanics using this geometrization in a higher-dimensional space. As to your question, I have not yet looked into Gleason's trouble with regard to the present theory.

Aparna Basu: The two lifetimes of particle escaping from magnetic mirrors be explained in the following terms. One lifetime for those particles which already have enough parallel energy to escape another for those which acquire this through collision processes.

R K Varma: The two lifetimes cannot be explained as you have suggested. First of all, if particles have enough parallel energy to escape, their lifetime of confinement is zero. Further, at the gas pressures used in the experiment, the collision times are much larger than any of the two lifetimes observed

Rhama Venkataraman: (1) The slow variation of g imposed introduces, or rather, may manifest as some limitations on the system itself. What is the justification? There is rather strong indication that the formulation is more specific and not general.

(2) Why the departures from the adiabatic conditions identified with quantum effect? Of what condition specifically? And how does it compare with the quantum effect, quantitatively?

R K Varma: The slow variation of g is assumed to enable one to define an adiabatic invariant and to recover the classical equations of motion with appropriate identification of potential, etc. When the condition of adiabaticity is violated, departures from classical behaviour would naturally occur and since such departures are, by our derivation, described by the Schrödinger equations they are inevitably to be identified with quantum behaviour.

Statistical Physics and Photoelectric Effect

Fluctuations, Critical Opalescence and Critical Point Phenomena

E.S.R. GOPAL

1. Historical introduction

The topic of fluctuations and critical opalescence is firmly associated with the names of Einstein and Smoluchowski, starting with their great work in the first decade of this century. The first significant paper by Einstein in 1905 on Brownian motion had a long title, 'Über die von der molekularkinetischen Theorie der Wärme geforderte Bewegung von in ruhenden Flüssigkeiten suspendierten Teilchen' ('On the motion of particles, suspended in stationary liquids and necessitated by the molecular kinetic theory of heat'). *Ann. Phys.* **17**, 549–560 (1905) (received on 11 May 1905; Published on 18 July 1905). This was followed a year later by another significant article 'Zur Theorie der Brownschen Bewegung' ('Towards the theory of Brownian motion') *Ann. Phys.* **19,** 371–381 (1906). In this paper Einstein showed that the Boltzmann relation between entropy S and probability W, viz., $S = k \ln W + S_0$, could be re-framed to give the probability of a fluctuating state x as

$$W(x)\,dx = (A/2\pi)^{1/2} \exp(-Ax^2/2)\,dx \tag{1}$$

A significant paper on Brownian motion by Smoluchowski appeared in the same year 'Zur kinetischen Theorie der Brownschen Molekularbewegung und der Suspensionen' ('Towards the kinetic theory of Brownian molecular motion and of suspensions') *Ann. Phys.* **21,** 756–780 (1906). Though Smoluchowski's contributions to the subject are excellent, as documented well in Chandrasekhar's review ('Stochastic problems in physics and astronomy'), *Rev. Mod. Phys.* **15,** 1–89 (1943), Smoluchowski generously refers to Einstein's work on p. 772 of this paper as "eine Diskussion der von Einstein befolgten, sehr sinnreichen Methoden einzugehen" ("go into a discussion of the very ingenious method followed by Einstein").

The crucial step of explaining the molecular scattering of light and critical opalescence was taken by Smoluchowski in 1908 with his paper 'Molecular-kinetischen Theorie der Opaleszenz von Gasen im kritischen Zustande, sowie einiger verwandter Erscheinungen' ('Molecular kinetic theory of the opalescence of gases in the critical region and of some allied phenomena') *Ann. Phys.* **25,** 205–226 (1908). In an apparently homogeneous medium, there are fluctuations of density around the mean value, caused by the thermal energy of the medium. Light is scattered by these fluctuations in density and hence by an apparently homogeneous medium.

The early observations of the light by material particles were done by Tyndall in 1869. Rayleigh calculated the scattering of light by a small dielectric sphere in 1871 and found it varying as λ^{-4}, thus accounting for the blue of the sky. In a gas, one has scattering by the molecules without any coherence from the various particles. However, in a solid or liquid, the contributions from adjacent sites are to be added coherently. Because of the definite phase relations among the light scattered from neighbouring regions, the light scattering from a perfectly homogeneous medium cancels cut in every direction other than the forward direction.

The phenomenon of critical opalescence noticed towards the end of the last century was a dramatic demonstration of the scattering of light from homogeneous media. When a fluid is taken near the critical point P_c, ρ_c, T_c, the light scattered from it increases markedly. The scattering becomes so intense near T_c that almost all the light comes out from the sides of the cell containing the fluid. The scattering medium appears to glimmer or 'opalesce' because of the intense secondary scattering. Yet the theory states that if a medium were homogeneous there should be no scattering of light.

It was this dilemma which was resolved by Smoluchowski in 1908. Very soon in 1910 a detailed calculation was completed by Einstein in his paper 'Theorie der Opaleszenz von homogenen Flüssigkeiten und Flüssigkeitsgemischen in der Nähe des kritischen Zustandes' ('Theory of the opalescence of homogeneous fluids and fluid mixtures in the neighbourhood of the critical state') *Ann.* phys. **33**, 1275–1298 (1910). On p. 1275 he starts his paper as 'Smoluchowski hat in einer wichtigen theoretischen Arbeit gezeigt...' ('Smoluchowski has shown in an important theoretical work that ...'). Einstein took into account the coherence among the wavelets scattered from adjacent media (which are essentially the Laue equations for light scattering). He then calculated the mean square amplitude of the fluctuations of density in the case of pure fluids and of compositions in the case of fluid mixtures. He thus obtained the expression for the intensity of the light scattered.

The differential scattering cross-section $(1/V)\ d\sigma/d\Omega$ is a convenient quantity to compare with the experiments. It is the flux of light or power scattered into a unit angle per unit scattering volume per unit incident intensity and is often called the Rayleigh ratio. The expression obtained by Einstein for the light scattering from a pure fluid was

$$\frac{1}{V}\frac{d\sigma}{d\Omega} = \frac{\pi^2}{\lambda^4}\left(\frac{\partial\varepsilon}{\partial_\rho}\right)^2 \rho^2 kT\,\chi_T \tag{2}$$

where ε is the dielectric constant and χ_T is the isothermal compressibllity. Near T_c, the compressibility χ_T becomes infinitely large and along with it the magnitude of the fluctuations. These make the light scattering increase enormously in the critical region.

The calculation of the light scattering intensity was completed by Ornstein and Zernike (*Proc. Roy. Acad. Sci.* Amsterdam, **17**, 793-806 (1914)). Einstein had assumed in his calculation no spatial correlations among the fluctuations. Ornstein and Zernike showed that one must include this if one were to calculate the scattering at finite values of $\mathbf{q} = \mathbf{k}_0 - \mathbf{k}$, $\mathbf{k}_0$ being the incident wave vector and $\mathbf{k}$ scattered wave vector. Thus $q \approx 2k_0 \sin(\theta/2)$ where θ is the angle of scattering. Ornstein and Zernike showed that

$$\frac{1}{V}\frac{d\sigma}{d\Omega} = \left[\frac{\pi}{\lambda^2}\left(\frac{\partial \varepsilon}{\partial \rho}\right)\right]^2 \rho S(q) \tag{3a}$$

$$\text{where } S(q) = \frac{1}{N}\left\langle \sum_j \sum_1 \exp\{i\mathbf{q}\cdot[\mathbf{r}_j(t) - \mathbf{r}_l(t)]\}\right\rangle \tag{3b}$$

is the structure factor of the fluid. The structure factor may also be written as

$$S(q) = 1 + \rho \int \exp(i\mathbf{q}\cdot\mathbf{r})\, g(r)\, d^3\mathbf{r} \tag{4}$$

where $g(r)$ is the radial distribution function, or the conditional probability per unit volume of finding an atom at $\mathbf{r}$ given that there is an atom at the origin.

For many purposes we can replace $g(r)$ by $g(r) - 1$ without affecting the integral and one can note that

$$1 + \rho \int [g(r) - 1]\, d^3\mathbf{r} = \rho\, kT \chi_T$$

a result obtained in standard statistical mechanics. The fact that χ_T diverges in the critical region implies that the tail portion of $g(r)$ is significant. Ornstein and Zernike assumed that asymptotically

$$g(r) - 1 \sim (1/r) \exp(-r/\xi) \tag{5}$$

where ξ is the correlation length which diverges as $T \to T_c$. Using this, or by a number of other equivalent procedures, one gets

$$\frac{1}{V}\frac{d\sigma}{d\Omega} = \frac{\pi}{\lambda^4}\left(\frac{\partial \varepsilon}{\partial \rho}\right)^2 \rho^2\, kT\, \chi_T (1 + q^2\xi^2)^{-1} \tag{6}$$

which is the famous Ornstein-Zernike equation for critical opalescence.

The theory of molecular scattering of liquid, thus pioneered by Einstein and Smoluchowski, has undergone many significant developments. The spectrum of the scattered light was studied by Brillouin to show that there must be doublets arising from the propagation of the density fluctuations as high frequency sound waves. Landau and Placzek showed that a central peak arises from the entropy fluctuations which diffuse away rather than propagate as waves. The discovery of the Raman process altered the very nature and object of the light scattering studies. These topics are reviewed elsewhere (Benedek, 1968; Fabelinski, 1968).

2. Critical opalescence: recent studies

Without considering these developments, let us get back to the critical opalescence problem. The development of lasers and the refinements in the experimental techniques of light scattering has made it possible to study several different aspects of the critical opalescence not only as regards the intensities but also regarding the spectral features of the scattered light. The main features studied are the following:

(a) Cross-section of the scattering $(1/V)\, d\sigma/d\Omega$ which has been mentioned earlier in connection with the Ornstein-Zernike form.
(b) Landau-Placzek ratio of the central peak to the Brillouin line intensities. Normally this is given by $(C_p/C_v) - 1$.
(c) Rayleigh line width of the central peak.
(d) Brillouin shift of the satellite lines arising from the Doppler shift of the scattered light from the propagating hypersonic waves.
(e) Line width of the Brillouin component.

A rich variety of experimental results have become available (Cummins, 1971; Chu, 1972; Hohenberg and Halperin, 1977). Instead of discussing all of them, let us see one or two illustrative features.

For example, the comparison of the experiments with the Ornstein-Zernike theory is commonly done by noting that

$$\left(\frac{1}{V}\frac{d\sigma}{d\Omega}\right)^{-1} = B(1 + q^2\xi^2) \tag{7}$$

where

$$B = \frac{\lambda^4}{\pi^2}\left(\frac{\partial \varepsilon}{\partial \rho}\right)^2 \rho^2\, kT\, \chi_T \tag{8}$$

Thus near T_c the plots of $\{(1/V)\, d\sigma/d\Omega\}^{-1}$ against q^2 should be straight lines. The slope is $B\xi^2$ and the $(q^2 = 0)$ intercept B. The slopes of the different isotherms should be the same. Thus one should be able to get the temperature-dependent compressibility $\chi(T)$ and the correlation length $\xi(T)$.

Modern laser techniques have made these determinations quite reliable, though the estimation of the absolute scattering cross-sections is still difficult. Thus it is customary to write the measured intensity at a scattering angle θ, corrected for geometrical factors but otherwise in arbitrary units, as

$$I(\theta) = I(0)\,(1 + q^2\xi^2)^{-1} \tag{9}$$

Figure 1 shows the plot of $[I(\theta)]^{-1}$ against $\sin^2(\theta/2)$, for the binary liquid system 2, 6-lutidine + water (Gulari et al, 1972), which has a critical separation temperature 307.013°K and a critical composition of 28.70 weight % of lutidine. Recall that $q = (4\pi\mu/\lambda)\sin(\theta/2)$ where μ is the refractive index of the medium and λ the wavelength of the laser light. There is indeed remarkable agreement with the theoretical prediction of parallel straight lines. This qualitative feature is true for all the systems studied so far.

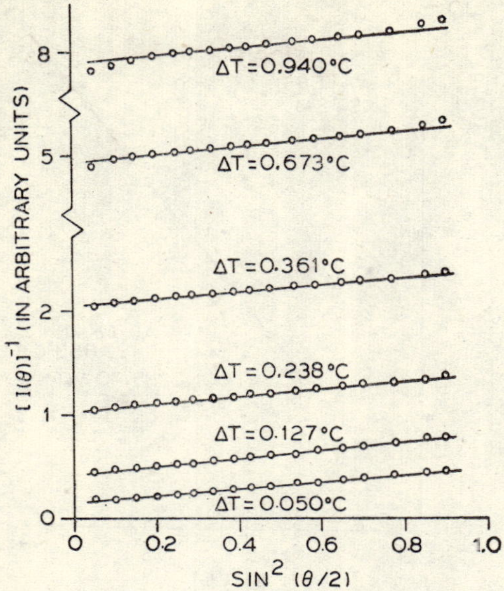

Fig. 1 Plot of reciprocal scattered intensity $[I(\theta)]^{-1}$, in arbitrary units, against $\sin^2(\theta/2)$ for a few temperatures in the critical region of the system 2, 6-lutidine+ water (after Gulari et al. 1972).

However, the precise temperature dependences of the divergences in the compressibility χ_T and the correlation length ξ deviate from the classical predictions and this has become the focal point of modern studies of critical phenomena (Stanley, 1971; Rowlinson, 1975). Based on the analytic expansion of the free energy or by using the Van der Waals equation of state, one can show that the compressibility should vary as $(T-T_c)^{-1}$ near T_c. Thus asymptotically as $T \to T_c$

$$\chi_T \sim (T-T_c)^{-1} \tag{10}$$

This requires that
$$\xi(T) \sim (T-T_c)^{-1/2} \tag{11}$$

which was also the functional dependence used by Ornstein and Zernike.

The experiments on the other hand give a different functional form for the asymptotic behaviour. For example, from eqns. (7-11) one should expect $[I(0)]^{-1}$ plotted as a function of $\Delta T = T - T_c$ on a log-log scale to have a slope 1. The experiments fit rather accurately a straight line with a slope 1.25 as shown in Fig. 2. This means that as $T \to T_c$

$$\chi_T \sim (T-T_c)^{-\gamma},\ \gamma \approx 1.25 \tag{12}$$

The critical exponent γ has a value 1.25 instead of 1 given by eqn. (10). Fig. 3 shows the behaviour of the correlation length as a function of $\Delta T = (T-T_c)$ on a log-log scale. The data fit accurately in equation $\xi(T) = 2.0\,[(T-T_c)/T_c]^{-0.62}$ Å. This means that

$$\xi(T) \sim (T-T_c)^{-\nu},\ \nu \approx 0.62 \tag{13}$$

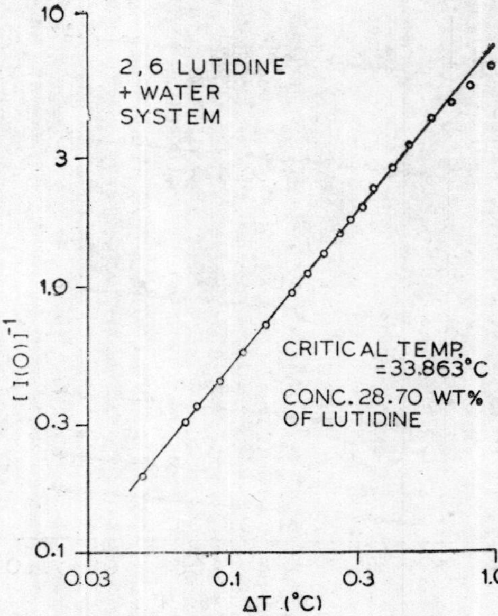

Fig. 2 Log-log plot of $[I(\theta)]^{-1}$ against $\Delta T = T - T_c$ for the lutidine + water system (after Gulari et al. 1972). The slope gives the exponent γ as 1.25.

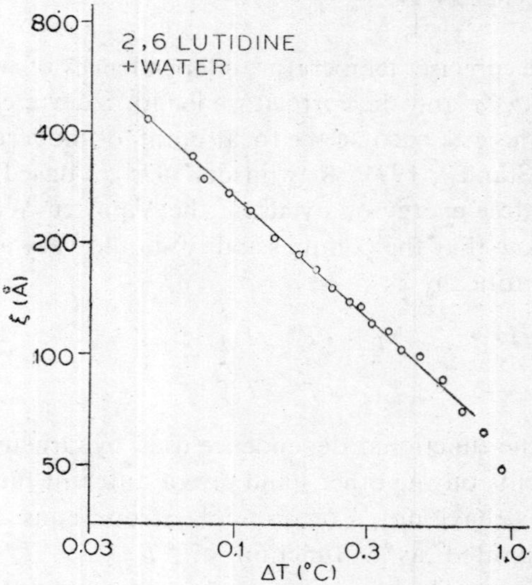

Fig. 3 Log-log plot of coherence length ξ versus ΔT for the system 2, 6-lutidine + water (after Gulari et al. 1972). The slope yields the exponent ν as 0.62.

which defines the critical exponent ν characterizing the rate at which the correlation length becomes infinitely large at T_c. Classically ν should be 0.5 (eqn. 11) which is not observed in practice.

The study of the critical exponents has become a very exciting area of statistical physics (Domb and Green, 1972; Ma, 1976). The statistical ordering processes rather than the specific strengths of the interatomic forces seem to determine the values of the critical exponents. For example, the magnetic susceptibility of a ferromagnet diverges near T_c with exactly the same exponent γ as the susceptibility. The order parameters, namely, the magnetization of a ferromagnet or the density difference of a liquid-gas system or the composition difference of a binary liquid system, all go to zero at T_c as $(T_c - T)^\beta$ with the same $\beta \approx 0.32$. The specific heat in all the systems behaves as $\Delta T^{-\alpha}$ with α of the order of zero. This universality of the critical phenomena has been responsible for the considerable interest in this field.

Without going into the various other details of light scattering studies of critical phenomena, let us consider only one more aspect. One can consider $\mathbf{q}$ as the momentum transfer in the scattering process $\mathbf{q} = \mathbf{k}_0 - \mathbf{k}$. If the correlation length is such that $\mathbf{q}\xi \ll 1$, then one may consider that local thermodynamic equilibrium is established in the process. If on the other hand the correlation length is much larger than the inverse of the wave vector transfer, i.e. $\mathbf{q}\xi \gg 1$, then the assumption of local thermodynamic equilibrium is not valid. Thus for a light scattering experiment, one has

$$\begin{aligned} &\mathbf{q}\xi \ll 1, \text{ hydrodynamic region} \\ &\mathbf{q}\xi > 1, \text{ critical region} \end{aligned} \tag{14}$$

and both the regions are experimentally accessible. The quantity most convenient for this discussion is the width of the central Rayleigh line in the scattered beam. This has a Lorentzian shape with a half width Γ. A detailed study of the coupling among the various fluctuating modes by Kawasaki (1976) lead to the following relation

$$\Gamma = \frac{kT}{16\eta^*}\left(\frac{2\mathbf{q}^3}{\pi}\right)\{(\mathbf{q}\xi)^{-1} + (\mathbf{q}\xi)^{-3} + [1 - (\mathbf{q}\xi)^{-4}] \tan^{-1}(\mathbf{q}\xi)\} \tag{15}$$

where η^* is the high frequency part of shear viscosity and is a constant. In the limiting cases of $\mathbf{q}\xi \ll 1$, the hydrodynamic region,

$$\Gamma \sim \mathbf{q}^3(1 + \frac{3}{5}\mathbf{q}^2\xi^2),$$

whereas in the critical limit of $\mathbf{q}\xi \gg 1$, one has $\Gamma \sim \mathbf{q}^3$. Normally $\Gamma/\mathbf{q}^3$ is plotted against $(\mathbf{q}\xi)$ to check this relation, as shown in Fig. 4. There is indeed a remarkable agreement with the Kawasaki relation.

One may say in a general way that the modern theory of critical opalescence phenomena agrees with the experiments in a very good manner. Admittedly experimental difficulties are considerable in checking some aspects of the theory, for example in the estimation of the critical exponent $\eta \approx 0.05$ which defines the asymptotic behaviour of the correlation function $g(r) - 1 \approx \exp(-r/\xi)/r^{1+\eta}$. The study of critical opalescence seems to be a good way of studying some aspects of the dynamics of the system.

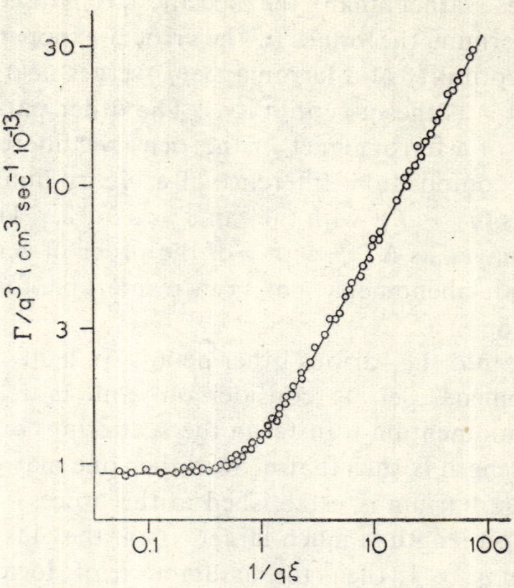

Fig. 4 Log-log plot of Γ/q^3 versus $(q\xi)^{-1}$ for the lutidine + water system (after Gulari et al. 1972). The line given by the Kawasaki equation is also shown.

3. Behaviour of electrical resistivity in the critical region

Let us now consider a different measurement, namely that of electrical resistivity in the critical region. One knows in a general way, for example from the Kubo relations, that a transport property is related to the correlation among the fluctuations. Since these fluctuations become large in the critical region one expects an anomalous behaviour of the electrical resistivity.

A calculation of the critical behaviour of electrical resistivity of a ferromagnet, using the recent ideas of critical phenomena, was done first by Fisher and Langer in 1968. They essentially started from the simple equation for conductivity $\sigma = ne^2\tau/m$. The conventional scattering theory can be formulated to give

$$\frac{1}{\tau} \sim b^2 \int S(q)\,(1-\cos\theta)\sin\theta\, d\theta$$

where $S(q)$ is again the structure factor or the Fourier transform of the correlation function. Fisher and Langer pointed out that the critical part of the resistivity arises largely from the $q \approx 2k_f$ values, which are also responsible for the thermal energies. Thus they obtained the main result that the critical electrical resistivity should be proportional to the thermal energy so that dR/dT should have the same α-divergence like the specific heat of the substance. The experiments on a number of ferromagnets and alloys showing order-disorder phenomena show the α-exponent for the singularity in dR/dT. For a review of the various theoretical calculation see Alexander et al. (1976) and Balberg et al. (1978). Thus the theory seems to be essentially correct.

However, in the case of binary liquid systems of the polar + non-polar type the experiments give a dR/dT singularity with an exponent $b \approx 0.35$, very different from the specific heat exponent $\alpha \approx 0.1$.

As an example one may discuss the behaviour of the methanol+n-heptane system which has $T_c \approx 323.248$ °K (critical composition 61.35 mole % of methanol) (Tiwari, 1978; Tiwari et al. 1979). The electrical resistivity (or more accurately the ac resistance at 1 kHz between a pair of electrodes nearly 2 cm in diameter and 1 mm apart) is as shown in Fig. 5. In the

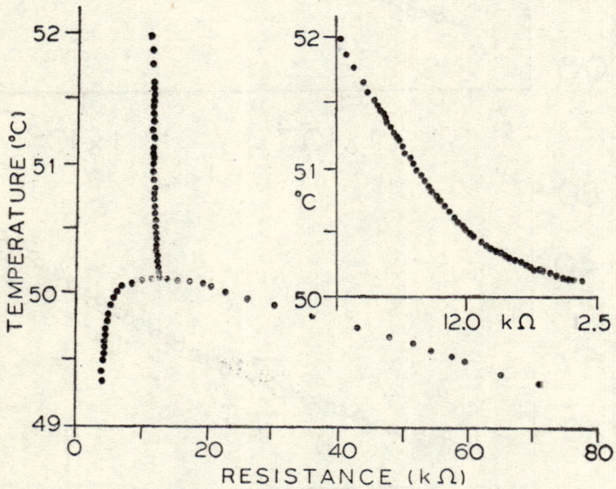

Fig. 5 Electrical resistivity of the binary liquid system methanol + n.heptane in the one phase (temperature $> 50.09_8$°C) and in the two phase regions. The inset shows the one phase resistance in an expanded scale. (critical composition 61.35 mole per cent of methanol).

one-phase region the resistance shows a slight increase as the temperature is lowered. The inset of Fig. 5 shows that there is a linear increase far away from T_c and that the increase becomes rapid as T_c is approached. In the two phase region, the resistance in the polar phase of methanol, which being heavier at the bottom, comes down to a few $k\Omega$, whereas the resistance of the upper non-polar n-heptane increases very rapidly. The behaviour in the two phase region is analysed in Fig. 6. The lower curve is a log-log plot of the resistivity difference $R_u - R_1$ against the reduced temperature $(T_c - T)/T_c$. The upper curve gives the analogous plot for the conductivity difference. It is seen that over the entire temperature interval the conductivity difference follows a relation

$$\sigma_1 - \sigma_u \sim [T_c - T)/T_c]^\beta \text{ with } \beta \approx 0.35 \tag{16}$$

The difference in resistance has the β exponent close to T_c but the plot deviates from linearity rather quickly. The magnitude of the exponent $\beta \approx 0.35$ and the fact that $\sigma_1 - \sigma_u$ is linear over a larger range of temperature need to be explained thoretically. In the one-phase region, one finds that away from T_c the resistance varies linearly with temperature.

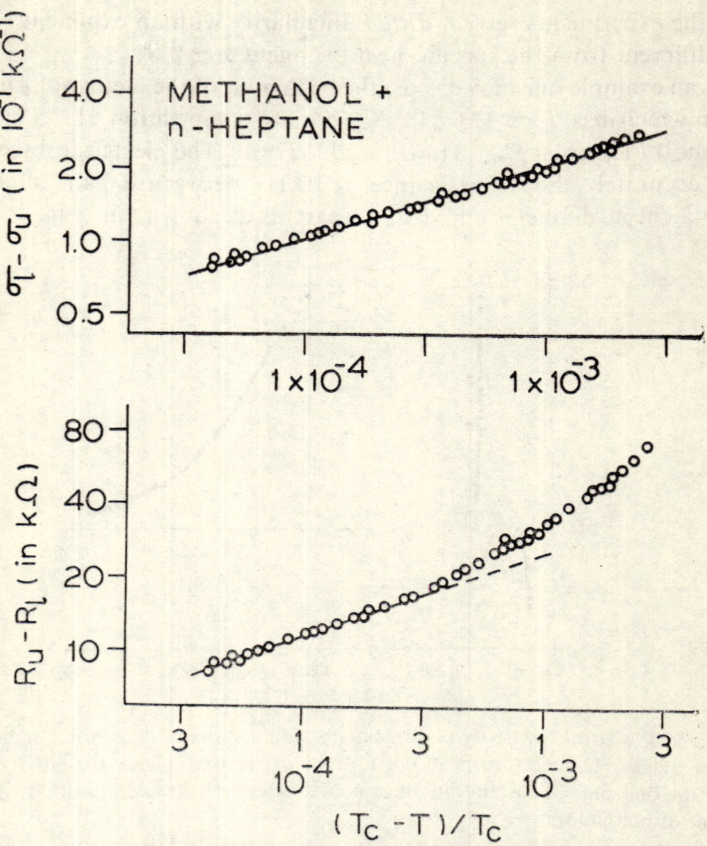

Fig. 6 Behaviour of the electrical resistance of the $CH_3OH + n\text{-}C_7H_{16}$ system in the two phase region. Conductivity difference between the two phases (upper curve) and the resistivity dlfference (lower curve) are plotted against $(T_c - T)/T_c$ on a log-log scale. The conductivity difference has a linearity over a larger temperature interval.

Thus one fits a linear equation to the R–T data away from T_c and calculates the deviation ΔR from the straight line fit. Fig. 7 shows a plot of $(\Delta R - \Delta R_c)/\Delta R_c$ against $(T - T_c)/T_c$ on a log-log scale with $\Delta R_c \approx 0.525\ k\Omega$. The curve plotted in the figure is a representation of

$$(\Delta R - \Delta R_c)/\Delta R_c = -\ 114.94\left(\frac{T - T_c}{T_c}\right)^{-0.63} - 724.98\left(\frac{T - T_c}{T_c}\right)$$

which is seen to be a good fit. It gives

$$dR/dT \sim [(T - T_c)/T_c]^{-0.37} \tag{17}$$

The electrical resistivity has been studied in a number of polar + non-polar binary liquids showing phase separation critical phenomena. The behaviour of $dR/dT \sim (T - T_c)^{-b}$, $b \approx 0.35$, seems to be nearly universal

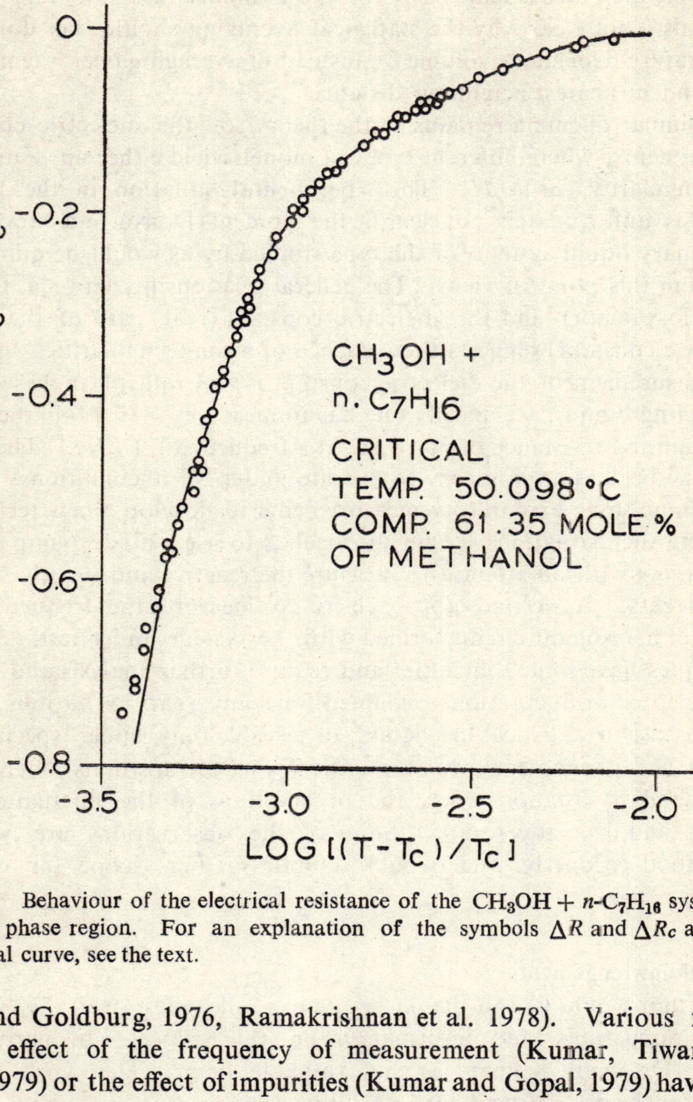

Fig. 7 Behaviour of the electrical resistance of the $CH_3OH + n\text{-}C_7H_{16}$ system in the one phase region. For an explanation of the symbols ΔR and ΔR_c and the theoretical curve, see the text.

(Shaw and Goldburg, 1976, Ramakrishnan et al. 1978). Various factors like the effect of the frequency of measurement (Kumar, Tiwari and Gopal, 1979) or the effect of impurities (Kumar and Gopal, 1979) have been studied. The exponent $b \approx 0.35$ seems to be unaffected and must be considered genuine.

The understanding of the discrepancy between the theoretical prediction of a specific heat like α-singularity for dR/dT and the experimental observation of a stronger $b \approx 0.35$ singularity is not complete as yet. The nearest neighbour scattering mechanism, as in the case of the scattering of the conduction electrons by the adjacent spins in a ferromagnet, or the analogous nearest neighbour hopping model give the same α-singularity. On the other hand models, where the averaging is done over the temperature dependent values ξ^3 of the correlation region, give a $1 - \nu$ singularity. Since $\nu \approx 0.65$, this is in good agreement with the experiments.

However, the theoretical basis for these models are not very rigorous. It is also not clear why the statistical averaging should be done over a temperature-dependent volume ξ^3, instead of averaging over a temperature-independent nearest neighbour distance.

A similar dilemma remains in the theory of the dielectric constant of these systems, where different types of models yield either an α or a $1 - \nu$ type singularity for $d\varepsilon/dT$. The experimental situation in the liquid-gas system is unfortunately not clear at the moment (Doiron and Meyer, 1978). The binary liquid systems of the type studied by us would be quite interesting from this point of view. The general relationship between the resistivity (loss factor) and the dielectric constant (real part of the complex dielectric constant) suggests the presence of strong singularities. However, the measurement of the dielectric constant is very difficult in these partially conducting liquids. Typically one has to measure $\sim$ 10 PF in the presence of a shunting resistance of $\sim 10\ k\Omega$ at a frequecy of 1 kHz. The conventional ac bridges are not easy to operate under such conditions. We are performing some exploratory measurements to develop a new technique of handling such situations. One approach is to use a diode pump charging-discharging with an attempt to measure the reactive and resistive parts of the currents. A second approach is to measure the frequency and Q factor of a resonant circuit formed with the system under test. Both the techniques have some difficulties and require further analysis and work.

The topic of fluctuations remained for many years a curious intellectual exercise in statistical mechanics, observable only under special conditions. The phenomena near the critical phase transitions offer, on the other hand, a situation where the observations of the fluctuations yield striking and dramatic results. Some of the observations are yet to be understood properly and would doubtless offer scope for continued work.

4. Acknowledgements

The author would like to thank his many colleagues over the years for useful discussions and information on this subject. In particular he thanks Dr. Anil Kumar for his valuable help. The work was also supported by PL 480 and DST (NTPP) grants.

References

1. Alexander, S., Helmen, J.S. and Balberg I., Phys. Rev. **B 13**, 304 (1976)
2. Balberg, I., and Helman, J.S., *Phys. Rev.* **B 18**, 303 (1978)
3. Benedek, G.B., 'Statistical Physics', Brandeis University Summer Institute in Theoretical Physics, Chretien, M. et al. (Eds.) p. 1–98. Gordon and Breach, New York (1968).
4. Chu, B., *Berichte Bunsen Gessel. Phys. Chem.* **76**, 202–214 (1972)

5. Cummins, H.Z., 'Critical Phenomena', Enrico Fermi International School, Course LI (Ed. M.S. Green), Academic Press, New York, p. 380–444 (1971)
6. Doiron, T. and Meyer, H., *Phys. Rev.* **B 17**, 2141 (1978)
7. Domb, C. and Green, M.S. (Eds.) *Phase Transitions and Critical Phenomena*, Academic Press, New York, Vol. I and later volumes (1972)
8. Fabelinski, I.L., *Molecular Scattering of Light* (Trans. R.T. Beyer), Plenum Press, New York (1968)
9. Gulari, E., Collings, A.F., Schmidt, R.L. and Pings, C.J., *Jour. Chem. Phys.* **56**, 6169–6179.
10. Hohenberg, P.C. and Halperin, B.I., *Rev. Mod. Phys.* **49**, 535–479 (1977)
11. Kawasaki, I., *Phase Transitions and Critical Phenomena* (Eds. C. Domb and M.S. Green), Academic Press, New York, Vol. 5A, 166–203 (1975)
12. Kumar, A., and Gopal E.S.R , *Pramana* **12**, 764 (1979)
13. Kumar, A., Tiwari, M.K. and Gopal, E.S.R., *Chem. Phys.* **39**, 118 (1979)
14. Ma, S.K., *Modern Theory of Critical Point Phenomena*, Benjamin, New York (1976)
15. Ramakrishnan, J., et al. *Jour. Chem. Phys.* **68**, 4098 (1978)
16. Rowlinson, J.S., *Liquids and Liquid Mixtures*, Butterworths, London (1975)
17. Shaw C.H., and Goldburg, W.I., *Jour. Chem. Phys.* **65**, 4906 (1976)
18. Stanley, H.E., *Introduction to Phase Transitions and Critical Phenomena*, Oxford University Press, Oxford (1971)
19. Tiwari, M.K., Ph. D. thesis, Indian Institute of Science, Bangalore, unpublished. (1978)
20. Tiwari, M.K., Kumar, A, and Gopal, E.S.R., preprint (1979)

Photo-emission from Solids*

MUKUNDA P. DAS

1. Introduction

The experiment demonstrating the phenomenon of photoelectron emission that reminds us of the importance of the development of the quantum theory was accidentally discovered by Hertz in 1887 and was correctly interpreted by Einstein in 1905.[1] Einstein was awarded the 1921 Nobel prize in physics under the citation "for the photoelectric law and his work in the domain of theoretical physics." Since then Einstein's law has become the basis of quantitative photochemistry in the same way as Faraday's law is the basis of electrochemistry. More recently the photoelectric effect has been largely used in the solid state devices bringing a new era in the electronic industry.

Einstein's simple interpretation of photo-emission has posed considerable challenges to the theorists over the last seven decades to understand the

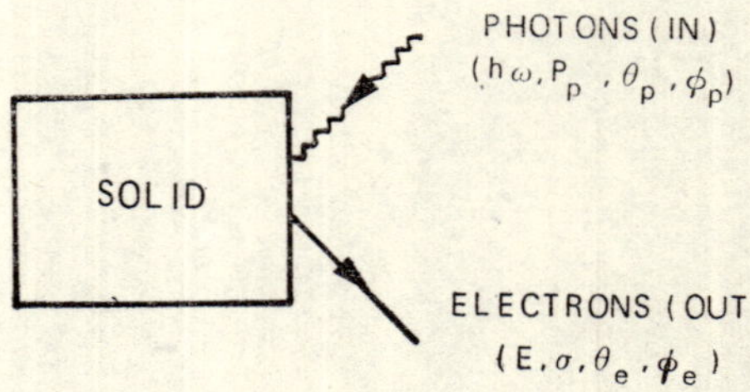

Fig. 1. ω, $\overline{P}_p$ and (θ_p, ϕ_p) are the frequency, polarization and direction of incoming photons. E, σ and (θ_e, ϕ_e) are the energy, spin and direction of outgoing electrons.

details of this phenomenon. What is essentially necessary is a description of the equilibrium and the excited electron structure of a matter with which the photofield is interacting resulting in the expulsion of the electron from the matter. This past few years have witnessed the development of various aspects of photo-emission in view of the availability of good vacuum and increasingly sophisticated experimental techniques. There are several independent variables as indicated in Fig. 1 which can now be controlled to reveal or supress, to a limited extent, some features one may wish to study or avoid. One can choose the energy, angle and the polarization of

*The work is supported by a reserch grant from the University Grants Commission, New Delhi.

the incident photons so as to analyse the escaping electrons from the matter with respect to their initial and/or final energy, angle of emission and spin polarization.

Here, it will not be possible to provide a complete historical review of the subject and so we refer the interested reader to some recent interesting review articles.[2-11] The aim of this work is simply to elucidate the current state of understanding of the phenomenon and to discuss how much the electron structure of solids is understood via photo-emission. The details could be found from original papers as referred below.

2. Photo-emission: theoretical aspects

As mentioned in the 'Introduction' and in Fig. 1 the photons with known energy, polarization and the angle of incidence expel electrons from the sample giving rise to an electric current which is measured as a function of energy or angle of emission. Sometimes the spin polarization of emerging electrons is also measured. Now the basic aim is to understand how does the current depend on these parameters.

There are two distinct formal approaches which we discuss below. One is known as scattering theory formalism based on the steady-state scattering theory of Lippmann and Schwinger, and Goldberger and Watson. This theory was developed by Adawi[12] and more recently by Mahan.[13] The other one is known as the quadratic response formalism based on the time-dependent perturbation theory developed due to Ashcroft and Schaich[14], Hermeking[15] and Caroli et al.[16]. Supressing the technicalities we shall discuss below both the approaches and show that both will yield the same result for the photo-electric current.

Scattering theory formalism

In this approach since one has a steady beam of photons shining the sample and a corresponding steady emission of escaping electrons the process is viewed as an inelastic scattering of photons. Assuming for simplicity the electrons in the sample to be independent, except for their interaction with photons and using a fully quantized description of radiation we write the Hamiltonian of the electron-photon coupled system as

$$\mathcal{H} = H_0 + H_I \tag{1}$$

where

$$H_0 = h + \sum_{\beta} \hbar\omega_\beta \, b_\beta^\dagger \, b_\beta \tag{2}$$

and

$$H_I = \frac{1}{\sqrt{\Omega}} \sum_{\beta} \gamma_\beta b_\beta \, (\bar{\varepsilon}_\beta . \bar{\nabla}) \tag{3}$$

h is the one-electron Hamiltonian of the sample involving electron lattice potential $V(\bar{r})$. $b^\dagger$ (b) is the creation (annihilation) operator for the radiation mode β whose frequency is ω_β and polarization $\bar{\varepsilon}_\beta$. The operators satisfy the usual commutation relation $[b_\beta, b_{\beta'}^\dagger] = \delta_{\beta\beta'}$ and $[b_\beta, b_{\beta'}] = 0$. γ_β is the electron-photon coupling constant given by

$$\gamma_\beta = \frac{ie\hbar}{m}\left(\frac{2\pi\hbar}{\omega_\beta}\right)^{1/2} \tag{4}$$

Ω is the volume of the quantized photon field. The initial unperturbed state of the system is a direct product of photon and electron state (here we ignore electron spin),

$$\psi_0 = | n_\beta, m\rangle \tag{5}$$

where n_β is the occupational quanta in the mode β with none elsewere and the electron is in the state $|m\rangle$. The eigenstate of the coupled system is written with the help of Lippmann-Schwinger equation

$$\Psi^+ = \Psi_0 + \frac{1}{\varepsilon_0 - H_0 + io^+} H_I \Psi^+ \tag{6}$$

where $\varepsilon_0 = \hbar\omega_\beta n_\beta + E_m$, with E_m as the eigenvalue for the mth electronic state. In first order to H_I,

$$\Psi^+ = \Psi_0 + \Psi_1 \tag{7}$$

where
$$\Psi_1 = \gamma_\beta \left(\frac{n_\beta}{\Omega}\right)^{1/2} | n_\beta - 1, \Phi_1\rangle \tag{8}$$

$$|\phi_1\rangle = G(\bar{\varepsilon}_\beta \cdot \bar{\nabla})/m\rangle \tag{9}$$

and
$$G = (E_m + \hbar\omega_\beta - h + i0^+)^{-1} \tag{10}$$

The electronic state $|\Phi_1\rangle$ in the vacuum far away from the sample which is semi-infinite is to be found out. To do this we write the Dyson equation

$$G = G_0 + G_0 V G \tag{11}$$

where G_0 is the Green function in the absence of the lattice (semi-infinite sample) electron potential V as

$$G = \left(E_m + \hbar\omega_\beta - \frac{p^2}{2m} + i0^+\right)^{-1} \tag{12}$$

G_0 is expanded as

$$\langle \bar{q}_t, R_z | G_0 | \bar{r}\rangle = e^{-i\bar{q}_t \cdot \bar{r}} \int \frac{dk_z}{2\pi} e^{ik_z(R_z - r_z)} \left[E_z - \frac{\hbar^2 k_z^2}{2m} + i0^+ \right]^{-1}$$

$$= -i \frac{m}{\hbar^2 q_z} e^{iq_z(R_z - r_z)} e^{-i\bar{q}_t \cdot \bar{r}} \tag{13}$$

which describes a Fourier component $\bar{q}_t$ of G_0 in a plane parallel to the surface of the semi-infinite sample. The quantities E_Z and q_Z are given by the energy conservation.

$$E_z = E - \frac{\hbar^2 q_t^2}{2m} = \frac{\hbar^2 q_z^2}{2m} \tag{14}$$

with $$E = E_m + \hbar\omega_\beta \tag{15}$$

and q_Z is a positive real or positive imaginary number. For $\bar{R}$ far outside the sample and $\bar{r}$ near it, we write

$$\langle \bar{q}_t, R_z \mid G \mid \bar{r} \rangle = -i \left(\frac{m}{\hbar^2 q_z} \right) e^{i q_z R_z} \langle f \mid \bar{r} \rangle \tag{16}$$

where $$\langle f \mid \bar{r} \rangle = e^{i\bar{q}\cdot\bar{r}} + \int d\bar{r}' \, \langle \bar{r} \mid G \mid \bar{r}' \rangle \, V(\bar{r}') \, e^{i\bar{q}\cdot\bar{r}} \tag{17}$$

is the final outgoing state.

Now using the above equation

$$\begin{aligned} \Phi_1(\bar{R}) &= - \int\limits_{q_z > 0} \frac{d\bar{q}_t}{(2\pi)^2} e^{i\bar{q}\cdot\bar{R}} \left(\frac{im}{\hbar^2 q_z} \right) \langle f \mid \bar{\varepsilon}_\beta . \bar{\nabla} \mid m \rangle \\ &= -(2\pi i) \int\limits_{q_z > 0} \frac{d\bar{q}}{(2\pi)^3} e^{i\bar{q}\cdot\bar{R}} \, \delta \left(E_m + \hbar\omega_\beta - \frac{\hbar^2 q^2}{2m} \right) \langle f \mid \bar{\varepsilon}_\beta . \bar{\nabla} \mid m \rangle \end{aligned} \tag{18}$$

Next if we calculate the current flowing away from the sample averaged over a plane at R_C with the vacuum incoming state as the appropriate final states of photo-emission we get

$$\langle J_z(\bar{R}) \rangle = \frac{e}{A} \left(\frac{m}{2\pi\hbar^2} \right) \int \sin\theta \, d\theta \int d\varphi \times \sum_m n(E_m) \frac{e\hbar q}{m} \langle m \mid H' \mid f \rangle \mid^2 \tag{19}$$

where A is the area perpendicular to Z and $H' = (n_c/\Omega)^{1/2} \, \gamma_\beta \, (\bar{\varepsilon}_\beta . \bar{\nabla})$. The integrand on the right hand side is the current emitted in the direction θ, φ, per unit area of the sample for the polarization mode $\beta . n(E_m)$ is the initial occupied density of states.

Quadratic response formalism

From experimental observation one knows that the photocurrent is directly proportional to the intensity of incident light. Hence, the steady state current response of the system is quadratic in the electromagnetic field amplitude. This is the basis of the formalism. The total Hamiltonian of the system has two parts as before

$$H = H_0 + H_I$$

H_0 is the Hamiltonian in absence of any applied radiation field. In the previous approach the electromagnetic field is fully quantized which was first suggested by Einstein while here a gauge is so chosen in which the scalar potential is identically zero.

$$H = -\frac{1}{c} \int d\bar{r} \, \bar{A}(\bar{r}) . \bar{J}(\bar{r}) \tag{20}$$

c is the velocity of light, $\bar{A}$ is the space-time dependent vector potential of the radiation field, $\bar{J}(\bar{r})$ is the current density operator for the electrons.

In the absence of any static field where no diamagnetic current exists, the only paramagnetic current that is written for its μth cartesian component:

$$J_\mu(\bar{r}) = \frac{e}{2m} \sum_i [p^i_\mu \delta(\bar{r} - \bar{r}_i) + \delta(\bar{r} - \bar{r}_i) p^i_\mu] \quad (21)$$

where $\bar{r}_i$ is the position of the ith electron. After the radiation falls on the sample and a steady state of emission is attained the canonically averaged current far away from the sample at a later time is given by

$$\langle J_\alpha(\bar{R}, t)\rangle = \sum_n \left(\frac{1}{Q^0} e^{-\beta E^0_n}\right) \langle n | e^{iHt/\hbar} J_\alpha(\bar{R}) e^{-iHt/\hbar} | n\rangle \quad (22)$$

where the quantity inside the parahtheses is the initial thermal weighting function of the many particle eigen state of H_0 whose eigenvalues are E^0_n, and Q^0 is its canonical partion function. These states evolve according to H as H_I is adiabatically turned on from the infinite past. Using the interaction representation

$$e^{iHt/\hbar} = e^{iH_0t/\hbar}\left[1 + \frac{1}{i\hbar}\int_{-\infty}^{t} d\tau\, H_I(\tau) + \ldots\right] \quad (23)$$

we obtain the second order in $\bar{A}$

$$\langle J_\alpha(\bar{R}, t)\rangle = \frac{1}{(\hbar c)^2}\int_{-\infty}^{t} d\tau_1 \int_{-\infty}^{t} d\tau_2 \sum_{\mu\nu} \int d\bar{r}_1 \int d\bar{r}_2$$

$$\times A_\mu(\bar{r}_1, \tau_1) A_\nu(\bar{r}_2, \tau_2) \langle\langle J_\mu(\bar{r}_1, \tau_1) J_\alpha(\bar{R}, t) J_\nu(\bar{r}_2, \tau_2)\rangle\rangle \quad (24)$$

The double angular bracket is ensemble average with respect to H_0. In this derivation the zero-order term represents the thermionic emision. The linear terms do not contribute to the dc photo-effect. It is very important to note that this form of the current has also been obtained by Wehrum and Hermeking[17] using the many-body Green function technique. Another alternative method employing Keldysh formalism for non-equilibrium many-body system has also yielded the same expression due to Caroli et al.[16]

The full many-body evaluation of this formula is unfortunately a formidbble task. Therefore, an independent particle reduction of this formula is necessary to calculate the current. This is carried out by incorporating an effective single particle Hami tonian. Using the single particle paramagnetic current the above equation giving the normal component of the outgoing current to the sample surface will reduce to[14]

$$\langle J_\alpha(\bar{R})\rangle = \frac{e}{A}\frac{2\pi}{\hbar}\sum_{m,f} n(E_m)\, |\langle m | H' | f\rangle|^2\, \delta(E_m + \hbar\omega - E_f) \quad (25)$$

where

$$H' = \frac{1}{2c}\int d\bar{r}\, \bar{A}(\bar{r}) \cdot \bar{j}(\bar{r}) \quad (26)$$

One notes here that the golden rule form of the photocurrent represented as a transition rate above demands the knowledge of the final state $|f\rangle$.

Expressions (19) and (25) are equivalent where

$$\bar{j} = \frac{e\hbar}{m}\bar{\nabla}$$

and

$$\bar{A} = \left(\frac{8\pi\hbar c^2 n_\beta}{\omega_\beta \Omega}\right)^{1/2} \bar{\varepsilon}_\beta$$

It is worth mentioning here that a complete evaluation of the expression (19) or (25) has not yet been attempted from first principle. It is partly because one does not know how to treat satisfactorily all ingredients and even if they are given, immense computational effects will be necessary to calculate (25). We indicate below a few problems involving this equation.

1. The final state $|f\rangle$ being the vacuum incoming state particularly for the single particle model it appears physically identical to the state involved in low energy electron diffraction (LEED). Once this is acceptable the damping of the final electron state is taken care of by the self energy of the electron by making it complex. This is a phenomenological attempt while more theoretical work is necessary to know the effective form of final state and its energy.
2. The knowledge of the electronic states and its energy in the sample is reasonably understood. If in the sample the initial states are extended, then the single particle model is appropriate. This works well in the ultraviolet photo-emission for the valence band of the sample. But for the local initial states like core level or adsorbate level excitations, one has to consider the many-body effects like relaxation shifts, shake up and shake off processes.
3. It is necessary to know the electromagnetic field that excites the photo-electrons. We have, in fact, used above the bare field which needs to be screened appropriately by the presence of the sample. There are interesting discussions on this problem in several recent papers[18–21].

3. Electron structure of solids

Now after qualitatively understanding the phenomenon of photo-emission we turn our attention towards the sample. At the expense of photon we have pulled the electrons out of it and measure the current as a function of its energy, angle of emission and sometimes of spin polarization. Here we address ourselves to only one important question: What information does the current give about the sample (particularly on its electronic structure, or what does the quantity, the electron distribution curve

(EDC), the measured quantity in the photo-emission experiments, corresponds to so far as the (solid) sample is concerned?

A key function in the electron theory of solids is the density of states which is defined as,

$$n(E) = \sum_m \delta(E - E_m) \qquad (27)$$

where E_m is the energy of the electron in the state m. For a crystalline solid one takes the help of canonical energy band theory to write the density of state function as

$$n(E) = \frac{1}{(2\pi)^3} \sum_i \int d\bar{k}\, \delta(E - E_l(\bar{k})) \qquad (28)$$

where $E_l(\bar{k})$ is the energy of the electron in the band i having the wave-vector (reduced) $\bar{k}$ in the Bloch state. On shining light on the solid one optically excites the electrons either by a direct or by a non-direct process. In the direct process it is essential to consider the law of conservation of momentum. The electron goes from i band to f band, which is above the Fermi level E_F, by an optical transition with photon energy $\hbar\omega$ restricted to a surface of $\bar{k}$ space of constant inter-band energy difference given by

$$\Omega_{fl}(\bar{k}) \equiv E_f(\bar{k}) - E_l(\bar{k}) - \hbar\omega = 0 \qquad (29)$$

The total number of direct transitions at photon energy is given by a quantity known as joint density of states.

$$\mathcal{J}(\hbar\omega) = \frac{1}{(2\pi)^3} \sum_f \int' d\bar{k}\, \delta(\Omega_{fi}(\bar{k})) \qquad (30)$$

where prime in the integration sign denotes that $E_i < E_F < E_f$. We note here that $\mathcal{J}(\hbar\omega)$ gives the imaginary part of dielectric constant, with an appropriate matrix element, which is measured in the experiment of optical absorption. Now we represent another quantity by calling it as the energy distribution of joint density of state (EDJDS) defined by

$$\mathcal{D}(\hbar\omega, E) = \frac{1}{(2\pi)^3} \sum_{fi} \int' d\bar{k}\, \delta(\hbar_{fi}(\bar{k}))\, \delta(E - E_l(\bar{k}))\; | M_{fi} |^2 \qquad (31)$$

The first δ-function selects those transitions with $\bar{k}$-conservation and the second with a particular initial energy. M_{fl} is the appropriate matrix element that weights the optical transitions. Now EDJDS is the quantity representing the internal energy distribution of photo-excited electrons. This differs from density of states because it contains extra quantities $\hbar_{fl}$ and $| M_{lf} |^2$.

Another process of optical excitation mentioned above is known as non-direct transition. In this process the momentum conservation selection rule is completely abandoned and only energy conservation is considered to be important. The EDJDS is taken as the product of initial and final density of state $(n(E - \hbar\omega)\cdot n(E))$. This process is thought to be suitable for localized states of electrons and for the non-crystalline materials. In this process other complications arise from many-body considerations.[22]

The electron distribution curve (EDC) of the photo-emission which one gets in an experiment is believed to be the EDJDS multiplied by a quantity T, known as the probability factor that takes care of the transport and the escape of the electrons to the vacuum. T is a function of $\bar{k}$, the momentum of the electron and Φ, the work function of the solid.

Now we make here a contact with the so-called "Three-step model.[23-25] This is a phenomenological description of photo-emission constituted of (i) optical excitation in the solid followed by subsequent (ii) electron transport to the surface including the possibility of inelastic scattering by the electrons and/or photons and (iii) finally escape into the vacuum. The model has been very widely used in interpreting experimental data. The basic idea of this step picture is to analyse the phenomenon in a sequence and one can very ably discuss each sequence to a great detail.[26] The step (i) is involved in finding out EDJDS as discussed above and (ii) and (iii) together involve T.[11] In the formal expression[25] we find out that the quantity $| H'_{mf} |^2$ contains information about the step model while the final state $| f \rangle$ contains the information about the last two steps. We have noted before that accurate evaluation of[25] is a hard task, therefore specific model calculations have been attempted[27] resulting in the same golden rule form. The new quantities to appear in the formula for current are (i) the effective escape depth due to the screening of the electromagnetic field and inelastic collision of the escaping electrons and (ii) the transmission probability. *In this compromise a detailed calculation have now been made on a realistic scale*[28] *where the three step model is a special case of the general formal theory of sec.* 2.

Now in this framework the interpretation of EDC spectra involves calculation of EDJDS. The steps needed to calculate are (i) calculation of the band structure $E_i(\bar{k})$, (ii) evaluation of the interband matrix element $M_{if}(\bar{k})$, (iii) mometum space interpolation of the quantities to permit accurate integration and (iv) evaluation of the momentum space integral that defines the energy distribution. As mentioned before, the final state used is a time-reversed LEED wave function.[29] The comparison of EDC with calculated EDJDS reveals certain consistency in the structures of the spectra which vary in shape and location that depends very much on the materials. Here we do not review the various experimental and computational results showing agreements or otherwise. An interested reader will find ample evidence for the usefulness of this approach in identifying the various origin of features in EDC. [11, 30]

At this point we focus our attention on two important sophisticated developments involving the origin and the angular dependence of photo-emission. First, by origin we mean whether photo-emission is from the bulk of the materials, what is known as volume effect, or is it from the surface of the materials known as surface effect. The appearance of Bloch wave energy in an infinite crystal reveals the information about the bulk electronic structure. In this limit the photon energy is high enough and the escape length is large. In constrast to the volume effect the first step of the three-step process is enough to explain the mechanism of surface photo-emission. Historically, this effect was explored in detail[31-33] by considering only the surface potential. The optical transitions are made possible by considering two-dimensional surface lattice network. Here, the surface state theory is totally incorporated.[34, 35] It is possible to have emission from surface states which may be either intrinsic or associated with adsorbed species. However, now that there is general awareness that both mechanisms are operative, which (if either) is dominant depends on the details of the experiments.

Secondly, the angle-resolved electron energy distribution curve[36] informs not only about the energy but also the direction of electrons emitted through the sample surface (θ_e, ϕ_e of Fig. 1). The motivation of this study may be stated as follows. By measuring the kinetic energy and by knowing the direction of propagation, one knows the momentum of the escaping electron simply from $E = \hbar^2 k^2/2m$. In principle, therefore, one has the exciting possibility of having a single point in Espace rather than having the integrated quantities as in eqn. (31). From the knowledge of energy and direction one deduces the dispersion relation $E(\bar{k})$, particularly the parallel component of momentum. Mahan[13] has drawn a strong similarity between the physics of photo-emission and that of LEED. Now people believe that the angle-resolved photo-emission may be used as alternative to LEED, particularly for studying the topography and surface electronic structure of solids.

4. Summary

We have discussed above the photo-electron spectroscopy, its conceptual foundation and the recent theoretical developments. It is a method used like other spectroscopic techniques such as light absorption and scattering, characteristic electron energy loss and X-ray fluorescence. The potential application of photo-emission is now well-recognised for the investigation of electronic structure of solids. Other than its commercial interest in the development of photo cathodes and its extensive application in electronic devices, it is now used for the study of all kinds of solids, metals, semi-conductors, insulators, magnetic materials, glasses, etc. We have not separately discussed the photo-emission from the core levels (what is known as XPS) and emission from the valence bands of solids (UPS). We have also not discussed anything about the experimental systems. With the availability of sources like synchrotron radiation one gets reliable data for careful theoretical analysis.

We conclude by mentioning the theoretical development of the constant "Austrittsarbeit" that appears in Einstein's (commonly known as work function). Its detailed calculation with the help of many-electron theory particularly for simple metals[37-38], has provided in important check of the photo-threshold.

References

1. Einstein, A., *Ann. Physik* **17**, 132 (1905) (Its English translation in A.B. Arons, M.B. Peppard, *Am. J. Phys.* **33**, 367 (1965).
2. Koch, E., Haensel, R., Kunz, C., (Eds.) *Vacuum Ultraviolet Radiation Physics*, Pergamon (1975).
3. Bennett, L.H. (Ed.) Proc. of 3rd Materials Research Symposium on Electronic Density of States, NBS Publication No. 323 (1970).
4. Spicer, W.F., in Optical properties of Solids (Ed.) B.O. Seraphin, North Holland (1976) p. 631.
5. Eastman, D.E., in *Technology of Metals Research*, Vol. **6**, E. Passaglia, Wiley, p. 413 (1972).
6. Feuerbacher, B., Willis, R.F., *J. Phys.* (C) **2**, 169 (1976).
7. Feuerbacher, B., Fitton, B., Wills, R.F., (Eds.) *Photoemission and the Electronic Properties of Surfaces*, Wiley, (1977).
8. Glasser. M.L., Bagchi, A., *Prog. Sur. Sc.* **7**, 113 (1976).
9. Cardona, M., Ley, L. (Eds.) *Photoemission in Solids*, Springer-Verlag (1978).
10. Siegmann, H.C., *Phys. Rep.* **17**, 37 (1975).
11. Smith, N.V., CRC, Critical Reviews in Solid Sate Science, **2**, 45 (1971).
12. Adawi, I., *Phys. Rev.* **134 A**, 788 (1964).
13. Mahan, G.D., *Phys. Rev.* **B 2**, 4334 (1970).
14. Ashcroft, N.W., Schaich, W., in Ref. 3, Schaich, W., Ashcroft, N.W., *Phys. Rev.* **B 3**, 2452 (1971).
15. Hermeking, H., *Zeits. Phys.* **253**, 379 (1972), *J. Phys.* (C) **6**, 2898 (1973).
16. Caroli, C., Lederer-Rozenblatt, D., Roulet, B., Saint-ames, D., *Phys. Rev.* **B 8**, 4552 (1973).
17. Wohrum, R.P., Hermeking, H., *J. Phys.* (C) **7**, L16 (1974).
18. Kliewer, K.L., *Phys. Rev.* **B 14**, 1412 (1976); *Phys. Rev.* **B 15**, 3759 (1977).
19. Bagchi, A., *Phys. Rev.* **B 15**, 3060 (1977).
20. Feibelman, P.J., *Phys. Rev.* **B 14**, 762 (1976).
21. Mukhopadhyay, G., Lundqvist, S., *Phys. Script*, **17**, 69 (1978).
22. Gadzuk in Ref. 7.
23. Fan, H.Y., *Phys. Rev.* **68**, 43 (1945); Kane, E.O., *Phys. Rev.* **127**, 131 (1972).
24. Spicer, W.E., *Phys. Rev.* **112**, 114 (1958); Burgland, C.N., Spicer, W.E., *Phys. Rev.* **136**, 1030, 1044 (1964).
25. Thomas, H., *Zeits Phys.* **147**, 395 (1957).
26. Spicer, W.E., *Comm. Solid State Phys.* **5**, 105 (1973).
27. Schaich, W.L., *Phys. St. Sol.* 66b., 527 (1975). (In this work the effective escape depth and the transmission probability have been described and are introduced into the expression for the current).
28. Pendry, J.B., *Surface Science* **57**, 679 (1976) and in Ref. 7.
29. Feibelman, P.J., Eastman, D.E., *Phys. Rev.* **B 10**, 4932 (1972).
30. Moruzzi, V.L., Williams, A.R., Janak, J.F., *Phys. Rev.* **B 8**, 2546 (1973); Grobman, W.D., Eastman, D.E., Freeouf, J.L., *Phys. Rev.* **B 12**, 4405 (1975). See also Ref. 3 and 11.
31. Tamm, I., Shubin, S., *Zeits. Phys.* **68**, 97 (1931).
32. Frolich, H., *Ann. Phys.* **7**, 103 (1930).
33. Mitchell, K., *Proc. Roy. Soc.* **146 A**, 442 (1934).
34. Gadzuk, J.W., *Phys. Rev.* **B 10**. 5030 (1974).

35. Liebsch, A., *Phys. Rev.* **B** 13, 544 (1976).
36. Smith, N.V., in Ref. 9 and Dietz, E., Eastman, D.E., *Phys. Rev. Lett*, **41**, 1674 (1978) (Erratum PRL 42, 133 (1979).
37. Lang, N.D., Kohn, W., *Phys. Rev.* **B** 3, 1215 (1971).
38. Pant, M.M., Das, M.P., *J. Phys.* (F) **5**, 1301 (1975); Monnimer, R., Perdue, J.P., Langreth, D.C., Wilkins, J.W., *Phys. Rev.* **B 18**, 656 (1978).

Discussions

E S R Gopal: If one uses X-ray incident energies, they would be larger than the work function details. So it should be possible to check the details of the mechanism of emission. Is it true?

M P Das: The X-ray photo-electron spectroscopy (XPS, which is also known as electron spectroscopy for chemical analysis, ESCA) employing X-rays provides a powerful tool for the measurement of core states as well as the valent states of materials. When chemical binding takes place, this can detect the chemical shifts induced in the core levels. XPS is largely used for the absorption of gases on solid surfaces. However, its resolution is less than that of UPS.

S C Tiwari: What are the specific and special uses of photo-electron spectroscopy as compared to other methods of band structure calculation?

M P Das: Photo-electron spectroscopy (PES) provides information which is similar and somewhat complimentary to that of a technique like optical absorption. The optical method is insensitive to surface studies of materials since the penetration depth of light is over 100 Å. But in PES the penetration depth is 5 Å (see W.E. Spicer in *Chemistry and Physics of Solid Surfaces*; Ed. by R. Vanselow and S.Y. Tong. CRC Press, 1977). The important point is the measurement of the dielectric constant which involves the summation over the final states. It is necessary to deconvolute the data in order to interpret them in terms of the electronic states of solid, which is a formidable task while the EDC of PES with certain approximation is equated with the final state density of states. This avoids the difficulty of deconvolution. For band structure information from PES, see Ref. 2 above.

S C Tiwari: Is there any attempt to determine band structure of GaAs and Si, etc.?

M P Das: Yes, some of the relevant references are:
for GaAs: *J. Electron Spectro. and Related Phenomea*, **5**, 603 (1974); *Phys. Rev. Lett.* **33**, 1601 (1974).
for Si: *Phys. Rev. Lett.* **28**, 1378 (1972).

Abstracts of Contributed Papers

Relativistic and Statistical Foundations of Quantum Mechanics

C.K. Raju

It is pointed out that the notion of a particle as a mass point or a point singularity in a field is incompatible with the Schwarzschild solution and the Heisenberg principle. The proposed model for extended particles is based only on gravitational forces to avoid the introduction of unknown forces. Specifically, the gravitational action of universe balances the gravitational self action of the particle mass, leading to a random harmonic oscillation.

It is contended that quantum mechanics describes the dynamics of such extended particles. A new interpretation of the wave function in this context and the corresponding derivation of the Schrödinger equation, are given. The usual probability interpretation appears in this theory as an excellent approximation, valid only for regions which are large compared to the characteristic dimensions of the particle. The classical probabilistic framework used, leads naturally to a representation of observables in terms of positive operator valued measures. The classical correspondence is tentatively explored.

Using the Schwarzschild-Tetrode-Fokker action it is shown that charged particles in this theory are stable if they are surrounded by a large number of charged particles in a similar energy range. Further, the model gives rise to some of the qualitative features of nuclear forces.

It would be possible to test this theory by:

(i) comparison of Planck's constant with the theoretical value obtained in terms of cosmological constants,

(ii) failure of the probability interpretation in regions comparable to the characteristic dimension of the particle,

(iii) measurement of the lifetimes of unstable charged particles in different environments.

A Quantum-like Structure in Classical Mechanics

R.K. Varma

The problem of non-adiabatic escape of charged particles from adiabatic magnetic traps is discussed. It is shown that the ensemble behaviour of the system of particles confined in the adiabatic magnetic traps can be described through a set a probability amplitudes obeying Schrödinger-like equations. Recent experimental results confirm the predictions of the theory that the particles in the adiabatic traps injected with a given total energy and 'first action invariant' μ reside in the trap with a multiplicity of lifetimes which are constant multiples of the lowest. This system, it is argued, thus constitutes an example of the existence of a quantum-like structure in classical mechanics, where the role of $\hbar$ is played by μ.

'Pauli Effect' in Relativity

S. Banerji and M.K. Ray

There seems to be a lot of controversy about the correct formulation of the electromagnetic energy-momentum tensor for a material medium, two different forms of which were proposed by Abraham and Minkowski towards the beginning of this century. The former tensor is symmetric in its indices while the latter is non-symmetric so that the latter cannot be directly substituted on the right hand side of Einstein's field equations of general relativity. Pauli remarked that for an isotropic and homogeneous material medium the force density derived from Abraham's energy-momentum tensor contains an extra term proportional to the rate of change of the Poynting vector in addition to that obtained from Minkowski's expression. This extra term is referred to in the literature as 'Abraham term' or 'Pauli effect'. The presence of this Pauli effect has been experimentally verified by Walker and Lahoz. Israel showed that Minkowski's energy-momentum tensor when slightly modified also predicts this effect. Recently Taub derived a third expression for the energy momentum tensor from a general variational principle, which is also symmetric. We show here that this also predicts the 'Pauli effect'.

Ether as Advocated by Einstein and Others

P. Achuthan, T.S. Shankara and K. Venkatesan

The paper points out that the ether hypothesis had an unqualified approval of Einstein, contrary to the general belief that he discarded it; it also deals with an important corollary of this hypothesis.

The ether hypothesis played an important role in the evolution of physics till the beginning of this century. Various models of it were envisaged by Descarres, Huygens, Euler, Cauchy, Faraday, Maxwell and many others. But when the Michelson-Morley experiment which supposedly settled the fate of ether gave the well-known null result. Lorentz suggested a suitable mechanical property for ether to retain the hypothesis. However, Einstein asserted that what is relevant is only relative motion and the hypothesis of ether was not necessary. This notion was later misinterpreted to mean that Special Relativity had killed the concept of ether. Einstein in his lecture "Ether and Relativity" has made it amply clear that what Special Relativity does is to make ether superfluous rather than disprove its existence. In fact, he gives weighty reasons in favour of ether on the basis of General Relativity: In order to look upon rotation as real, ether is also to be looked upon as real. General Relativity recognizes that 'empty space' is neither homogeneous nor isotropic compelling us to describe its state by 10 functions $g_{\mu\nu}$.

As a corollary to the existence of ether, we have a unit time vector orthogonal to the hypersurface of ether. This at once ensures causality even in interactions involving spacelike and time-like particles. Indeed the appropriate momentum for the spacelike particles will have the form

$$p_\mu = \frac{m}{\sqrt{(u^2/c^2) - 1}} (u \hat{n}, c) = \frac{m}{\sqrt{(1 - c^2/u^2)}} \left(c\hat{n}, \frac{c^2}{u} \right)$$

with positive root, the metric being $(-, -, -, +)$. This suggests the interchange of the roles of space and time so that the ordinary space becomes hyperbolic, which in turn, implies an associated Mach cone.

Any experiment to detect these spacelike particles also reveals the existence of ether by exhibiting an anisotropy of the latter in their wake. This crucial property of spacelike particles is not envisaged in any of the tachyon theories and, therefore, to stress their distinction, we call them Rugtigons (*Ruch*: light in Sanskrit), which literally means particles travelling faster than light.

Finally, it appears worthwhile to develop the concept of such a non-mechanical ether along the lines of the vacuum of quantum field theory.

On the Physical Constancy of some Tensors in Relativistic Hookean Elastic Systems

L. Radhakrishna and R.V. Virkar

A tensor is known to be physically constant (p.c.) if its convected derivative vanishes with respect to flow of the medium. The necessary and sufficient conditions for the stress tensor T_{ab} of a Hookcan Elastic system to be p.c. are obtained. The existence of p.c. cubically symmetric elastic body is established. For an isotropic elastic body in special relativistic approximation the p.c. of deformation tensor θ_{ab} exists in the absence of initial strain and the conservation of energy density as well as expansion with respect to free flow. In the Newtonian approximation for an isotropic body it is found that the vorticity tensor (or θ_{ab}) could not be p.c. in the absence of initial strain. The characterization p.c. of Einstein Curvature tens or E_{abc^2} is accomplished.

Collineations in Null Electromagnetic Field in General Relativity

Zafar Ahsan and Saiyed Izhar Husain

A geometric structure on V_4 characterizing the null electromagnetic field has been defined and different types of collineations along the propagation and Polarization vectors have been proved. It is shown that $\mathcal{L}_s F_{ij} = 0$

implies $F_j^k \mathcal{L}_s g_{ik} = 0$, which does not necessarily imply that S (propagation vector) is killing. Thus even under certain geometric conditions, it is not possible that the symmetry of the space-time is equivalent to the existence of certain killing vectors, although this happens for non-null electromagnetic fields. Finally, a conservation law, first given by Katzin et al. for null fields, has been given for the present structure.

Embedding Class of the Plane Symmetric Space-time

S.N. Pandey, S.P. Sharma and Pramila Goyal

The plane-symmetric space-time of Taub, admitting three parameter group of motion in the Euclidean plane, is studied with reference to its embedding in a higher dimensional flat space. It has been established that its embedding class cannot exceed three. A set of sufficient conditions for this space-time to be of class-2 has been obtained. In the static case three distinct embeddings in a 6-flat space have been worked-out. A necessary and sufficient condition is given for the plane-symmetric space-time to be of class-1.

Tachyonic Lorentz Transformations and Transformation of Relativistic Wave Functions

K.D. Purohit and B.S. Rajput

Transformation equations for the relativistic wave function has been derived under the operations of Poincare group for imaginary mass system by using the extended Lorentz transformations for tachyons and it has been shown that interchange of roles of space and time co-ordinates of a tachyon automatically incorporates the interchange of roles of energy and momentum.

Dirichlet Problem for Maximal Submanifolds of Lorentzian Manifold

Daniel Bancel

The existence of a maximal submanifold (with respect to area) is an important property for a space-time solution of the initial value problem, positivity conjecture for the gravitational mass of an asymptotic flat space-time.

In this paper we shall consider the problem of determining the existence of maximal spacelike submanifolds with prescribed $(n-2)$ dimensional boundary for a space-time $V_n = R \times V_{n-1}$ where V_{n-1} is a complete Riemannian manifold with metric ds^2 the metric on V is $ds^2 = dt^2 - d\sigma^2$. We shall give sufficient restrictions on the boundary in order for this Dirichlet-problem to be solvable. The method is based on a priori bounds for solutions in view of an application of Leray-Schauder fixed point theorem. Some theorems of this paper, with a sketch of the proofs have been announced in Comptes Rendus série A (1978).

Causality Violations and Cuts in a Space-time

P.S. Joshi

Various causality assumptions have been considered for a space-time. The relation between the violation of these conditions and cuts in a space-time has been examined. It is shown that for a space-time without any cuts the causality and the distinguishing conditions turn out to be equivalent. In conclusion it is stated that for a causal space-time only cuts are responsible for any type of higher degree causality violations.

Global Definition of Path-integrals of Riemannian Manifolds and their Evaluation in the WKB Approximation

A.N. Maheshwari

Using a generalized Feynman-Kac relation the path integrals for propagation kernels are defined on configuration spaces which are Riemannian manifolds. Techniques for computing the path integrals in a semi-classical expansion are developed.

Complex Dynamics in Hamiltonian Particle Motion

R.L. Clerc and I. Gumowski

The motion of a charged relativistic particle in a periodically varying force field is considered. It is assumed that this motion is described by an autonomous Hamiltonian of order four or by a non-autonomous one of order two, leading to a system of canonical equations which are not completely integrable. It is further assumed that these equations permit the construction of an unambiguously defined two-dimensional real-valued recurrence (point-mapping of two-dimensional surface onto itself). The singularities of the recurrence are studied by means of a mixed analytico-numerical method. A particular case, originating in astrophysics and for which a certain number of discrete trajectories of a 'stochastic' nature are already known, is examined for singularities. As expected, a set of classical singularities is found: cycles (periodic points) of various order, and invariant curves intersecting at homoclinic and heteroclinic points. It appears, however, that this particular non-integrable case involves also an unusual singularity, related to a degenerate non-uniqueness of implicitly defined functions.

Electromagnetic Fields in Space-times with Local Rotational Symmetry

S.V. Dhurandhar, C.V. Vishveshwara and J.M. Cohen

A major cause of difficulty of integrating the electromagnetic field equations in a curved space-time is their coupled structure. Standard methods which decouple the equations in flat space-time fail in curved space-time since the space-time curvature leads to a stronger coupled system. Cohen and Kegeles have shown that the Hertz potential formalism which is useful in flat space-time may be extended to all curved space-times and the Debye formalism (two component Hertz potential) may be extended to a wide class of space-times, the generalized Goldberg-Sachs class; which consists of those space-times which admit a shear-free congruence of null goedesics along the repeated principal null direction of the Weyl tensor. In this paper we confine our investigations to perfect-fluid space-times with local rotational symmetry given by Ellis and Ellis and Stewart, which is a subclass of the generalized Goldberg-Sachs class of space-times. This class consists of of a wide range of interesting space-times: Friedmann models, Schwarzschild, Kerr, Godel, Kantowski and Sachs universes, Taub-NUT, anisotropic spatially homogeneous cosmological models, etc.

We adopt the Newmann-Penrose formalism for our study. We first treat the problem at a general level, that is, we make use of the general form of the line-element of space-times with local rotational symmetry and set up the governing equation for the Debye potentials. It is seen that a considerable amount of information may be derived from this equation without specializing to any particular space-time. We then discuss some particular space-times, namely, Kantowski and Sachs universes, Taub-NUT, anisotropic spatially homogeneous cosmological models.

Axially Symmetric Stationary Electrovac Solutions

S. Chatterjee and S. Banerji

The equilibrium shape of a rotating star being an oblate spheroid, axially symmetric solutions of Einstein's field equations are of interest in astrophysics. The only astronomical objects discovered so far where general relativistic effects are not negligible are pulsars, which are known to possess high magnetic field. This has made stationary solutions of Einstein-Maxwell equations relevant to astrophysics. Perjes and Israel and Wilson have given independently a new class of solutions of source-free Einstein-Maxwell equations which can be interpreted as the external gravitational and electromagnetic fields of a spinning source with unit specific charge. Bonnor and Ward found a solution corresponding to the exterior field of two charged spinning sources. Starting from Zipoy's solutions in oblate and prolate spheroidal co-ordinates for the source-free gravitational field we generate some axially symmetric stationary solutions of source-free Einstein-Maxwell equations by using Perjes' method. All these solutions became Euclidean at infinity. The asymptotic behaviour and the singularity of the solutions are studied in order to gain some insight into the nature of the source. The solution in prolate spheroidal co-ordinates is found to contain closed time like lines.

In the case of the simplest solution considered by Perjes and Bonnor and Ward both the specific charge and the gyromagnetic ratio, i.e., magnetic moment/angular momentum are unity. But in our solution although the specific charge again becomes unity, the gyromagnetic ratio becomes 2.

Static Spherically Symmetric Solution of Einstein-Klein-Gordon Equations

Manisha Bhattacharyya and Bidyut Kumar Datta

The Einstein-Klein-Gordon equations for scalar meson cloud are set up along with certain consistency condition. The geometry inside an atomic nucleus is discussed and the existence of a static spherically symmetric solution is demonstrated.

Thin Cylindrical Shell of Dust under Rigid Rotation in General Relativity

A. Papapetrou, A. Macedo and M. M. Som

Following the method developed by Papapetrou and Hamoui, a solution of the Einstein equations corresponding to a thin shell of dust under rigid rotation solution is obtained. The region interior to the shell is flat and the exterior vacuum region is chosen as a Lewis manifold. An essential limitation of this type of shell is that it does not allow the limit M_1 and $L_1 \rightarrow 0$, where M_1 and L_1 are the mass and the angular momentum per unit length. It is shown that the limitation is a consequence of the fact that the Lewis metric does not contain the Minkowski metric as a special case.

Spiral Density Waves in an Inhomogeneous Disc Galaxy

A. Ambastha and R.K. Varma

We have considered a differentially rotating, razor-thin disc galaxy in hydrodynamic framework. Various dynamical models indicate that the total mass density projected on the galactic plane decreases fairly fast as the galactocentric distance is increased, so as to make the gradient effects important enough. We find that Lin's neutral density waves become overstable in such a system, and thus infinitesimally small signals can grow to observational levels. It is found that such a system distingnishes between trailing and leading spirals; trailing ones being preferred over the leading ones. The amplification rates are also calculated. The energy loss due to the radial propagation of the density waves is compensated since they are generated continuously in the system and hence the overstability of the density waves could provide one possible explanation for the maintenance of the spiral structure in many disc galaxies.

Galaxy Formation in Eddington-Lemaitre Universe

V.B. Johri and M.P. Pathak

Bondi (1961) and Rindler (1969) have pointed out that the general solution of the Friedmann equation

$$R^2 = C/R - k + (1/3)\,\Lambda\, R^2$$

is quite intricate as it involves elliptic functions; however, we have obtained non-elliptic solutions which reveal certain interesting features. In the first part of this paper, we have found three non-elliptic solutions of the Friedmann equation out of which the one with $C \leqslant +\, 2k^{3/2}/3\Lambda^{1/2}$ corresponds to Eddington-Lemaitre universe. As is well known, Einstein static model is unstable against random density fluctuations and eventually transforms to Eddigton-Lemaitre universe. In the second part of the paper, we have shown by using tetrad technique (Johri, 1969) that the density fiuctuations grow exponentially in the Eddington-Lemaitre universe as

$$K = B_1 \exp\,(\sqrt{C_1/2}\,t) + B_2 \exp\,(\sqrt{C_1/2}\,t)$$

and that the statistical density fluctuations might eventually grow into galactic clusters within suitable time scale, i.e., 3.5×10^9 years. Simon (1970) has got similar results by using a classical technique due to Landau and Lifshiftz.

Neutrinos in Compact Objects

C.V. Vishveshwara and A.K. Kembhavi

We consider in this paper the possibility of neutrinos being bound in the strong gravitational field of a compact object. For this purpose we have examined the massless Dirac equation for curved space-time following the formalism of Brill and Wheeler, and Unruh. The background geometry is provided by a spherically symmetric object of uniform density. The wave equation is separated in Unruh's representation of the curved space-time γ matrices. In the usual notation the radial equations are

$$e^{\nu/2-\lambda/2}\frac{dR_1}{dr} - iwR_1 = \frac{ke^{\nu/2}}{r}R_2 \qquad (1)$$

$$e^{\nu/2-\lambda/2}\frac{dR_2}{dr} + iwR_2 = \frac{ke^{\nu/2}}{r}R_1 \qquad (2)$$

There are only two independent equations because of the zero mass of the neutrino. Introducing functions F and G through the relations

$$F^1 = \tfrac{1}{2}(D + A)\,G + \omega F \qquad (3)$$

$$G^1 = \tfrac{1}{2}(D - A)\,F + \omega G \qquad (4)$$

where $F^1 = R_1 + R_2,\; G^1 = (R_1 - R_2)/i,\; D = d/dR = e^{\lambda/2-\nu/2}\, d/dr$

$A = \frac{ke^{\nu/2}}{r}$, it can be shown that in the approximation where terms k are neglected relative to terms k^2, the functions F and G satisfy the second order equations

$$D^2 + (w^2 - A^2)\,F = 0,\; D^2 + (w^2 - A^2)\,G = 0 \qquad (5)$$

These are the equations used by Brill and Wheeler.

The A^2 term in the second order equations acts as an effective potential which $\to \infty$ as $r \to 0$, has a minimum in the interior of the star, a maximum in the empty region outside the star and $\to 0$ as $r \to \infty$, for suitable values of the parameter $2M/R$, where M is the mass of the star and R its radius. The second order equations can be treated in the W.K.B. approximation. In this manner we have obtained the energy levels of a neutrino effectively trapped in the potential, for various values of the free parameters, neglecting the slow leakage that necessarily takes place. Having obtained the energy levels, the lifetime of a neutrino in these levels is estimated using suitable generalizations of the usual W.K.B. formulae. We discuss the astrophysical implications of a 'sea' of neutrinos trapped in a strong gravitatienal field.

Relativistic Equation for Axisymmetric Gravitational Collapse with Escaping Neutrinos

M.D. Patel

Relativistic field equations for the gravitational collapse with escaping neutrinos of the rotating oblate spheroid of perfect fluid are obtained. As each element of the fluid will cool off by emission of neutrinos at some rate depending on temperature and density, the stress energy tensor fails to satisfy a local conservation law. The field equations are the generalization of the equations presented by Chandrasekhar and Friedman. The collapsing core emits neutrinos and the surrounding region is an axially-symmetric radiation zone. It is shown that on the boundary of the radiating axi-symmetric collapsing source the interior metric matches with the metric of the axisymmetric radiation zone, presented by the author.

Kerr Metric in an Expanding Universe

S.N. Guha Thakurta

While there has been a good deal of discussion on the fitting of the Schwarzschild metric with the cosmological line element, the corresponding problem for the Kerr metric has so far received little attention. After a review of earlier works of McVittee, G.C., (1933), *Mon. Not. R. astron. Soc.* **93**, 325), Einstein, A., and Straus, E.G., (1945), *Rev. Mod. Phys.* **17**, 120); and a brief discussion on recent work by Vaidya, P.C., (1977), *Pramana* **8,** 512), on a metric claimed to represent a body with angular momentum in the Einstein static universe and another in the expanding universe, the present paper goes on to consider a metric which is conformal to the Kerr metric.

It turns out that although in a sense it may represent a Kerr metric in Einstein-de Sitter universe (the space curvature $k = 0$) with the conformal factor a function of time alone, the energy momentum tensor is not that of a perfect fluid, although the expansion is everywhere isotropic.

Apparently, a similar procedure does not work for universes with non-flat space sections ($k = 0$), the conformal factor in this case turns out to be a function of $(t^2 - r^2)$.

At the non-rotating limit ($a = 0$), in the Einstein-de Sitter universe ($k = 0$), when the metric is conformal to the Schwarzschild metric, the energy momentum tensor may be interpreted as a perfect fluid energy tensor with an energy flux in the radial direction.

Frequency Shift of Radiation from Spherical Objects in Brans-Dicke Theory of Gravitation

A. Banerjee and D. Bhattacharya

An analytic expression for the frequency shift of radiation from an exploding or collapsing object in the Brans-Dicke theory of gravitation is obtained. It is shown that there is a significant deviation from the results obtained in Einstein's theory due to the presence of the scalar field. The radiation emitted from the source near the singularity in the Brans Dicke theory of gravitation exhibits only a finite frequency shift for the case of explosion, while we observe an infinite magnitude of blueshift in the radiation emitted from a Schwarzschild white hole at the moment of explosion from the singular state. For collapse, however, one always finds redshift, which may even amount to an infinite magnitude in the extreme case of approaching the singularity.

Hawking Radiation of Scalar and Dirac Quanta from a Kerr Black Hole

B.R. Iyer and Arvind Kumar

Unruh's technique of replacing the time-dependent problem of collapse by a choice of appropriate boundary conditions on the past horizon of the black hole (the ξ-quantization scheme) for derivation of the well-known Hawking radiation is extended to the Kerr black hole for the scalar and especially for the massive spin half field. The extension to the massive spin half field is made possible following the recent separation of Dirac equation in Kerr metric due to Chandrasekhar. Orthonormal sets of scalar and Dirac ξ-modes with respect to a generalized inner product defined in the extended manifold are constructed and the consistency of the new quantization scheme is exhibited. The more conventional η-scheme, which for Dirac field was developed by us in an earlier work, is reviewed and contrasted with the ξ-scheme. The expectation value of the energy-momentum tensor is evaluated asymptotically in the ξ-vacuum state, yielding explicitly the new Hawking flux of scalar and Dirac quanta, which in the limit of very large mass of the black hole reduces to the Unruh-Starobinsky effect obtained in the η-scheme. The most satisfying feature of the work is that the Fermi-Dirac distribution emerges as naturally for the Dirac quanta as the Bose-Einstein distribution for the scalar case, both corresponding to the same temperature. This establishes the validity of the ξ-scheme for fermions and confirms the association of temperature with a Kerr black hole.

On the Black Hole Energetics Relative to Tachyons

Naresh Dadhich

By considering the tachyon trajectories we establish the following:

(1) The second law of black hole physics breaks down for the tachyon-black hole interaction.

(2) It is possible to extract energy through a reversible process from a charged or a rotating source, that has not yet turned black.

These are demonstrated by giving specific examples.

Completeness of Tests of Local Hidden Variable Theories

S.M. Roy and Virendra Singh

Quantum mechanically for the Bohm-Aharonov system of two spin $\frac{1}{2}$ particles in a single state, the polarization correlation parameter is

$$P(\hat{a}, \hat{b})\Big|_{Q.M.} = \langle \vec{\sigma}_1 \cdot \hat{a}\, \vec{\sigma}_2 \cdot \hat{b} \rangle = -\hat{a}.\hat{b}$$

For a class of hidden variable theories, including those with $P(\hat{a},\hat{a}) = -1$, we formulate the Einstein-Bell locality condition, and obtain the following generalization of Bell's inequality. For $N = 2, 3, 4, \ldots,$

$$2 \sum_{i<j}^{N} n_i n_j P_{ij} \leqslant \sum_{i=1}^{N} n_i^2 + \tfrac{1}{2}(-1) \sum_{1}^{N} n_i$$

Here $P_{ij} \equiv P(\hat{a}_i, \hat{a}_j)$ and the n_i are integers, positive, negative or zero. For $P(\hat{a}, \hat{a}) = -1$, the inequalities with $n_1^2 = 0$ and 1 are shown to be the complete content of locality for $N = 3$, 4 or 5, and not to be so for $N \geqslant 6$.

Physical Reality and Quantum Mechanics

S.B. Khadkikar

An attempt is made to review the present status of the completeness of the quantum mechanical description of physical reality. A critical evaluation of the definitions of physical reality is made and the assertion about the same by Einstein, Podolsky and Rosen is examined. The probability in the interpretation of a wave function arises through the irreducible simultaneous indeterminacy between pairs of canonically conjugate variables and not because of unknown initial conditions. Thus the description given by quantum mechanics is complete. A justification for the conventional attitude or the Copenhagen interpretation is provided by experimental results as well as the deeper understanding of various paradoxes arising in several thought experiments. The onus of proving the existence of the possible hidden variables which may or may not lead to a deterministic mechanics lies with hidden-variable-theorists. However, the present conventional Copenhagen point of view has an aesthetic appeal in that 'particles' and 'waves' are unified into a single entity which represents all nature at its most microscopic and fundamental level. I expand on this viewpoint as follows:

An experimental set up may probe either (i) particle or (ii) wave (field) nature of the physical system. An example of (i) is Wigner's spectator in a Stern-Gerlach experiment or an example of (ii) is the Stern-Gerlach experiment without spectator where the two beams are recombined. The measurement of any individual particle attribute (position, momentum or spin) naturally quenches all other possibilities including the wave nature and one gets statistical answers. The final state then is a classical statistical mixture of states. If one does not attempt to measure number of particle by excluding measurement apparatus for particle attributes, the final state retains its wave (field) nature and one can set up experiments, to observe, for example, interference or diffraction phenomena. However, any ultimate measurement on say, electron field, involves probing of the particle nature and the result is necessarily statistical. Thus the Nature manifests itself only in terms of quantum fields which have both particle and wave (field) properties and it is the experimental set-up that decides which aspect is seen. The duality (Adwait) of the particle and wave-field nature of the elementary systems is the physical reality.

A Modified Set of Feynman Postulates in Quantum Mechanics

V.K. Thankappan

Feynmann has proposed the following postulates as the foundation of non-relativistic quantum mechanics:

F1. Associated with every path a of a particle from location A to location Q in space is a complex function $\phi_a\,(A,\,t_Q{:}A,\,t_A)$ given by $\phi_a\,(Q,\,t_Q{:}A,\,t_A) = a_a \exp\,(i/\hbar)\,S\,(Q,\,t_Q{:}A,\,t_{A_t})$, where S is the action associated with the path a;

$$S_a = \int_{t_A}^{t_Q} L\,dt,$$

L being the Lagrangian associated with the particle while t_A and t_Q are the time of departure from A and the time of arrival at Q, respectively. a_a is a 'normalizing constant' and $\hbar$ the Planck's constant upon 2π.

F2. The probability amplitude for a particle to go from A at time t_A

to Q at time t_Q is

$$\Phi\,(Q,\, t_Q\!:\!A,\, t_A) = \sum_a \phi\,(Q,\, t_Q\!:\!A,\, t_A)$$

where the summation is over all paths between A and Q.

F3. The total probability amplitude for a particle to arrive at the location Q from anywhere in space is given by the integral equation,

$$\Psi(Q,\, t_Q) = \int \Phi(Q,\, t_Q\!:\, A,\, t_A)\; (A,\, t_A)\; d\tau_A,$$

where $d\tau_A$ is a volume element containing the location A.

The following modifications are proposed for these postulates:

F1. Replace ϕ_α by

$$\Psi_a(Q,\, t_Q,\!:\, A,\, t_a) = \phi_a\,(Q,\, t_Q:\, A,\, t_a)\; \exp\; [(i/\hbar)\, J_S(Q,\, A)]$$

where, J_S is a function that depends on the spin s of the particle. t_a has the same meaning as t_A except that it could be different for the different paths.

F2. The probability amplitude for a particle to go from A at *some* time to Q at time t_Q is

$$\Psi_A(Q,\, t_Q) = \sum_a \Psi_a(Q,\, t_Q:\, A,\, t_a)$$

F2a. Only those paths contribute to the summation in **F2** that differ from a_c (the classical path) by less than $h/2$ in action.

F3. If $A, B, C, \ldots$ are locations corresponding to *similarly-prepared states* of a particle in an experimental set up, the number of particles arriving at a point of observation Q at time t_Q from the above locations, is proportional to

$$|\,\Psi(Q,\, t_Q)\,|^2, \text{ where } \Psi(Q,\, t_Q) = \textstyle\sum_A c_A\, \Psi_A(Q,\, t_Q)$$

These modifications are shown to have the following consequences:

(i) A shorter and simpler derivation of the Schrödinger equation.

(ii) Possible incorporation of the spin-dependence of the statistics applicable to a system of identical particles.

(iii) Derivation of the position-momentum uncertainty relationship, with the suggestion that the time-energy uncertainty relationship has a likely place only in the relativistic theory.

(iv) Derivation of a validity-criterion, in the form of a ratio involving the linear dimension of the particle, for the classical mechanics approximation.

(v) Explanation of the interference phenomenon as a purely statistical process.

Generalized Kasner Model and Its Counter Part in Brans-Dicke Theory

V.B. Johri and G.K. Goswami

Solutions of *BD*-field equations correponding to the Bianchi-type I metric and perfect fluid with equation of state $p = \rho$ (stiff matter) have been obtained. *BD*-field equations give rise to two solutions in which one corresponds to relativistic case (vanishing scalar field) and other is a proper *BD*-model. Various physical and geometrical properties of the models are discussed.

A Microscopic Theory of Gravitation Arising Due to the Interaction of Photons with Neutrinos

A. Das and P. Bandyopadhyay

This paper deals with a theory of gravitation based on the assumption that the whole universe is submerged into a sea of degenerate neutrinos and antineutrinos and photons can interact weakly with them. So it has been possible to design a dielectric constant prevailing throughout the space which is higher near massive body. The most elegant features of the model are (a) here the gravitational phenomenon is described through photon-neutrino weak coupling and thus has an explicit quantum-mechanical picture, and (b) Mach principle is found to be automatically implemented into the picture, and as a consequence of this principle cosmical numeric coincidences are found to be evolving naturally. The question of the violation of discrete symmetries and the interpretation of null result of Eötvös-Dicke-Braginski experiment have also been treated. It is found that the former cannot be experimentally detected and the later arises due to the local constancy of the velocity of light.

The 'Screen Effect' of the Earth in ETTG

I.A. Adamuti

In the Electro-thermodynamical Theory of Gravitation (ETTG), was obtained a formula of the interaction force between two bodies similar to that of Newton's. The ETTG takes into consideration the existence of a universal stationary isotropic 'gravitational' radiation having an intensity $K_{0g} = 2.46 \times 10^7$ Jm^{-2} s^{-1}, and, upon passing through matter, it has an attenuation coefficient reported to the mass unity 2.71×10^{-18} m^2 Kg^{-1}. Both quantities are fundamental physical constants, their product being equal to Newton's constant in the *MKS* systems

$$k_0 K_{0g}\, \varkappa = f = 6.668 \times 10^{-11}\ \mathrm{Nm^2\ kg^{-2}} \tag{1}$$

Due to this small attenuation of radiation K_{0g}, generated by all the bodies of the universe as in the Mach's idea, has a spherical symmetry in all the places, which are far from the massive bodies Its stationary value in a point P is formed by radiations from different ages, between 8 minutes until millions of years.

Considering a point A on the earth's equator, a day when the moon is in the first or the last quarter, the expressions of the difference of gravitational fleld to midnight and to midday in Newtonian (without screening) and in ETTG (with screening) are:

$$\Delta\{\delta[B = A(t=24)] - \delta[A(t=12)]\}$$

$$= \begin{cases} -fM_S\left(\dfrac{1}{r_{SA}^2}+\dfrac{1}{r_{SB}^2}\right) = -0.01184758\ mS^{-2} \\ -fM_S\left(\dfrac{1}{r_{SA}^2}+\dfrac{1}{r_{SB}^2}\right)k_{fS} \end{cases} \tag{2}$$

$$-\pi f \mu_S a_S \mu_E a_E \varkappa k_{fSBK_{0g}}$$

$$+ k_0 \frac{2\pi}{3} a_S^2 K_{Sg}\left(\frac{1}{r_{SA}^2}+\frac{1}{r_{SB}^2}\right)$$

$$= -0.01184756 - 0.00001953 + 0.00000002$$

$$-0.01186707\ mS^{-2} \tag{3}$$

In the relation (3)

$$k_{fS} = 1 - \frac{3}{4}\mu_S a_S \varkappa + \frac{2}{5}\mu_S^2 a_S^2 \varkappa^2 - \ldots \varkappa = 0.9999980063 \tag{4}$$

$$k_{fSBK_{0g}} = 1 - \alpha - \frac{(1+2\alpha)}{3} J_{SmB} - \frac{10-4\alpha}{3} J_{SmB}^2 + \frac{11+4\alpha}{15} J_{SmB}^3$$

$$+\left(\frac{2}{15}+\alpha\right) J_{SmB}^4 = 0.9966945638 \tag{5}$$

$$J_{SmB} = \frac{a_S}{r_{ES}} = 4.65188 \times 10^{-3} \tag{6}$$

$$\alpha = \frac{a_E - C_2}{a_E} = 1.6775972 \times 10^{-3} \tag{7}$$

$$C_2 = \sqrt{a_E C_E} = 6.36747 \times 10^6\ m \tag{8}$$

The relation (3) differs from the relation (2) in that:

(a) the factor k_{fS} in the first term: $k_{fS} \leqslant 1$,

(b) the third term is due to the 'gravitational' radiation of the sun K_{Sg}. Its contribution is very small, $2 \times 10^{-8}\ ms^2$ only,

(c) the second term represents the screen-effect term; one can see that it depends on the densities of the two bodies and on their radii.

The centrifugal effect due to the daily rotation of the earth around its axis is eliminated, having the same sign at both moments: $t = 12h$ and $t = 24h$. The centrifugal effect due to the yearly rotation of the earth around the sun is negative at the $12h$ and positive at the $24h$. The difference between the two moments is given by the relation

$$\Delta [\delta (w_R, B) - \delta (w_R, A)] = w^2 (r_{SB} + r_{SA}) = 0.01186180 \, mS^{-2} \qquad (9)$$

The total difference between the values of the gravitational accelerations at the two moments (the relations (9) and (2) and (3) is:

$$\Delta [\delta(B) - \delta(A)] = \begin{cases} 0.00001422 \, mS^{-2} & \text{(Newton's formula)} \\ -0.00000527 \, mS^{-2} & \text{(ETTG formula)} \end{cases} \qquad (10)$$

Such a difference would be easily observed with a pendulum ($1 = 1m$) having the values $0.5\mu S$ in the ETTG and $1.464\mu S$ in the Newtonian formula. The fact that this difference has not been observed yet is due probably to the oceanic tides, which modifies the form of the earth even in those moments, and whose effect is greater than the screen effect of the earth.

But the analysis of these perturbations will be made in a future work.

Strong Gravity and Elementary Particles

C. Sivaram

A possible role for general relativity in determining the mass spectrum of elementary particles is explored through the short-range strong gravitational field of massive self-interacting spin-2 mesons. The space-time structure in the immediate vicinity of a hadron is expected to be determined by strong gravity. Various models of strong gravity are considered and it is postulated that both strong and weak interactions of elementary particles may have their basis in strong gravity. Convincing numerical support for the conjecture is also provided.

LIST OF PARTICIPANTS

1. Achuthan, P
Department of Mathematics
IIT, Madras

2. Adamut, I
Institute de Cercetari P T
Industrial Electrotechni
EC J Vladimirescu 45–47
Sector 6, Bucharest
Romania

3. Addy, Indira
23, Mahandra Sarkar Street
Calcutta

4. Ahalpara, Dilip
Physical Researh Laboratory
Navrangpura
Ahmedabad

5. Akabari, Rasiklal P
Department of Mathematics
Gujarat University
Ahmedabad

6. Akolia, GM
P. G. Department of Mathematics
Saurashtra University
Bhavnagar

7. Alurkar, SK
Physical Research Laboratory
Navrangpura
Ahmedabad

8. Ambastha, AK
Physical Research Laboratory
Navrangpura
Ahmedabad

9. Angreji, PD
Vedhsahala
Near Naranpura Railway Crossing
Ahmedabad

10. Appa Rao, MVK
Tata Institute of Fundamental Research
Homi Bhabha Road
Colaba, Bombay

11. Arvind, Kumar
Department of Physics
University of Bombay
Kalina, Bombay

12. Ballabh, GM
Vedhshala
Ahmedabad

13. Bancel, D
Department de Mathematiques
Universite Paul Sabatier
118, Route de Narbonne
31077 Toulouse Cedex
France

14. Bandyopadhyay, N
c/o Professor A K Raychaudhuri
Physics Department
Presidency College
Calcutta

15. Banerjee, Asit
Department of Physics
Jadavpur University
Calcutta

16. Banerji, S
Department of Physics
University of Burdwan
Burdwan

17. Basu, Aparna
Physical Research Laboratory
Navrangpura
Ahmedabad

18. Basu, Rahul
J-1960, Chittaranjan Park
New Delhi

19. Basu, Uma
Department of Applied Bathematics
Calcutta University
92, Acharya Prafulla Chandra Road
Calcutta

20. Bhandari, N
Physical Research Laboratory
Navrangpura
Ahmedabad

21. Bhatnagar, KP
IA/47C Ashok Vihar
Delhi

22. Bhatt, HC
Physical Research Laboratory
Navrangpura
Ahmedabad

23. Bhatt KH
Physical Research Laboratory
Navrangpura
Ahmedabad

24. Bhatt PV
Department of Mathematics
South Gujarat University
Surat

25. Bhattacharya, D
Department of Physics
Jadavpur University
Calcutta

26. Bhattacharya, Manisha
9/1 Balaram Bose Ghat Road
Calcutta

27. Bhavsar, PD
Space Application Centre
Jodhpur Tekra
Ahmedabad

28. Bhonsle, RV
Physical Research Laboratory
Navrangpura
Ahmedabad

29. Biswas, AC
Tata Institute of Fundamental Research
Homi Bhabha Road
Colaba, Bombay

30. Borikar, SM
Department of Physics
Institute of Science
Nagpur

31. Brill, D
Department of Physics and Astronomy
University of Maryland
College Park
Maryland 20742, USA

32. Buti, Bimla
Physical Research Laboratory
Navrangpura
Ahmedabad

33. Calla, OPN
Space Application Centre
Jodhpur Tekra
Ahmedabad

34. Canuto, V
NASA
2880 Broadway
New York, NY 10025
USA

35. Chakraborty, DK
Physical Research Laboratory
Navrangpura
Ahmedabad

36. Chakravarti, SP
Makhan Chatterjee Lane
Suksanatantala
Chandernagore
Dt-Hooghly

37. Chakravarty, N
'Manjari'
Lesimai Lane
Shillong

38. Chandratreya, GL
'Seetaram'
1233 B, Apte Road
Pune

39. Chatterjee, SK
Department of Physics
University of Burdwan
Burdwan

40. Chatterjee, TK
Department of Astronomy
Osmania University
Hyderabad

41. Chitnis, EV
Space Application Centre
Jodhpur Tekra
Ahmedabad

42. Chitre, DM
Raman Research Institute
Hebbal P.O.
Bangalore

43. Chitre, SM
Tata Institute of Fundamental Research
Homi Bhabha Road
Colaba, Bombay

44. Clerc, RL
U.E.R. de Mathematiques
Universite Paul Sabatier
118, Route de Narbonne
31077 Toulouse Cedex
France

45. Cowsik, R
Tata Institute of Fundamental Research
Homi Bhabha Road
Colaba, Bombay

46. Dadhich, Naresh
Department of Mathematics
University of Poona
Pune

47. Dalal, Hara Gopal
S.N. Bose Institute of Physical Sciences
92, A.P.C. Road
Calcutta

48. Das Ashesh
22-A, Garcha First Lane
Calcutta

49. Das, MP
P.G. Department of Physics
Sambalpur University
Jyoti Vihar, Burla
Orissa

50. Das, Sumit Ranjan
15/6 Jamir Lane
Calcutta

51. Das, T
Tata Institute of Fundamental Research
Homi Bhabha Road
Colaba, Bombay

52. De, UK
Department of Physics
Jadavpur University
Calcutta

53. Deshpande, MR
Physical Research Laboratory
Navrangpura
Ahmedabad

54. Dewangan, DP
Physical Research Laboratory
Navrangpura
Ahmedabad

55. Dhar, Deepak
Tata Institute of Fundamental Research
Homi Bhabha Road
Colaba, Bombay

56. Dhurandhar SV
Tata Institute of Fundamental Research
Homi Bhabha Road
Colaba, Bombay

57. Durgapal, MC
Department of Physics
DSB College
Nainital

58. George, Abraham
Madras Christian College
Madras

59 Ghosh, SK
Tata Institute of Fundamental Research
Homi Bhabha Road
Colaba, Bombay

60. Gopalan, K
Physical Research Laboratory
Navrangpura
Ahmedabad

61. Goswami, GK
Department of Mathematics
University of Gorakhpur
Gorakhpur

62. Gunzig, E
Universite Libra de Bruxelles
Pool de Physique
Campus Plaine
Code Postal No. 225
Bid du Triumphe
1050 Bruxelles, Belgium

63. Gupta, BK
Department of Physics
Osmania University
Hyderabad

64. Gupta, Rajeev
Staff Quarters
Bal Bhavan
Kolla Road
New Delhi

65. Hari Dass, ND
Raman Research Institute
Bangalore

66. Husain, S. Izhar
Professor of Mathematics
Aligarh Muslim University
Aligarh

67. Iyengar, KVK
Tata Institute of Fundamental Research
Homi Bhabha Road
Colaba, Bombay 400005

68. Iyer, BR
Department of Physics
University of Bombay
Vidyanagari
Bombay

69. Jain, Sanjeewani A
Deptartment of Physics
IIT, Powai
Bombay

70. Johri, VB
Department of Mathematics
University of Gorakhpur
Gorakhpur

71. Joshi, Pankaj
Department of Mathematics
Saurashtra University
Bhavnagar

72. Kaohot, KR
Department of Mathematics
Lukhdhirji Engineering College
Morvi, Gujarat

73. Kanavi, Shivanand
Department of Physics
IIT, Powai
Bombay

74. Karade, TM
Department of Mathematics
Nagpur University
Nagpur

75. Kaul, CL
NRL-BARC
Barzulla
Srinagar

76. Kembhavi, AK
Tata Institute of Fundamental Research
Homi Bhabha Road
Colaba, Bombay

77. Keshavamurthy, RN
Physical Research Laboratory
Navrangpura
Ahmedabad

78. Khadkikar, SB
Physical Research Laboratory
Navrangpura
Ahmedabad

79. Khan, MQ
St. Andrew's Post-Graduate College
Gorakhpur

80. Khare, Avinash
Physical Research Laboratory
Navrangpura
Ahmedabad

81. Kothari, DS
101 Bhupalpura
Udaipur
Rajasthan

82. Krishnarao, J
Department of Mathematics
Saurashtra University
Bhavnagar

83. Krishnaswamy, KS
Tata Institute of Fundamental Research
Homi Bhabha Road
Colaba, Bombay

84. Maheshwari, AN
Professor of Physics
RCE-Mysore
Mysore

85. Maity, Satyaranjan
Department of Physics
Presidency College
Calcutta

86. Mathew, PM
Sr. Professor of Theoretical Physics
University of Madras
AC College Building
Madras

87. Mathur, UNL
Nuclear Theory Group
Department of Physics
IIT, Kanpur

88. Mehendale, DP
220, Bhagirath Society
Sahakar Nagar 1
Pune

89. Mehra, AL
Government College
Tonk
Rajasthan

90. Mohan, M
Physical Research Laboratory
Navrangpura
Ahmedabad

91. Mohanty, G
Department of Mathematics
IIT, Kharagpur

92. Morrison, Phil
Department of Physics
Massachusetts Institute of Technology
Boston, Massachusetts
USA

93. Mukherjee A
Tata Institute of Fundamental Research
Homi Bhabha Road
Colaba, Bombay

94. Nagaraj, M
Department of Mathematics
Central College
Bangalore

95. Nagesha Rao, N
Physical Research Laboratory
Navrangpura
Ahmedabad

96. Naik, Prahagllad C
Institute of Physics
A/105, Sahead Nagar
Bhubaneswar

97. Narlikar, JV
Tata Institute of Fundamental Research
Homi Bhabha Road
Colaba, Bombay

98. Narlikar, VV
701 Colaba Housing Colony
Homi Bhabha Road
Bombay

99. Nayak, Pratibindhya
Institute of Physics
A/105 Saheed Nagar
Bhubaneswar

100. Neeraj Chaubey
Department of Mathematics & Astronomy
Lucknow University
Lucknow

101. Newman, ET
University of Pittsburg
Faculty of Arts & Science
Department of Physics and Astronomy
Pittsburg, PA 15260
USA

102. Nitsure, Nitin
34/11, Erandavana
Prabhat Road
Lane No. 8
Pune

103. Nuthakki, P
Physics Department
Andhra University
Waltair

104. O Costa de Beauregard
University de Paris VI
Laboratoire de Physique Theorique
Institut Henri Poincare
11 Rue Pierre et Marie Curie
75231 Paris Cedex 05
France

105. Oza, Ashwin
Physical Research Laboratory
Navrangpura
Ahmedabad 380009

106. Pachner, J
University of Regina
Regina
Canada S4S OA2

107. Padmanabhan, T
T. C. 3, —130
First Puthan Street
Manacad
Trivandrum

108. Pal, A
Tata Institute of Fundamental Research
Homi Bhabha Road
Colaba, Bombay

109. Panchapakesan, N
Department of Physics
University of Delhi
Delhi

110. Pande, AK
Nuclear Physics Section
Department of Physics
Banaras Hindu University
Varanasi

111. Pande, Atul Kumar
Department of Physics
DSB College
Nainital

112. Pandey, SK
Theoretical Astrophysics Group
Tata Institute of Fundamental Research
Bombay

113. Pandey, SN
Department of Mathematics
University of Roorkee
Roorkee (U.P.)

114. Pandey, US
Research Scholar
Department of Physics
Gorakhpur University
Gorakhpur

115. Pandit, SN
Tata Institute of Fundamental Research
Homi Bhabha Road
Colaba, Bombay

116. Papapetrou, A
Universite de Paris VI
Laboratoire de Physique Theorique
Institut Henri Poincare
11 Rue Pierre et Marie Curie
75231 Paris Cedex 05
France

117. Paranjape, S
Tata Institute of Fundamental Research
Homi Bhabha Road
Colaba, Bombay

118. Parekh, JH
Vedhsahala
Near Naranpura Rly Crossing
Ahmedabad

119. Parikh, JC
Physical Research Laboratory
Navrangpura
Ahmedabad

120. Patel, LK
Department of Mathematics
Gujarat University
Ahmedabad

121. Patel, MD
Department of Mathematics
Sardar Patel University
Vallabh Vidyanagar
Dist. Kaira

122. Pathak MP
c/o Dr. V. B. Johri
20 D Hirapuri
Gorakhpur University
Gorakhpur (U. P.)

123. Perry, Judith
Max Planck Institut Fur
Physik and Astrophysik
Fohringer Ring 6
D 8000 Munchen 40
Fed. Rep. of Germany

124. Prabhakar Rao, P
Department of Mathematics
IIT, Kharagpur

125. Pradhan, A
Department of Mathematics
Hindu Degree College
Zamania R. S.
Ghazipur

126. Prasad, G
Department of Mathematics
Faculty of Science
Banaras Hindu University
Varanasi

127. Prasad, SS
Department of Physics
University of Gorakhpur
Gorakhpur

128. Prasanna, AR
Physical Research Laboratory
Navrangpura
Ahmedabad

129. Pratap, RV
Physical Research Laboratory
Navrangpura
Ahmedabad

130. Purohit, KD
Lecturer in Physics
G.G.I. College
Srinagar
Garhwal (U.P.)

131. Puroshotham, S
Department of Mathematics
Saurashtra University
Bhavnagar

132. Radhakrishna, L
Reader in Mathematics
Shivaji University
Kolhapur

133. Raghava Rao, R
Physical Research Laboratory
Navrangpura
Ahmedabad

134. Rai, LN
59, Dalmia Hostel
Banaras Hindu University
Varanasi

135. Rajgopal, AS
Assistant Professor of Mechanical Engineering
Coimbatore Institute of Technology
Aerodrome P.O.
Coimbatore

136. Rajagopal, ES
Professor of Physics
Indian Institute of Science
Bangalore

137. Rajasekharan, G
Head, Department of Theoretical Physics
University of Madras
AC College Building
Madras

138. Rajeev, SG
III Yr. B.Sc.
Department of Physics
University College
Trivandrum

139. Rajput, BS
Head, Physics Department
Garhwal University
Srinagar (U.P.)

140. Raju, CK
Indian Statistical Institute
T.S.J.S. Sansanwahl Marg
New Delhi

141. Ramadurai, S
Tata Institute of Fundamental Research
Homi Bhabha Road
Colaba, Bombay

142. Ramamurthy, S
Department of Nuclear Physics
Andhra University
Waltair

143. Ramamurthy, VL
Department of Physics
S.V. University College
Tirupati (A.P.)

144. Ramaswami, S
Space Application Centre
Jodhpur Tekra
Ahmedabad

145. Ramchandran, R
Tata Institute of Fundamental Research
Homi Bhabha Road
Colaba, Bombay

146. Ramesh Chandra
Department of Mathematics
St Andrew's College
Gorakhpur

147. Rana, NC
Tata Institute of Fundamental Research
Homi Bhabha Road
Colaba, Bombay

148. Rao, ABP
1/8 Shobhanagar
Sholapur

149. Rao, JR
Department of Mathematics
IIT, Kharagpur

150. Rao, MN
Physical Research Laboratory
Navrangpura
Ahmedabad

151. Rawal, JJ
Nehru Centre
Sterling Centre, 7th Floor
Dr Annie Basant Road
Bombay

152. Ray, Mrinal Kanti
Department of Physics
University of Burdwan
Burdwan

153. Raychaudhuri, AK
Presidency College
Calcutta

154. Razdan, H
NRL-BARC
Baghat Barzulla
Srinagar

155. Reddy, DRK
Department of Applied Mathematics
Andhra University
Waltair

156. Rohtagi, Rajeev
FLAT 11B
317 BG Kher Marg
Bombay

157. Roy, P
Tata Institute of Fundamental Research
Homi Bhabha Road
Colaba, Bombay

158. Roy, SM
Tata Institute of Fundamental Research
Homi Bhaba Road
Colaba, Bombay

159. Saikia, DJ
Radio Astronomy Centre
NTIFRI PO Box 8
Ootacamund

160. Saraykar, RV
Department of Mathematics
Nagpur University Campus
Nagpur

161. Sarkar, S
Tata Institute of Fundamental Research
Homi Bhabha Road
Colaba, Bombay 400005

162. Satya Prakash
Physical Research Laboratory
Navrangpura
Ahmedabad

163. Seifert, HJ
Hochschule der Bundeswehr
Hamburg Fachbereich
Mas chinenbau-Mathematik
Postfach 700822
D-2000 Hamburg 70
Fed. Rep. of Germany

164. Sen, Abhijit
Physical Research Laboratory
Navrangpura
Ahmedabad

165. Shah Atulkumar, R.
499/C Rasta Peth
'Gunjan' Block No. 3
Pune

166. Shankara Sastry, K
Department of Astronomy
Osmania University
Hyderabad

167. Shankara, TS
Mathematics Department
IIT, Madras
Madras

168. Sharma, DN
Department of Mathematics
Kashmir University
Srinagar

169. Sharma, SP
Mathematics Department
University of Roorkee
Roorkee

170. Sheorey, VB
Physical Research Laboratory
Navrangpura
Ahmedabad

171. Sheth, YI
9, Lecturers' Quarters
Opp. University G House
Gujarat University
Ahmedabad 380009

172. Shirke, JS
Physical Research Laboratory
Navrangpura
Ahmedabad

173. Shivaram, C
Department of Physics
Indian Institute of Science
Bangalore

174. Sidharth, BG
75, Raja Basarta Roy Road
Calcutta

175. Singh, Gulab
Department of Mathematics
Banaras Hindu University
Varanasi

176. Singh, KP
Tata Institute of Fundamental Research
Homi Bhabha Road
Colaba, Bombay

177. Singh, Onkar Nath
Department of Mathematics
Faculty of Science
Banaras Hindu University
Varanasi

178. Singh, T
Applied Mathematics Section
Institute of Technology
Banaras Hindu University
Varanasi

179. Singh, V
Tata Institute of Fundamental Research
Homi Bhabha Road
Colaba, Bombay

180. Sohnius, M
Max Planck Institut Fur
Physik und Astrophysik
Fohringerring 6
D 8000 Munchen 40
Fed. Rep. of Germany

181. Som Murari, M
Departmento de Fisica
Mathematica Instituto
de Fisica, U.F.R.J.
Rio de Janeiro
Brazil

182. Srinivasa, BR
Department of Mathematics
Central College
Bangalore

183. Srivastava, DC
Department of Physics
University of Gorakhpur
Gorakhpur

184. Srivastava, SK
Department of Mathematics
Gorakhpur University
Gorakhpur

185. Stephens, SA
Tata Institute of Fundamental Research
Homi Bhabha Road
Colaba, Bombay

186. Subbaraya, BH
Physical Research Laboratory
Navrangpura
Ahmedabad

187. Subramanian, K
16, Emerald Building
Plot 41/45
Jai Shankar Co-operative Society
Chedda Nagar
Bombay

188. Sundaram, AK
Physical Research Laboratory
Navarangpura
Ahmedabad

189. Swami SP
1552/13, Parijat
Subhashnagar
Barshi

190. Tarachand Singh, RK
Department of Mathematics
IIT, Kharagpur

191. Thakar, SC
Anandnagar LIG-680
Bhavnagar

192. Thakore, SR
Space Application Centre
Jodhpur Tekra
Ahmedabad

193. Thakurta, SNG
Department of Physics
Presidency College
Calcutta

194. Thankappan, VK
Physics Department
University of Calicut
Calicut

195. Tikekar, Ramesh
Department of Mathematics
Sardar Patel University
Vallabh Vidyanagar

196. Tiwari, SC
SSC Div. CEERI
Pilani

197. Tiwari, RN
Department of Mathematics
IIT, Kharagpur
Kharagpur

198. Trivedi, SP
'Gandhikutir'
2125 Krishnanagar
Bhavnagar

199. Trivedi, VM
Physics Department
Gujarat University
Ahmedabad

200. Tulsani, V
Sindhu Mahavidalaya
Panchpaoli
Nagpur

201. Vaidya, PC
Vice Chancellor
Gujarat University
Ahmedabad

202. Vardya, MS
Tata Institute of Fundamental Research
Homi Bhabha Road
Colaba, Bombay 400005

203. Varma RK
Physical Research Laboratory
Navrangpura
Ahmedabad

204. Varshneya, MC
Professor of Physics
MPKV, Rahuri

205. Venkataraman, Rhama
Physical Research Laboratory
Navrangpura
Ahmedabad

206. Venkatesan, TR
Physical Research Laboratory
Navrangpura
Ahmedabad

207. Venugopal, R
'Rema Vihar'
Diamond Hill
Vellayambalam
Trivandrum

208. Verma, SD
Physics Department
Gujarat University
Ahmedabad

209. Virkar, KV
Staff Quarters
Walchand College of Engineering
Sangli

210. Vishveshwara, CV
Raman Research Institute
Hebbal P.O.
Bangalore

211. Vyas, UD
Sir PP Institute of Science
Bhavnagar

212. Walker, M
Max Planck Institut Fur
Physik und Astrophysik
Fohringerring 6
D 8000 Munchen 40
Fed. Rep. of Germany

213. Walwadkar, BB
Plot No. 36 S.T Colony
Karre's Bungalow
Rajarampuri
Kolhapur

214. Zafar, Ahsan
Mathematics Department
Aligarh Muslim University
Aligarh

215. Radhakrishna, V
Raman Research Institute
Hebbal P.O.
Bangalore

216. Yash Pal
Space Application Centre
Jodhpur Tekra
Ahmedabad

217. Ramanathan, KR
Physical Research Laboratory
Navrangpura
Ahmedabad

218. Pisharoty, PR
Physical Research Laboratory
Navrangpura
Ahmedabad

219. Lal, Devendra
Physical Research Laboratory
Navrangpura
Ahmedabad

220. Pandya, SP
Physical Research Laboratory
Navrangpura
Ahmedabad

221. Sreekantan, BV
Tata Institute of Fundamental Research
Colaba, Bombay

222. Williams, Ruth M.
Girton College
Cambridge CB3 OJG
England

223. Subbarao, BK
D-22 Naval Officers' Flats
Colaba
Bombay

224. Pandey, SN
Department of Applied Sciences
M.M.M. Engineering College
Gorakhpur (U.P.)

225. Tripathi, SN
Department of Physics
University of Gorakhpur
Gorakhpur (U.P.)

226. Katkar, LN
Department of Physics
Shivaji University
Kolhapur

227. Roghalia, NN
Dahod
Gujarat

228. Datta Mazumdar, A
Pachhunga College
Aizawl
Mizoram